Language Acquisition and Development

Language Acquisition and Development
A Generative Introduction

Misha Becker and Kamil Ud Deen

The MIT Press
Cambridge, Massachusetts
London, England

This book was set in Times New Roman by Westchester Publishing Services, Danbury, CT. Printed and bound in the United States of America.

Library of Congress Cataloging-in-Publication Data

Names: Becker, Misha Karen, 1973- author. | Ud Deen, Kamil, author.
Title: Language acquisition and development : a generative introduction / Misha Becker, Kamil Ud Deen.
Description: Cambridge : MIT Press, 2020. | Includes bibliographical references and index.
Identifiers: LCCN 2019019584 | ISBN 9780262043588 (hardcover)
Subjects: LCSH: Language acquisition. | Children--Language. | Communicative competence in children.
Classification: LCC P118 .B423 2020 | DDC 401/.93--dc23
LC record available at https://lccn.loc.gov/2019019584

10 9 8 7 6 5 4 3 2 1

Contents

Acknowledgments xi

I MODULE 1: LANGUAGE ACQUISITION IN THEORETICAL CONTEXT 1

1 Introduction: What Is Language Acquisition? 3
1.1 The Logical Problem of Language Acquisition 3
1.2 The Developmental Problem of Language Acquisition 9
1.3 Overview of Chapters 10
1.4 Further Reading 13
1.5 References 13

2 Theoretical Approaches to Studying Language Acquisition 15
2.1 Universal Grammar 16
2.1.1 Motivations for Universal Grammar 16
2.1.1.1 The Problem of Induction 17
2.1.1.2 An Example from Language 21
2.1.1.3 The Poverty of the Stimulus 24
2.1.2 The UG-Based View of Language: A Computational System 29
2.1.3 Competence versus Performance 30
2.1.4 Flavors of UG Approaches 31
2.1.4.1 Continuity 31
2.1.4.2 Principles and Parameters 32
2.2 Statistical Tracking 35
2.3 Modern Constructivist Approaches 39
2.3.1 What Is a Domain-General Mechanism? 39
2.3.2 The Constructivist View of Language: Form-Meaning Pairings 40
2.3.3 How Constructivism Works 41
2.4 How Does Constructivism Differ from the UG-Based Approach? 45
2.5 Summary 47
2.6 Further Reading 48
2.7 Exercises 48
2.8 References 50

II MODULE 2: BUILDING A SOUND SYSTEM 53

3 Early Speech Perception 55
3.1 Speech Sound Discrimination 57
3.2 Perceiving Phonemic Contrasts 63
3.3 Finding Word Boundaries: Speech Segmentation 66
3.3.1 Infant-Directed Speech 67
3.3.2 The Importance of Prosody and Rhythm 70
3.3.3 Phonotactic Constraints 72
3.4 Summary 73
3.5 Further Reading 73
3.6 Exercises 74
3.7 References 75

4 Speech Production and Phonological Development 79
4.1 When Are Vocalizations Part of Language? 79
4.2 Building a Sound System 82
4.2.1 What Is a Phoneme? 82
4.2.2 Early Phoneme Inventory 83
4.3 Common Phonological Processes 85
4.3.1 Substitutions 86
4.3.2 Assimilations 87
4.3.3 Syllabic Processes 88
4.3.4 Covert Contrasts 90
4.4 Accounting for Patterns: Phonological Rules 90
4.5 Accounting for Patterns: Constraints 94
4.6 Summary 98
4.7 Further Reading 99
4.8 Exercises 99
4.9 References 101

III MODULE 3: WORD MEANING AND WORD STRUCTURE 103

5 Word Learning 105
5.1 Characteristics of Early Word Production 105
5.1.1 The Vocabulary Spurt 107
5.1.2 Content of Early Vocabulary 107
5.1.3 Early Uses of Nouns: Overextension and Underextension 108
5.1.4 Early Vocabulary Comprehension and Fast Mapping 110
5.2 The Problems of Word Learning, and the Limitations of Ostension 111
5.3 Principles That Guide Word Learning 116
5.3.1 Principle 1: The Principle of Reference 116
5.3.2 Principle 2: The Whole Object Constraint 117
5.3.3 Principle 3: The Principle of Mutual Exclusivity 118
5.4 Learning Verbs via Syntactic Bootstrapping 120
5.5 Summary 128

5.6 Further Reading 129
5.7 Exercises 129
5.8 References 131

6 **The Acquisition of Morphology 135**
6.0 Introduction 135
6.1 The Foundation: Roger Brown 137
6.1.1 Brown's Method for Establishing When a Morpheme Has Been Acquired 138
6.1.1.1 Obligatory Contexts 138
6.1.1.2 90% Criterion 139
6.1.1.3 Consistency 139
6.1.1.4 Mean Length of Utterance 141
6.1.2 Brown's Findings 142
6.2 Acquisition of a Rule, or Memorized Chunk: Jean Berko (Gleason) 143
6.3 General Properties of the Acquisition of Inflection 145
6.3.1 Rapidity and Accuracy 145
6.3.2 Prefixation versus Suffixation 146
6.3.3 Rich versus Impoverished Morphology 147
6.3.4 Kinds of Morphological Errors 148
6.3.4.1 Errors of Commission versus Errors of Omission 148
6.3.4.2 Overregularization and U-Shaped Development 151
6.4 The Role of Input 153
6.5 Summary 155
6.6 Further Reading 155
6.7 Exercises 155
6.8 References 157

IV **MODULE 4: THE SENTENCE LEVEL 165**

7 **Syntactic Development 167**
7.0 Introduction 167
7.1 Bootstrapping into Syntax: Semantic Bootstrapping 167
7.2 Functional Structure and Optional Infinitives 170
7.2.1 Functional Categories and Structure 171
7.2.2 Telegraphic Speech 177
7.2.3 Optional Infinitives 178
7.2.3.1 The Truncation Hypothesis 180
7.2.3.2 Form-Position Contingencies in Optional Infinitives 184
7.2.3.3 Null Subject Contingencies in Optional Infinitives 187
7.2.3.4 Wh-question Contingencies with Optional Infinitives 188
7.2.3.5 Summary of Optional Infinitives and Truncation 188
7.3 Other Aspects of Functional Structure 190
7.3.1 Negation 190
7.3.2 Questions 191
7.3.3 Passive Construction 197
7.3.4 Relative Clauses 201

7.4 The Problem of Variable Reference 203
7.4.1 The Binding Theory 204
7.4.2 Principle of Reflexives (Principle A) 206
7.4.3 Principle of Pronouns (Principle B) 211
7.4.4 Principle of R-Expressions (Principle C) 213
7.4.5 Summary of the Binding Principles 215
7.5 Summary 216
7.6 Further Reading 217
7.7 Exercises 217
7.8 References 218

V MODULE 5: BEYOND MONOLINGUAL AND TYPICAL LANGUAGE ACQUISITION 223

8 Language Acquisition under Nontypical Circumstances 225
8.1 Late First-Language Acquisition 225
8.1.1 Feral Children 226
8.1.2 Genie 227
8.1.3 Chelsea 230
8.2 Language Acquisition in Deaf Children 232
8.2.1 Acquisition of Sign Language in Deaf Children 233
8.2.2 Late Acquisition of ASL 238
8.2.3 Acquisition of Oral Language in Deaf Children 239
8.2.4 Summary 242
8.3 Language Acquisition in Blind Children 242
8.3.1 Lexical and Grammatical Development 242
8.3.2 Acquisition of Perception Verbs 243
8.3.3 Summary 245
8.4 Impaired Language Acquisition 245
8.4.1 Specific Language Impairment 245
8.4.1.1 Grammatical Characteristics of SLI 247
8.4.1.2 Causes of and Explanations for SLI 249
8.4.2 Autism and Autism Spectrum Disorder 251
8.4.3 Hemispherectomy 253
8.5 Summary 256
8.6 Further Reading 256
8.7 Exercises 256
8.8 References 259

9 Acquisition of More than One Language 265
9.0 Introduction 265
9.1 Bilingualism in Early Childhood: Simultaneous Bilingualism 266
9.1.1 The Single-System Hypothesis 267
9.1.2 The Separate-Systems Hypothesis 268
9.1.3 The Interdependent Development Hypothesis 270
9.1.4 Code-Switching 272

9.2 Successive Bilingual Acquisition 274
9.3 Language Attrition and Heritage Language 275
9.4 Language Endangerment and Language Revitalization 278
9.4.1 Is Language Revitalization Important? 279
9.4.1.1 Scientific Knowledge, Including Linguistic Knowledge 280
9.4.1.2 Cultural Knowledge 280
9.4.1.3 Identity 281
9.4.1.4 Autonomy 281
9.4.2 How Are Languages Revitalized? 282
9.5 Summary 284
9.6 Further Reading 285
9.7 Exercises 285
9.8 References 286

Appendix A: English IPA Symbols 289
Appendix B: Methods in Child Language Acquisition 293
Introduction 293
Module 1: Naturalistic Data 293
What Is Naturalistic Data? 294
How Naturalistic Data Is Collected 295
Module 2: Production Data 296
Elicited Production 296
Elicited Imitation 298
Priming 299
Module 3: Comprehension Data 300
Grammaticality/Acceptability Judgment 300
Truth Value Judgment Task (TVJT) 301
Picture Selection 303
Intermodal Preferential Looking Paradigm / Eye Tracking 304
Act Out Task 306
Methodologies for Infant Studies 307
Brain-Based Methods 309
Further Reading 310
References 310

Notes 313
Index 317

Acknowledgments

The idea for this book was born where many great ideas are born—during a heady coffee break at the Boston University Conference on Language Development. We were comparing notes on how we teach introductory acquisition courses, and we were both struck by the paucity of textbooks that address child language acquisition, framed within generativist linguistic theory, and aimed at students who have relatively little background in linguistics. We resolved to develop our own textbook, and several years later, here we are. But in many ways the roots of this book should be traced back to our graduate school days. We were both students of Nina Hyams—scholar, leader, and mentor extraordinaire. It was Nina's sage guidance and teaching, along with truly compelling subject matter, that hooked us. Under her tutelage we developed our passion for studying how children create grammars and our desire to share that passion with others. We gratefully dedicate this book to Nina, in appreciation of all she has done for her students, the field of linguistics, and the field of language acquisition. She has made an indelible mark, both on us as scholars and, more importantly, on the field as a whole.

Since our days at UCLA we have, of course, benefited from many other linguists who have taught and challenged us, served as sounding boards and collaborators, and generally shaped us as scholars. Among them are Stephen Crain, Susie Curtiss, Katherine Demuth, Jill de Villiers, Lila Gleitman, Maria Teresa Guasti, Jeff Lidz, Diane Lillo-Martin, William O'Grady, Colin Philips, Luigi Rizzi, Tom Roeper, Jeannette Schaeffer, Bonnie Schwartz, William Snyder, Rosalind Thornton, and Ken Wexler. We particularly wish to express thanks to Stephen Crain, Diane Lillo-Martin, Theo Marinis, Jason Rothman, Jennifer Smith, Anne Michele Tessier, Rosalind Thornton, and an anonymous reviewer for insightful comments on drafts of this book. Their input has undoubtedly increased the quality of this book many times over.

Numerous other linguists have contributed directly or indirectly to the content of this book and to the ideas as we present them. These include Adrianna Belletti, Elika Bergelson, João Costa, Cécile De Cat, Shin Fukuda, John Grinstead, Theres Grüter, Cornelia Hamann, Li (Julie) Jiang, Victoria Mateu, Reiko Mazuka, Keiko Murasugi, Akira Omaki, Robyn

Orfitelli, Ana Teresa Pérez-Leroux, Tetsuya Sano, Carson Schütze, Koji Sugisaki, Kristen Syrett, and Charles Yang. It is a pleasure and an honor to be part of a field made vibrant and exciting by these amazing scholars.

We are also grateful to our own students for challenging us to be better teachers, for asking good questions, and for feedback and corrections on earlier drafts of these materials as we developed them. We've been very lucky to have some superb graduate students who have driven our ideas and our research, and we are indebted to them. Among them we'd like to thank Jinsun Choe, Iyad Ghanim, Inmaculada Gomez Soler, Megan Gotowski, Raina Heaton, Ryan Henke, Kum Jeong Joo, Chae-Eun Kim, Soyoung Kim, Susannah Kirby, Hye-Young Kwak, Elaine Lau, Grant Muāgututi'a, Jun Nomura, Akari Ohba, Hiroko Sato, Nozomi Tanaka, Jennie Tran, and Sejung Yang.

We are also indebted to Amy Brand, Marc Lowenthal, and Anthony Zannino of MIT Press for encouraging us to pursue our idea, and to John Donohue of Westchester Publishing Services for his patience with us and his attention to detail. Finally, of course, we wish to thank our families for their extreme patience with our late-night and/or early-morning writing and editing marathons, and for providing sustained (and sustaining!) emotional support throughout this multiyear endeavor.

Module 1: Language Acquisition in Theoretical Context

1 Introduction: What Is Language Acquisition?

This book is about one of the great mysteries of the human experience—how we go from being small, cute, noisy blobs that don't understand or produce language to eloquent kindergartners who not only know the meanings of several thousand words but can rattle off stories (some true, some maybe not), commentaries, opinions, and questions seemingly without end. One parent came home from work one day to an exhausted partner who complained that their 3-year-old daughter had not stopped talking since eight o'clock that morning. How do children, who can't tie their shoes, who may or may not be potty-trained, and who can't compute basic mathematical operations, perform this feat?

In this book we'll walk through many of the stages that children go through on their way to gaining eloquence and provide some answers to the question of how they do it. One of the main answers we start with is that babies are not just passively experiencing their world. Instead, their brains are designed to anticipate that human language will have certain properties—for instance, that sentences have a hierarchical structure—and this predisposition allows them to rapidly assimilate important information about language from their environment.

This answer (that children are born predisposed to learn language) eases the problem, but doesn't lessen the mystery. We still want to explain how children begin to tease apart the fluid stream of sounds coming at them from different speakers and possibly from different languages. We still need to unravel the process by which children take an individual word (once they are able to isolate individual words) and figure out which of the infinite possible meanings or concepts in the universe that one little word is supposed to label. The question of how children figure out the rules of their particular language is still a puzzle to be solved. This book won't answer *all* of these mysteries—linguists still have a lot of work to do—but we'll answer many of them. So, let's get started!

1.1 The Logical Problem of Language Acquisition

A driving idea in the field of language acquisition is known as the **Logical Problem of Language Acquisition** (LPLA). Simplifying (for now), the LPLA notes that the manner in which children acquire language is not predicted by the kind of language that they hear.

There is a gap between what children hear and experience and what they are eventually able to do with language. A key issue in the field, and one that this book is centered around, is how children bridge this gap. How do they go from nothing to everything in the manner that they do? As we will see, children learn language universally (all normal children do so regardless of which language they are born into), quickly (by kindergarten, typically), easily, and relatively uniformly. Moreover, they do all this without any meaningful correction or instruction, and in the face of information that is ambiguous at best, and misleading at worst. This is a genuine puzzle, one that was instrumental in the formation of the field of modern linguistics, and one that continues to guide all modern linguistic and language acquisition theory.

We'll go into this in more detail in chapter 2, but to start, let's think about what ingredients are needed to make language acquisition happen. What do you think is necessary to acquire language? One very obvious ingredient is language itself—children need to be exposed to language. We call this **input**. We'll talk much later in the book about what happens if language input is not available to children in the early years of life (see chapter 8), but for now we can assume that if you don't have exposure to language, you won't acquire it.

So, we can say that language input is necessary. But is it sufficient? This is a very different question. What would it mean for language input to be sufficient?

One way to think about this question is to take stock of all that has to be learned when one learns a language: vocabulary, the sounds in the language, how words are ordered in phrases and sentences, how to turn a statement into a question, and so on. These words and rules vary from language to language, and children might be able to learn about them just by listening and "picking it up." After all, kids are like sponges, right? We hear this all the time. Looking at it this way, it seems like input alone might be sufficient.

But let's take it a step further. Is the language you know really just a catalog of all the words and sentences you've heard before? Sometimes, when you hear a new word, it is impossible to understand until someone gives you more information. But often you're able to figure out some aspect of the meaning of the new word using clues in the sentence. The point is that learning the meanings of words does not happen simply because you hear the word in your environment. Rather, there's some mental and linguistic work that happens that helps figure out the meaning. This points to a more complex process than simply being exposed to language: the child is an agent of this process of learning, and not a passive sponge.

Furthermore, if all there was to language acquisition was learning what you hear, then you would not be able to produce and understand brand-new sentences—but surely you do so every day. "The grass screamed danger to the foot soldiers in the Mighty Mouse army." Have you ever heard that sentence before? Likely not, but you know it is a sentence of English, and you understand what it means (even if that meaning is kind of weird). The ability to understand that sentence could not have come from a catalogue of previously heard sentences. That's because you know (implicitly) the rules of English syntax, and this allows you to understand the (admittedly bizarre) message in that novel sentence.

Similarly, your knowledge of grammatical structure helps you understand and create new words. Every year dozens, if not hundreds, of words enter the lexicon, and speakers are able to adapt to them quite easily. When new ways of communicating develop through different forms of technology, the language and its speakers adopt new verbs like *email*, *instant-message*, or *ping*, and we use them just like verbs that have been in the language for centuries ("I'm emailing you that file right now," "She just instant-messaged her friend," "Ping me if you want to meet up"). You can **generalize** using the rules of grammar you already know so that you can easily handle these newcomers to the language; typically you can do this the very first time you encounter one of them. You are not dependent on prior exposure through the input.

Another way to think about the sufficiency of language input, specifically with respect to children's language, is to notice all the neat things they say. Here are a few examples:

(1) a. I'm gonna fall this on you!
 b. Don't giggle me.
 c. My legpit hurts.
 d. I want you pick up me.

These are "errors" from the perspective of adult grammar. Children surely have not heard these kinds of sentences from their parents—children invent them on their own. The errors themselves might be funny, but when we look at them carefully we find that they are infrequent (most utterances are totally unremarkable—the cute, deviant ones stand out), highly systematic, and rule driven. Both 1a and 1b involve taking a verb that is intransitive (a verb that has a subject but no object), using it with a direct object (*this* in 1a; *me* in 1b), and expressing a meaning related to causation (1a means, roughly, "I'm gonna make this fall on you," and 1b means "Don't make me giggle"). But this is exactly how verbs that allow an alternation between intransitive and transitive uses in adult language work: the verb *break* denotes a change of state when it has only a subject (*The vase broke* ≈ The vase underwent a change from an unbroken to a broken state), but it has a causative meaning when it has both a subject and an object (*Mary broke the vase* ≈ Mary made the vase break). Children seem to generalize the *break* patterns to other verbs like *fall* and *giggle*, even though they never hear *fall* or *giggle* in the transitive pattern with a causative meaning. So children clearly go beyond what they are exposed to in their input.

The neologism in 1c, *legpit*, is a reasonable extension of *armpit* to a joint in the leg which, when closed, creates a hollow space. We have both *fingernails* and *toenails*, so why shouldn't we have both *armpits* and *legpits*? The statement in 1d takes a common construction in English that *sometimes* allows variable word orders, but *not always*. Notice that *up* and the object (*me*) can go in either order when the object is a full noun phrase (I want you to pick *the book up* / I want you to pick *up the book*), but not when the object is a pronoun

(I want you to pick *it up* / *I want you to pick *up it*).[1] Here, the child simply extended the rule to allow pronouns to vary as well as nouns. So you can see that children are creative with their language: to refute a popular myth about children, they are clearly not sponges, soaking up what they hear.

In the following chapters we'll detail other examples of productions that deviate from adult language, from pronunciation to word formation to sentence building. In these examples and the ones we've given already, we can see that children's apparent errors are logical and rule governed. They are also systematic in another sense: children growing up in very different families, different parts of the country, and even across the world are surprisingly uniform in the types of errors they make. Of course there are individual differences among children just as there are among adult speakers, but the uniformity is what stands out. Moreover, the general stages that children go through are quite predictable, both within languages and across languages. For example, we know that children across all languages will typically produce their first word just after the first birthday (with some minor variation, of course); children will go through a one-word stage until just before the second birthday; they go through a stage referred to as *telegraphic speech* until around age 4; by about age 4 they are fluent speakers of their language; and by around age 5 or 6, children across the globe have acquired the majority of the grammar of their language. This is why a first-grade classroom is one of the noisiest environments on Earth: it is filled with children making the most of their recently acquired grammars.

Not only is acquisition uniform, it is quick. As suggested above, by their sixth birthday children have put together not only the basic rules of their target language but even quite complex rules that allow them to ask questions, embed clauses inside other clauses, create passive sentences and endlessly long descriptions with adjectives, adverbs, and relative clauses. They take a bit longer to learn rules for discourse and pragmatics, like how to tell good jokes and how to know what is appropriate to say, and of course vocabulary words are learned throughout the lifespan. But to become a fluent speaker within three or four years and to acquire a complete grammar within six years is still fast. Compare that to how long it takes you to acquire a second language. With six solid years of learning and speaking a second language, most adults can learn a lot of vocabulary and become quite proficient speakers of a second language. But there will almost inevitably be grammatical structures that they don't quite master. Even proficient second-language learners of English, for example, commonly make errors with the use of the articles *a* and *the* (e.g., *It was very interesting journey* or *working on the similar problem as I*; cited in Dušková, 1983) or with the use of *-s* for verbs with third-person-singular subjects (e.g., *He have to finish his work*). Native English-speaking children of elementary school age do not make errors like this. Adults also often retain an accent in their second language, so some aspects of the phonology of the learned language are simply beyond their reach.

It is important to remember that motivation cannot fully explain differences in learning outcomes between children and adult language learners. Consider immigrants, for example.

They move to a new country and have to learn the language. This is in their economic and social interests—they are highly motivated—and yet very rarely do they fully master the new language. They often acquire some basic skills—enough to get by and function. But typically they do not gain fluency to the degree that they are indistinguishable from native speakers. But the vast majority of children do it. In fact, this accomplishment is so routine that we sometimes struggle to see why it is an impressive feat. Just because all children learn language within six years does not make it any less amazing.

Add to that the fact that learning language seems very easy for children—they just do it. They don't need to be forced to learn language, and they don't complain about it either. "Oh Mom, I don't want to learn new verbs today," said no child, ever. It is something that they simply seem equipped and ready to do. Again, compare this to how we learn a second language as adults. If you immerse an adult in a second language, they are simply not going to learn that language in the same way, as fast, or as easily as children do.

What's more, studies show that children very rarely get correction and explicit feedback about the errors they make. As an adult learning a second language, when you make a mistake, your teacher is likely to correct you and tell you what you did wrong. How else are you going to know that what you said was incorrect? And in writing it comes as the dreaded red ink, but such feedback is important for second-language learners. This is called **negative evidence**: evidence for what is not possible in a language. The opposite (**positive evidence**) is evidence for what is possible in a language.

Children hear lots of positive evidence: every sentence that they hear in their environment is positive evidence and constitutes one piece of data that they can use to learn a language. But we tend not to give children negative evidence (correction). Parents are busy people: they don't have the time or energy to give grammar lessons in between getting dinner ready, doing the laundry, cleaning up the bowl of cereal that just spilled, and so on. Children do not get much negative evidence to help them learn the language, but they get plenty of positive evidence. So how are they to know, when they say something ungrammatical, that this is not a possible sentence in their language? This adds to the puzzle of how children learn language.

Making the problem worse, when parents do make a conscious effort to correct children's errors, they invariably fail. A well-known example, attributed to Braine (1971), demonstrates this failure:

(2) Child: Want other one spoon, Daddy.
Father: You mean, you want the other spoon.
Child: Yes, I want other one spoon, please, Daddy.
Father: Can you say, "the other spoon"?
Child: Other … one … spoon.
Father: Say "other."
Child: Other.

Father: "Spoon."
Child: Spoon.
Father: Now say "other ... spoon."
Child: Other ... spoon. Now give me other one spoon?

The child here seems to be immune to the father's attempts to correct the child. Despite numerous, very explicit attempts to correct the child, the child simply adheres to their original pattern. Moreover, even when correction does have an effect on children, they often attend to the wrong thing or only a part of what the parent is trying to correct. The following example from McNeill (1966) shows this.

(3) Child: Nobody don't like me.
Mother: No, say "nobody likes me."
Child: Nobody don't like me.
[Eight repetitions of this dialogue follow.]
Mother: No, now listen carefully, say "NOBODY LIKES ME."
Child: Oh! Nobody don't LIKES me.

Here, the child produced an error (inserting *don't* and saying *like* instead of *likes*), and the adult tried to correct it by explicitly modeling the correct form for the child. Not only did the child not pay any attention for the first nine repetitions of this back-and-forth, but when they eventually did, they paid attention to the lesser of the two errors in the original sentence. The more obvious one (insertion of *don't*) was simply ignored.

Summarizing, children don't learn through drilling and correction. Their utterances are spontaneous, creative, and systematic, and their ability to spout brand-new sentences is infinite. All children across the globe, irrespective of the language they are born into, acquire language, and they do so quickly, uniformly, and with great ease. How does this happen? The key idea that we focus on in this book is that the task that children have before them in acquiring a language is not to acquire sentences, as people often think. Rather, their job is to acquire *a system*. What you "know" when you know a language is not a list of sentences, but rather an engine that lets you generate an infinite set of words and sentences. But you, as a child, figured out this engine in only a few short years, and with no correction or direct instruction. How did you do that? Without hearing all the words and sentences of your language, you have, by the time you start kindergarten, acquired the machinery that will allow you to create whatever sentences you want for the rest of your life. What this means is that you know some things that were never explained to you, which is the gap that we referenced earlier. The fact that children routinely overcome this gap presents us with a puzzle that linguists refer to as the Logical Problem of Language Acquisition, or LPLA (Chomsky, 1986; Hyams, 1986). The LPLA basically asks: How do you figure out the underlying machinery of language, without it being explained to you, in such a short amount of time, with such ease, so uniformly, and without negative evidence? This

Sidebar 1.1: Different Kinds of Negative Evidence

Children rarely hear explicit corrections (referred to as direct negative evidence). However, some researchers have argued that children make use of *indirect* negative evidence—something that is slightly more common in child-directed speech. This consists of things that indirectly indicate to the child that their utterance was incorrect. For example, if the child says, *The ball falled down*, the parent might say, *Oh, the ball fell down, did it?* This way, the parent did not directly correct the child (no direct negative evidence), but the parent did indirectly indicate to the child that the verb should be *fell*, not *falled*.

This may seem like excellent evidence for the child, but it isn't. It tells the child something, but not what exactly, is wrong. So the child gets very little useful information from a recast, except that there may be something wrong. Moreover, parents do not *always* (or even usually) provide recasts, so what happens to all those errors that are not met with a recast? Do they get learned as part of the grammar? All of this means that a child can't rely on recasts to correct any errors in their grammar, and so it is widely assumed that negative evidence (of any kind) is somewhat ineffectual as a mechanism to explain how children learn language.

issue is addressed in much more detail in the next chapter, but suffice it to say that this logical problem is the foundation for all of modern linguistics.

We're now in a position to go back to our question about whether the input is sufficient. If you end up knowing stuff that wasn't explained to you, and that perhaps wasn't even there in the input (like knowledge of underlying sentence structure), then the input is *not sufficient*. If it's not sufficient, what else helps you acquire language? We've already alluded to the answer, or part of it: babies are born with expectations about how human language works. We'll flesh out this idea in the next chapter, along with some competing ideas in the field.

1.2 The Developmental Problem of Language Acquisition

We have introduced a logical puzzle about how children acquire language: If language input is insufficient, by itself, for children to acquire all the rules of grammar, how is language acquired so quickly, and further, how is it possible at all? The partial answer we just provided is that children are born with certain preconceptions about how grammar *can* work. These preconceptions restrict children's hypothesis space and eliminate a lot of the guesswork that would otherwise be needed (e.g., children don't need to wonder if their language will have hierarchical structure if human language *must* have hierarchical structure). But this answer raises a new puzzle: If children have preconceptions about how language works, why does language acquisition take so long? Even though we argued above that three or four years is a short period of time for language acquisition to occur, if children are preequipped with expectations of how language works, we might ask why the process is not even quicker. This is known as the **Developmental Problem of Language Acquisition**, or DPLA. More specifically, the DPLA is concerned with how and why children go through the particular stages that they do. Why do they progress the way that they do, and not in the infinite other logically possible ways?

Most of this book is devoted to spelling out answers to the DPLA by explaining what children have to learn about language and the developmental stages they go through. We'll see how humans go from being tiny blobs (or, at least, blobs-with-a-predisposition-for-language) to thinking, speaking, articulating, comprehending chatterboxes. There are many facts and theories, and at times it can get overwhelming. If you find yourself wondering, as you read this book, whether there isn't some simple way to sum up language acquisition, remember this: *language acquisition is a process of grammar creation.* All children, almost no matter what circumstances they are born into, create a grammar for themselves. That is what language acquisition is.

1.3 Overview of Chapters

This book is divided into five modules. The first lays some of the theoretical groundwork for the rest of the book, as we have already begun to do in this chapter. Why start with the theoretical arguments, which are abstract and philosophical? We realize that many readers will be eager to get to the facts about children's language. But, as with any scientific inquiry, it is vitally important to be aware of one's theoretical framework before interpreting any observed data. This is because your theories about the world (e.g., about language, about how things move through space, about the structure of matter) form a kind of lens through which you observe stuff and interpret those observations. In fact, theories so thoroughly inform the way we interpret what we observe that we sometimes go to great lengths to explain our observations in a way that keeps them in line with those theories.

To see why this is important, let's take an example from astronomy. For centuries, astronomers' theory of celestial motion held that the planets moved at constant speeds in orbits of perfect circles. This theory enabled them to make reasonably good, though imprecise, predictions of where planets and stars would appear in the sky at different times of the year. Even as they shifted from an Earth-centered to a Sun-centered model of the solar system, they held fast to the old idea of uniform circular motion. This limited their ability to improve the quality and accuracy of their predictions about events like solar eclipses, and they were still unable to satisfactorily explain why planets do not appear to move at a constant speed as they orbit the Sun. Astronomers were forced to come up with models of greater and greater complexity to explain their observations because their theory told them that planets had to move in perfect circles. It was not until astronomers abandoned the idea of uniform circular motion that they were able to develop and accept a new theory of motion (planets, in fact, move in elliptical orbits at speeds that vary with their distance from the Sun) and move on to more successful approaches to interpretation, explanation, and prediction. This shows that the theory that you hold significantly impacts how you interpret observable facts. The facts about planetary motion have not changed, but our theories have.

New theories enable us to see more useful ways to organize and interpret those very same facts.

Numerous other examples could be taken from medicine, physics, and other scientific disciplines, including linguistics. Science is the endeavor to *know* and understand our world (the word *science* comes from the Latin word *scire*, 'to know'), and we understand the world as we do via our theories, which lead us to have expectations about how stuff behaves. Scientists may disagree with one another about which theory is right, and then they perform experiments to test the predictions each one makes—this is how science moves forward. But in order to develop testable predictions, it is essential to be clear about what your theory is in the first place.

The second module explores children's development of the sound system of language. Chapter 3 focuses on infants and young children's **perception** of speech sound, including their segmentation of the speech stream and how they begin to perceive speech sound in categories (these are **phonemes**). Chapter 4 then turns to speech **production**. We look at how children figure out which phonemes are found in their language and how they make regular, systematic alterations of adult words when they pronounce them. Finally, we look at two approaches to **explaining** children's phonological systems, one in terms of **rules** and the other in terms of **constraints**.

The third module proceeds to the word level: How are words learned, and how are words formed? Chapter 5 tackles the first question, namely, how children figure out what words mean (also called lexical **semantics**). We give some descriptions of how children's early vocabularies take shape, but we spend most of that chapter probing the constraints and **biases** that limit children's guesses about what a word might mean so that they aren't stuck going through the infinite universe of possibilities. After all, if you are going to learn upward of eight new words per day, while also learning to walk and feed yourself, you can't be combing through an infinite set of hypotheses for each word meaning.

Besides the question of how word meanings are learned, words have their own structure, known as **morphology**. In chapter 6 we begin with a description of how children acquiring English begin to piece together the subparts of words. The description is followed by some theories about how children form generalizations about morphological patterns, or **paradigms**.

The fourth module is concerned with sentence structure, or **syntax**. This large chapter has four subparts. The first, on **argument structure**, looks at how children figure out the basic skeleton of a sentence: how the constellation of arguments (or noun phrases) arrange to make up a basic sentence and what semantic roles those arguments typically play. The second and third subparts have to do with what is known as **functional structure**. We focus a good deal on children's patterns of marking (or not marking) **tense**. This phenomenon has consumed a great deal of theoretical debate in linguistics, but more importantly, it can potentially reveal that children's underlying knowledge of sentence structure goes far beyond what one hears on the surface when they talk.

The remaining sections of chapter 7 look at other aspects of functional structure (how children acquire **negation** and **questions**, for example, and some more **complex constructions**) and, finally, how children interpret **pronouns** (words like *him* and *her*).

While the first module sets the theoretical stage for analyzing children's language, and while modules two, three, and four give an empirical and theoretical overview of how language is acquired by monolingual children, the fifth and final module takes us beyond the basic case of a child acquiring a single language under typical circumstances. In chapter 8 we explore how language is acquired under a host of nontypical circumstances: from a nontypical age (**late first-language acquisition**—acquiring a first language outside of infancy and early childhood) to a nontypical set of sensory inputs (acquiring language without sound or without sight—i.e., in **deaf** or **blind** children) to a nontypical set of cognitive faculties (acquiring language with certain cognitive disorders, such as **specific language impairment** or **autism**, or even after a **hemispherectomy**—being without half of one's brain!).

Chapter 9 begins with a situation that is fairly typical but nonetheless falls outside the ordinary circumstance that the rest of the book assumes: acquisition of more than one language. We look at how **bilingual language acquisition** occurs, whether the languages are presented simultaneously or in sequence. In that chapter we also explore how languages can come and go, either across a lifespan within an individual, as in **heritage speakers**, or across generations within a language community, as happens when languages become **endangered**. But because this book is really about language acquisition, not language death, we end that chapter by looking at how endangered languages can become **revitalized**.

Some readers might find that the chapters go into more detail than they need; for these readers, the next-to-last section of each chapter (before the summary section) can be skipped. Some readers, on the other hand, might feel that the chapters do not go into quite enough depth; for these readers, each section ends with some suggestions for further reading. And for all readers, each chapter has some sidebars that are intended to provide additional information relevant to the chapter and to invite the reader to explore a "food for thought" kind of question. At the end of each chapter, there are some exercises designed to test your understanding of the material in that chapter and to offer some practice analyzing children's language.

Finally, there are two appendices: appendix A provides charts of the International Phonetic Alphabet, which we use throughout the book to indicate the pronunciation of a word. For any student who is a little rusty on all those symbols, the charts are there for your reference, and there are some links to webpages where you can hear the sounds associated with those symbols. Appendix B summarizes some of the experimental methodologies that are widely used in the field of language acquisition. In various chapters we will mention experiments that employ these methodologies, and appendix B is meant to offer additional detail about how those experiments are run, should you want to design some of your own.

We've written this book because, even after many years of probing children's knowledge of language, the mystery of this singular human achievement still fascinates us. We hope that in reading this book you'll not only learn some facts about how children become powerful wordsmiths but, perhaps more importantly, come to wonder at the process of it and appreciate all the unanswered questions. And maybe you'll think of new ways to answer them.

1.4 Further Reading

Crain, Stephen, and Diane Lillo-Martin. 1999. *An Introduction to Linguistic Theory and Language Acquisition*. Malden, MA: Blackwell Publishing.

1.5 References

Braine, Martin D. 1971. The acquisition of language in infant and child. In Carroll Reed (ed.), *The Learning of Language*, pp. 7–95. New York: Appleton-Century-Crofts for the National Council of Teachers of English.

Chomsky, Noam. 1986. *Knowledge of Language: Its Nature, Origin and Use*. New York: Praeger.

Dušková, L. 1983. On sources of errors in foreign language learning. In Betty Wallace Robinett and Jacquelyn Schachter (eds.), *Second Language Learning: Contrastive Analysis, Error Analysis, and Related Aspects*, pp. 215–233. Ann Arbor: University of Michigan Press.

Hyams, Nina. 1986. *Language Acquisition and the Theory of Parameters*. Boston: Reidel.

McNeill, David. 1966. Developmental psycholinguistics. In Frank Smith and George A. Miller (eds.), *The Genesis of Language: A Psycholinguistic Approach*, pp. 15–84. Cambridge, MA: MIT Press.

2 Theoretical Approaches to Studying Language Acquisition

In chapter 1 we laid out the basic Logical Problem of Language Acquisition (LPLA): children acquire language quickly, universally, with relatively few errors, showing remarkable uniformity, systematicity and adherence to rules, and not needing any kind of correction or grammar lessons. How do children do this? Researchers address this question in numerous ways, but the most basic division in the field centers on the question of whether language learning is primarily driven by the input that children hear or by innate knowledge about human language. In this book we adopt the point of view that emphasizes innate knowledge, and in the first part of this chapter, we flesh out that perspective more fully. This view is known as the Universal Grammar (UG), or "generative," approach to language acquisition, and it explains how children acquire language despite the LPLA and why they acquire language in the precise way that they do (the Developmental Problem of Language Acquisition, or DPLA). But it is important for students of language acquisition to know about other theoretical approaches, so in the second part of this chapter we address some theories that take a different perspective, referred to as input-based approaches. These approaches emphasize the regular patterns found in language input and explain language acquisition in terms of children's ability to recognize and make sense of these patterns. We will investigate two such theories of learning that invoke general learning mechanisms—learning mechanisms that apply not only to language but to other domains within human cognition such as vision, number, and spatial knowledge. Although we believe the input-based approaches are less suited to addressing the LPLA and the DPLA, we'd like to emphasize that they are legitimate and worthy of consideration. As we shall see, they provide an intuitive, novel, and extremely clever way in which children learn certain aspects of language.

It is important to recognize that among input-based approaches, there are many different viewpoints, specific claims, and assumptions about language learning, some more extreme than others. As an example of an extreme approach, the school of psychology known as behaviorism held that *all* language learning (and in fact, all learning) was based on the input. On this view, the learner was simply an empty vessel to be filled with the experience of hearing language in the input. Learning was considered nothing more than forming a stock of habits and drawing associations between a stimulus and a response—much like the stimulus appraisal that all animals can do. Think: Pavlov's dogs. This view is no longer taken seriously as an account of how language is learned, and many people who study animal cognition do not believe it accounts for animal learning either (see Gallistel, 1990).

Nowadays all researchers of language agree that there must be some structuring or organizing activity on the part of the child in order for language learning to be successful. But don't let this fool you: not everyone agrees on everything. There remain deep and important divisions in the field, and this is what makes the field of child language acquisition so interesting and vibrant. The debates and the opportunity for discovery are stimulating and enticing, which is precisely why so many researchers enter the field every year.

We begin with a discussion of the UG-based approach to language acquisition. We then introduce the first input-based approach, referred to as Statistical Tracking, and then the second input-based approach, which we label the Constructivist approach.

2.1 Universal Grammar

2.1.1 Motivations for Universal Grammar

Language input plays a critical role in language acquisition. For one thing, children acquire the language they hear rather than some other language—that's just common sense. On one level, the common explanation that children are "sponges" is totally true—they soak up the input in their environment, and it seems effortless. But is that really how children acquire language? Do they just memorize their input, simple as that? Well, as we showed in chapter 1, the answer to that question is categorically no. Children don't just mimic their input, and their knowledge of language is not just a catalog of all the sentences and words they've heard. Children go far beyond what they hear, and this is the key observation that motivates the theory of Universal Grammar (UG). More specifically, the primary observation that drives UG approaches to language acquisition is that there are certain aspects of language that simply *cannot* be learned from the input.

Children acquire language

with **N**o correction

Although input is impoverished
Through biases
Uniformly
Rapidly
Easily

To understand why this is the case, we need to first explain two very important concepts that form the basis for all generativist thinking:

(i) The Problem of Induction
(ii) The Poverty of the Stimulus

Once we have laid this foundation, we will go into some of the details of the UG-based approach, explaining the various camps within this approach, what they all have in common, and what some of the differences are. Like within any vibrant intellectual group, there is not complete unanimity in ideas. Rather, the UG-based approach is filled with new ideas and is constantly changing and adapting to new findings. It can be quite frustrating for newcomers and outsiders to get a handle on the nuances of the most modern theory (as it is constantly evolving), but this is also what makes it such a dynamic field. Nonetheless, the foundational ideas remain constant, which we now discuss.

2.1.1.1 The Problem of Induction

We've just noted that language input is essential for language acquisition, but we don't mean that children simply mimic their parents. Rather, what we mean is that the input provides words and other language-particular information about the language being acquired, and it serves as the evidence on which children test their innate hypotheses about language. The following analogy from Noam Chomsky may be helpful, where the "key" is language input and the "engine" is knowledge of language (in Piattelli-Palmarini, 1980, p. 172): "It's a little like turning on the ignition in an automobile: It's necessary that you turn the key for the car to run, but the structure of the internal combustion engine is not determined by this act." In other words, you need some input (the key, the gasoline), but the engine, the basic framework for language, is already there. The job of linguists is to make explicit what the engine is like and how exactly the key works to turn it on: *How* do children make use of the language they hear spoken around them to flesh out their innate biases? It is the job of linguists because these innate biases are considered to be specific to language—they don't apply to any other domain other than language. In section 2.3.1, we will see some examples of domain-general learning mechanisms (mechanisms that apply across multiple domains of cognition, not just language) that could take input and turn it into knowledge. But domain-general learning mechanisms are typically far too broad for language. Processes like analogy and association formation (usually considered to apply across domains of cognition) are so all-purpose, so overpowering, and so wide in scope that they create more problems than they solve. It's a little like trying to slice your sandwich in half using a chain saw: the instrument does the job (it cuts), but it does way more than you intend. The problem of induction is one such problem that we get with processes like analogy. We will explain this more below.

Sidebar 2.1: Induction and Universal Grammar

> The problem of induction is a problem only if we assume the child has none of the generative system in place to begin with. If the child already knows some of the system, then the problem of induction is mitigated to some degree. This is why the UG-based approach does not face the same problem of induction that input-based approaches do.

The language experience for a child consists of a series of communicative acts. Adults and others in their environment are constantly directing their intentions to the child, and this is accompanied by bits of language. One of the child's jobs is to pick up on that language and figure out what it means. But the child's most important job is to go beyond simply figuring out the meaning/intention of the adult's individual utterances. The child must figure out what the *system* behind those individual utterances is, so that they can eventually produce utterances like them (and any others that constitute well-formed language). It's a little like that old saying: *Give a man a fish and you feed him for the day; teach a man to fish and you feed him for a lifetime.* Likewise, if the child decodes individual utterances, they get immediate but short-term rewards (giving a man a fish, in this metaphor). But if and when the child masters the underlying system of language (learning to fish, in this metaphor), then the child has control over all of language, for life.

Moreover, we know that speakers have a creative, *generative* capacity for language—that is, after all, one of the hallmarks of human language. We saw in chapter 1 that children don't simply memorize all the sentences that they have heard and then reuse them through the rest of their lives. How do we know this? Well, we are able to understand sentences that we have never heard before—we do all the time, in fact. This means that language does not consist of a really long list of sentences that we all know. Rather, language consists of an abstract system that allows us to generate all possible grammatical sentences. So what the child is developing must be this abstract system underlying language. And importantly, the only real external evidence the child has for the nature of this abstract system is the individual examples that constitute the linguistic input. The child never gets to *see* the underlying system—it is abstract, and no one ever talks about it explicitly—in fact, adult speakers are not even aware of this abstract system.

Induction: the process of going from individual examples to a general rule

So if we accept that the task facing the child is to acquire an abstract system that can generate all the sentences of their language (and doesn't incorrectly generate impossible sentences) and that the only evidence before the child are individual examples of language, we need to ask how the child does this. This is precisely the process of **induction**: going from examples to a general rule. Induction has long been recognized as difficult, perhaps impossible, without the aid of some sophisticated help for the learner. Anticipating our discussion, constructivists argue that domain-general learning mechanisms suffice to overcome this problem of induction, whereas the UG-based approach claims that (i) domain-general mechanisms fail to solve this problem and (ii) children are born with biases that help avoid the very real pitfalls of induction.

So what is so difficult about induction anyway? At first blush, it seems pretty easy to deal with: you get examples, you observe what those examples have in common, and you postulate a generalization that captures your observations. But it is not so simple. Let's exemplify this with some nonlinguistic cases before we move on to the more pertinent linguistic cases. Consider the series of numbers in 1:

(1) 1 . . . 2 . . .

What is the next number in this series? Let's assume these two numbers have been put into this series on the basis of a generalization. Which generalization was used? It could be 'Add one to the previous number' (in which case the next number is 3), or it could be 'Double the previous number' (in which case the next number is 4), or it could even be 'Add 9 to the previous number, divide by 2, and subtract 3' (in which case the next number is 2.5). The point is that just by being exposed to individual examples, you can't immediately tell how those examples were intended to be related to each other. There are too many possibilities.

Here is a slightly more complex example. Imagine we have a deck of cards and we play a game: The dealer chooses five cards (not randomly, but she gets to pick the cards she wants), and she shows you four of them. Your job is to guess what the fifth card is. The only restriction is that she must pick all five cards on the basis of some rule. That is, you should be able to guess her fifth card if you can guess the rule she has in mind. So our dealer thinks for a moment, and then she picks out her five cards. She shows you the first four cards:

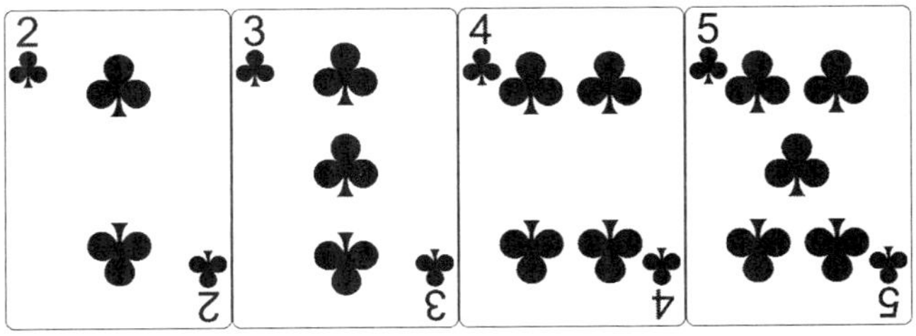

Looking at these four cards, the 2 of clubs, the 3 of clubs, the 4 of clubs, and the 5 of clubs, what would you guess is the fifth card? A 6 of clubs? That would be a good guess, but you'd be wrong (an ace of clubs would be an equally good guess, and equally incorrect). Your guess is likely based on the idea that the dealer picked a numerical sequence of cards that were all clubs. A 6 of clubs would be the highest in the sequence, thus completing the run. This makes sense because the numbers are sequentially organized, and the cards are all clubs. But sadly for you, here is the fifth card:

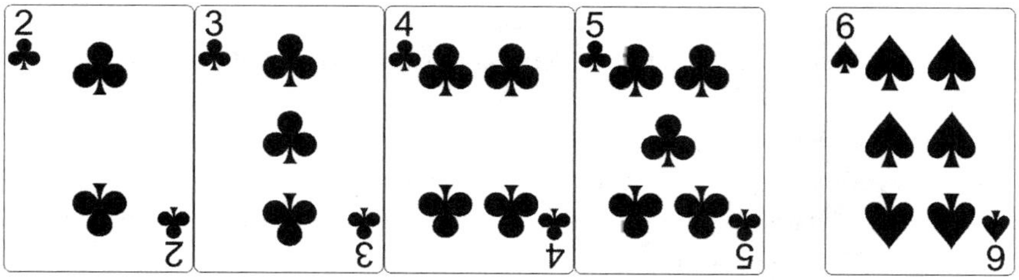

The rule the dealer had in mind when she picked her five cards was "a sequence of cards of any black suit." Notice that the original four cards that you saw were consistent with that rule, but you did not immediately notice that. (You might have, actually, but for the sake of this explanation, let's assume you didn't.) It just so happened, by chance, that the first four black cards the dealer picked were of the same suit, but that was not her intention. So while your guess may have been the most likely, it wasn't the one the dealer had in mind when she selected the five cards. For poker players, the incorrect rule you had in mind was the straight flush, while the one the dealer had in mind was the more likely straight (with the slight stipulation of the same color).

But how were you to know that, you ask? That's the point: there is no way for you to have known that the dealer had a different rule in her mind. It's in her mind, after all: you can only go by what's in front of you. The data you had in front of you (the first four cards) were consistent with at least two hypotheses, and you picked the most obvious one to you.

Hypothesis 1: sequential cards, all of one suit (incorrect hypothesis)

Hypothesis 2: sequential cards, all of one color (correct hypothesis)

This may not seem fair. It may seem like that rule was very arbitrary. And you'd be right—that was a totally rigged game! In language, the rules themselves are not completely arbitrary, but they do vary from language to language, and the child has to figure them out on the basis of what they hear in the input. That is, the child is faced with a series of examples, and they have to figure out how those examples are tied together. The data they get is necessarily ambiguous—it is consistent with multiple hypotheses, just like the cards above are, as we shall see shortly. The child needs to figure out not just *any hypothesis* that is consistent with the data, but the *correct hypothesis*—the abstract structure that generated the utterance. What is the underlying system that allowed the adult to use that particular form? Given the complexity of language and the variation that we see in the input to different children, this is (arguably) an impossible task without something else to help the child. Let's now take a look at a more relevant example of the problem of induction from language.

2.1.1.2 An Example from Language

Let's bring this into focus by looking at a classic linguistic example of the problem of induction, initially put forward by Chomsky (1971) and later taken up by Crain and Nakayama (1987). Chomsky noted that most (syntactic) constructions in human language are compatible with many, many possible analyses (at least to the naïve learner), only one of which is actually correct. In fact, the entire field of syntax is devoted to exactly that: figuring out what the correct underlying structures of various word sequences are. There are so many ways to analyze any given sequence of words that thousands of smart, obsessively hardworking people devote their professional lives to this endeavor. The example Chomsky used is yes-no question formation in English.

In order to form a yes-no question, one simply takes the auxiliary verb (*is* in 2a) and moves it to the front of sentence 2b.[1] Notice that other things like modals (in sentences with no auxiliaries, for example) might move to the front of the sentence, as in 3a–b.

(2) a. The man is in trouble.
 b. Is the man in trouble?

(3) a. I can go.
 b. Can I go?

From these examples, one might hypothesize that the rule in English for yes-no question formation is actually quite simple: move the first verbal element (auxiliary verb *is* or modal verb *can* in the examples above) to the front of the sentence and add question intonation.

Hypothesis 1: Linear Order Hypothesis

Move the first verbal element to the front of the sentence. This hypothesis works very well for the vast majority of yes-no questions in the child's environment, even sentences like those in 4, in which there are multiple verbal elements. Notice that 4b, in which the first auxiliary is fronted, is acceptable, but 4c, in which the second auxiliary is fronted, is unacceptable.

(4) a. The man is in trouble now that he is in custody.
 b. Is the man in trouble now that he is in custody?
 c. *Is the man is in trouble now that he in custody?

So Hypothesis 1 works very well. This hypothesis is structure *independent* in that it does not make reference to any syntactic structure. Rather, it is a hypothesis that makes reference to linear order. It states that in order to form yes-no questions, you need to identify the *linearly* first verbal element (linearly first = starting from the first word in the sentence, moving to the next) and move it to the front of the sentence. But there is a big problem with this hypothesis: human language syntax is not a linear system but rather a hierarchical structure (see figure 2.1). That is, it is organized into a stratified structure, typically

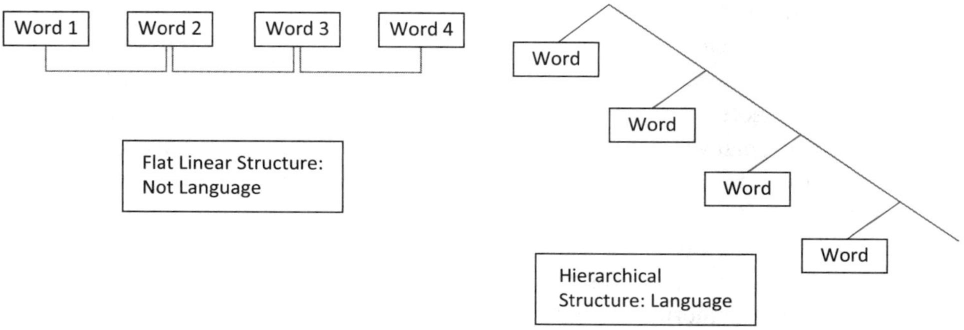

Figure 2.1
Linear vs. hierarchical sentence structure.

represented as syntax trees. This is simply how syntax works (any introductory course on syntax will teach you this), so a linear hypothesis is bound to be incorrect.

To see why this linear hypothesis is clearly incorrect, let's look at another example. Hypothesis 1 (linear order) works for examples 2–4, but it falls apart with examples like 5, which also has two auxiliary verbs but in a slightly different configuration. Applying the above linear principle faithfully (move the *first* auxiliary to the front of the sentence) to 5, we get the incorrect 5b. Clearly, something is wrong here. What we actually need is a principle that somehow gets us 5c.

Linearly first aux → | Linearly second aux →

(5) a. The man who is in trouble is now in custody.
b. *Is the man who in trouble is now in custody?
c. Is the man who is in trouble now in custody?

To account for the sentences in 5, we can't have a principle that says, 'Move the linearly second verbal element' or 'Move the linearly last verbal element' since that would directly conflict with Hypothesis 1 and generate the wrong result for sentences like 4. So we are in a bit of a quandary: How do we reconcile example 4 with example 5? Don't worry—there is a solution. And the solution is a rule that makes reference to the hierarchical structure of *syntax*; in other words, it is a **structure-dependent rule**.

The first thing to notice is that in sentence 5a, the first auxiliary verb is actually part of the subject of the sentence, which is *The man who is in trouble*. That means this first auxiliary is not the main auxiliary of the main sentence. The main sentence in 5a is *(Subject) is now in custody*, where (Subject) = *The man who is in trouble*.

(5) a. [The man who is in trouble] is now in custody.

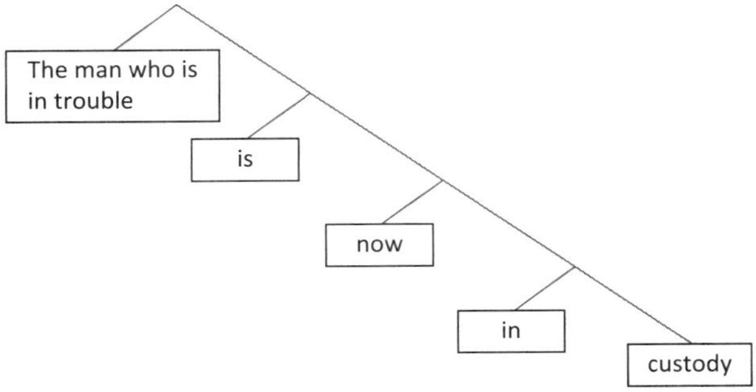

Looking at this sentence this way, the auxiliary that should be moved to the front of the sentence is the auxiliary verb that belongs to the main sentence. The linearly first auxiliary is utterly irrelevant for this purpose, since it occurs somewhere inside the subject of the sentence and so is invisible to the process of yes-no question formation when the rule makes reference to its structural position. Using linear order blindly (as in Hypothesis 1) does not allow for the observation that in this case, the important auxiliary is the linearly second one, while in example 4, it is the linearly first one. So if we reformulate our hypothesis as below, we get the correct yes-no question in 5c as well as every other yes-no question in the language, including 2–4.

Hypothesis 2: Structure Dependent Hypothesis

Move the verbal element in the main clause (i.e., after the subject) to the front of the sentence. This is called the structure dependent hypothesis because it makes reference to syntactic structure, not to linear order. It is vital to spell out the rule this way because the majority of yes-no questions that children hear are single-auxiliary questions and thus compatible with both hypotheses (and indeed other hypotheses, if you put your mind to it). So why don't (at least some) children make errors with questions like 5b, **Is the man who in trouble is now in custody?*, in which they front the wrong auxiliary? This kind of error has not been described in the literature, and we must ask why. If children are genuinely learning from the input and not guided by any predispositions, why don't they, at least some of the time, go down this incorrect path?

Before we go any further, let's ask if it is true that children do not actually make errors in which they front the wrong auxiliary. We suggested that they don't, but where's the evidence? Fortunately, this has been tested empirically. Crain and Nakayama (1987) tested thirty children aged 4–6 years old. They elicited yes-no questions from children using a very clever protocol. They prompted children to ask a Jabba the Hutt action figure questions, and Jabba then answered yes or no. The experimenter provided children with a prompt like 'Ask Jabba if the dog that is sleeping is on the blue bench.' This protocol is clever because it provides the child with a full model of a sentence containing multiple auxiliary verbs, but because the question is embedded in a conditional if-clause, the model sentence does not have a fronted auxiliary verb. So all the child has to do is take that model sentence and choose one auxiliary to move to the front of the sentence.

The short version of their results is that children produced many correct yes-no questions (fronting the main-sentence auxiliary). They also made some errors (as children this age typically do), but *never once* did any of the children produce errors consistent with the linear order hypothesis—that is, '*Is the dog that sleeping is on the blue bench?' This shows that children do not consider the linear order hypothesis, and so they seem to be constrained by structure from very early in development (as young as 3;2 in Crain and Nakayama's study).[2]

In sum, the example shows that simple constructions like yes-no questions are logically compatible with several different rules or analyses. There is no principled way to avoid adopting an incorrect hypothesis on the basis of input alone if all you have access to is the input and general learning mechanisms that make no reference to linguistic structure. What is needed is something that guides you to the correct hypothesis; in this case, that something is actually a very broad-level bias toward *structure*. The bias simply says, "When you hear language, think structurally, not linearly." That is the essence of UG—a learning bias that guides children in their analysis of the linguistic input.

Notice that structure dependence is a linguistic property. It is not a general learning mechanism—nowhere else in cognition does such a mechanism play any role. This is because the "structure" referred to by the principle of structure dependence is uniquely and purely linguistic in nature.

But there remains one other way that children could reach the correct hypothesis for yes-no questions without being guided by UG: perhaps the input provides sufficient evidence to solve this problem after all. As we shall see in the next section, the input is indeed rich in many ways, but it is severely impoverished in other ways, so the input is not going to save this particular situation. Instead, the input is so severely impoverished (in the relevant sense) that it actually forms the second motivating factor for the UG approach: the poverty of the stimulus.

2.1.1.3 The Poverty of the Stimulus

The term **poverty of the stimulus** refers to the idea that the input to children is missing certain important information. *Poverty* here means poor or impoverished, and *stimulus* refers to what children hear (the input to children). So the expression 'poverty of the stimulus' means that the input is poor or insufficient. Note that the stimulus being impoverished does not mean that parents fail to speak to their children with enough language. The

Sidebar 2.2: More Input Does Not Help!

How do hundreds of more examples like the following help disambiguate between a structure-dependent rule for yes-no question formation and one that depends on linear order?

- Is the cookie good?
- Did Daddy leave already?
- Is the TV on?
- Is that my phone ringing?
- Are you hungry?
- Did you eat all your carrots?

And so on. (Answer: They don't.)

poverty of the stimulus makes a claim about certain kinds of evidence being absent from the input, not really about the communicative richness of language.

Think back to the little card game we played earlier. The trouble you had with the game was that the four cards the dealer put in front of you were consistent with multiple hypotheses. And the dealer selected the cards such that the most obvious hypothesis was not the correct one. Unfair, true, but it was illustrative of the problem of induction and the process of inductive reasoning: drawing conclusions based on bits of evidence.

One reason this was difficult was that the dealer only gave you four cards from which to build a hypothesis. What if she gave you a dozen cards? Would that make it easier? Probably. So opponents of the UG approach often point out that the problem of induction may be solved by *more evidence*. The more evidence a child encounters, the easier it is to induce the correct properties of language.

Returning to the yes-no question issue, perhaps more evidence would allow the child to solve this problem without any innate bias toward structure. Possible? Sure, but let's think about this a little more. What exactly does "more evidence" amount to? Simply hearing more yes-no questions of any type would *not* help. If a child got a million yes-no questions, all with a single auxiliary, they would be no closer to avoiding the problem of induction than a child that gets a single yes-no question, because all single-auxiliary yes-no questions are compatible with at least two hypotheses. And, in fact, we know that children do hear lots of yes-no questions in the input. The problem is that these are overwhelmingly single-auxiliary questions. The argument from the poverty of the stimulus is not an argument about the *amount* of input that children get. Rather, it is about the absence of the precise kinds of evidence needed to overcome the problem of induction. As Chomsky (1971) points out, the only evidence that would inform the child that a linear order hypothesis is incorrect is questions of the kind in 5c:

(5) c. Is the man who is in trouble now in custody?

This sentence type, and this sentence type alone, would tell the child that using Hypothesis 1 (move the first verbal element) is incorrect. This would be akin to our batch of cards (the 2-3-4-5 of clubs sequence) containing one card that was a spade, thereby informing you that we were picking cards by color, not by suit. In the sample of cards below, the four of spades is the crucial evidence that tells you that the dealer did not have suit in mind but just a sequence of black cards.

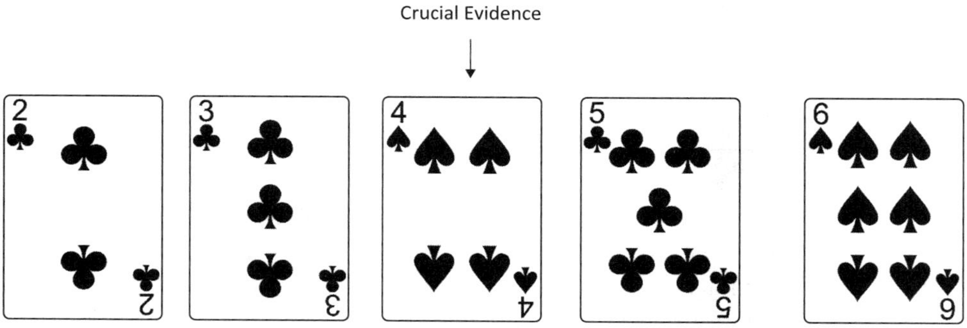

If the child gets disambiguating evidence of the kind in 5c, then they might be able to tell that the correct rule for yes-no question formation must make reference to syntactic structure. But how often do you think children hear questions like 5c? These are complicated patterns, and most parents don't talk to their toddlers like this (cf. Pullum and Scholz, 2002). In fact, Legate and Yang (2002) found that in the speech to two children, double-auxiliary yes-no questions in which the second auxiliary had been fronted over the first (i.e., examples like 5c) occurred exactly zero times out of a combined 66,871 utterances, of which there were a combined 29,540 questions. There were some wh-questions in which two verbal elements occurred, with one fronted over the other (e.g., *Where's the part that goes in between*), but these were exceedingly rare, with a combined rate of 0.06%.[3]

That's rare, to be sure, but the argument from the poverty of the stimulus is not really about the precise rate of rare evidence. With yes-no questions, sure, the evidence against Hypothesis 1 is rare, but the important point is that 99.94% of the data is consistent with multiple hypotheses. With the statistics stacked so overwhelmingly toward ambiguous patterns, surely at least some children, at least some of the time, will misanalyze their language. Crain and Nakayama show that this simply is not the case.

So the argument from the poverty of the stimulus says that (i) language is inherently ambiguous in terms of how one could analyze it and (ii) children simply never see this ambiguity because they are predisposed to analyze language in a structural manner. In a sense, children have blinders on that ensure that they only see the analyses that are consistent with how we know human language works. And because of that, they never even consider the hypotheses that are incompatible with UG properties, like a linear rule. This accounts for why children so expertly avoid misanalysis of their language and so unwaveringly adhere to principles of UG.

What this shows is that the kind of patterns that children consider is *restricted*, and not simply directed, by the input. In that sense, this is very much like certain other phenomena in animal cognition. Rats can learn to associate a flash of light with an electric shock and a funny taste in their water with an episode of sickness. But it turns out that not all types of associations can be learned. For example, while rats will learn to associate a flash of light with an electric shock and a funny taste in their water with an episode of sickness (and so they will avoid the funny-tasting water), they *cannot* learn to associate a flash of light with getting sick or funny-tasting water with electric shock, even if the two events are presented with a high degree of regularity. If learning were simply a matter of associating one thing with another, then rats should be able to learn that a flash of light could make them sick, just as it could herald an oncoming electric shock. But this does not seem to occur to the rats. In other words, the rats' set of hypotheses about what can make them sick is *restricted*.

We've seen that there is both strong empirical evidence and strong logical evidence for the argument from the poverty of the stimulus. But still, we might ask: How *plausible* is it that children are predisposed to see language in a particular way? Cutting right to the chase, this idea is *very* plausible. Not only do we have parallels in the cognition of other animals, but we also have countless examples of exactly that kind of thing in other domains of human cognition, as noted by many linguists (e.g., Fodor, 1983). Consider the most obvious one: optical illusions. Optical input to the eye, like linguistic input, is massively ambiguous in terms of how it could be interpreted. In terms of a physical specialization, we see the world in the colors of the rainbow because our eyes have been specialized to see only those wavelengths of the light spectrum. But more relevant to us are optical illusions. Take the famous "horizontal lines" optical illusion (figure 2.2). This figure consists of perfectly horizontal lines with unevenly aligned black and white squares. Our minds find vertical unevenness unnatural to parse, so we perceive the horizontal lines as if they are sloped at the edges. Let's emphasize this point: the horizontal lines in figure 2.2 are perfectly parallel—your mind is making them look loopy! This shows that the visual domain of our mind has preferences—it doesn't like vertical unevenness—and it imposes those preferences on what we see. Our minds are predisposed to interpret optical information in one particular way, and that analysis of the optical input just happens to be inaccurate, resulting in an optical illusion.

Vision, therefore, has a highly articulated internal structure: it is so specialized to see the world in one particular way that we can find hundreds of ways to trick our minds. And judging by the millions of internet hits for the term **optical illusion**, we get great pleasure out of it. Moreover, these specializations that result in optical illusions are completely unique to vision. They serve no purpose outside of vision and are a clear indicator of domain-specific architecture. This is also true in other domains of cognition—we are biased to interpret data in particular ways, ways that are suited to the needs of that particular domain of cognition. Moreover, none of this is learned behavior. We don't learn to see optical illusions—we simply do. Such optical illusions are universal, present from the earliest testable ages, and thus are widely assumed to be innate.

Figure 2.2
A common optical illusion. The horizontal lines don't look parallel (but they are). The black and white are not aligned, but our eyes naturally follow the vertical lines and compensate for the unevenness by interpreting the horizontal lines as sloping.

If it's the case that other domains of cognition have domain-specific biases innately built into them, then why would this be a strange thing for language to have? The answer is that *it isn't strange*. It is perfectly normal and is a reasonable basis on which to pursue a research program, which is what the generative approach to language acquisition does.

In sum, then, the arguments from the poverty of the stimulus and the problem of induction lead us to the conclusion that language acquisition is impossible if it is driven exclusively by what is heard in the input. But language is surely acquired by every typical child in typical circumstances (as well as many atypical circumstances). What's the solution to this puzzle? Well, the solution is that the mind is biased to interpret data in a manner that is consistent with the structural properties of language. Furthermore, since the biases themselves are not explicitly taught or derivable from experience of the world, and since all children have them, we make the further claim that they are innate—part of what it means to be born a human. This is the essence of the UG approach. As we will see below, the Constructivist approach to language acquisition (a variety of input-based approaches) either downplays the importance of the problems of induction and poverty of the stimulus or addresses them only partially.[4]

So far we have described UG in quite broad terms: there are innate biases that prevent children from considering analyses of their language that are incompatible with human grammar. This view of language has as its roots the theory proposed by Noam Chomsky in the 1950s and 1960s, which laid out the architecture of grammar, and the concepts of competence and performance, which play an important role in UG-based approaches to language acquisition. At this point let's make these ideas more explicit.

2.1.2 The UG-Based View of Language: A Computational System

On the UG-based approach, language consists of two major components: (i) a lexicon and (ii) a computational unit. The lexicon contains all the lexical entries for the language: nouns, verbs, inflectional morphemes, and so on. The computational unit consists of a series of procedures that combine those lexical entries (see figure 2.3). For each sentence, the desired lexical items are extracted from the lexicon and combined to form sentences according to a series of rules that will differ slightly from language to language. These procedures build a structure that adheres to certain fundamental properties that linguists have discovered about human language. For example, we know that human language is organized into hierarchical structures (represented as trees). The procedures within the computational unit will build structures of this sort. We also know that language involves various kinds of dependencies, sometimes local dependencies (such as gender agreement between an adjective and a noun, e.g., *una casa bella*, 'a-fem. house-fem. pretty-fem.') and sometimes lengthy dependencies (such as the one created when a wh-element is moved to the front of the sentence, across multiple clauses, e.g., *What did the man say that Mary thinks that John ate [what]?*). There are various restrictions on such dependencies, and the procedures of the computational unit of grammar are designed to obey those restrictions.

This system is highly appealing since it explains why language is infinite: it takes atomic lexical items and combines them in endless ways. It also allows for the creativity and fluidity of language while at the same time providing a structure that limits the kind of variation language might exhibit.

This view of human language, then, is based on a strong computational unit that essentially determines the structure of human language. Because of the very architecture of this computational unit, some things simply never occur in human language. For example, language is not linearly ordered because this computational unit does not work that way. Its architecture is such that all it can do is build hierarchical structures, so syntax is not ever going to make reference to a linear order rule. This is like the rat never considering that funny-tasting water might cause an electric shock or that a flash of light might make it vomit.

What other properties does the computational unit provide? This is an empirical question, and one that linguists are continually working to answer. Whatever those particular properties of the computational unit turn out to be, the UG hypothesis states that children already have a fully formed computational unit. They are born with the procedures that create well-formed linguistic units, and they do so without any difficulty whatsoever. But if that is the case, why don't children speak perfectly just as soon as they acquire some of the lexicon? There are several reasons for this, but before we can explore these reasons, we need to understand a foundational distinction within the UG-based approach to language: competence versus performance.

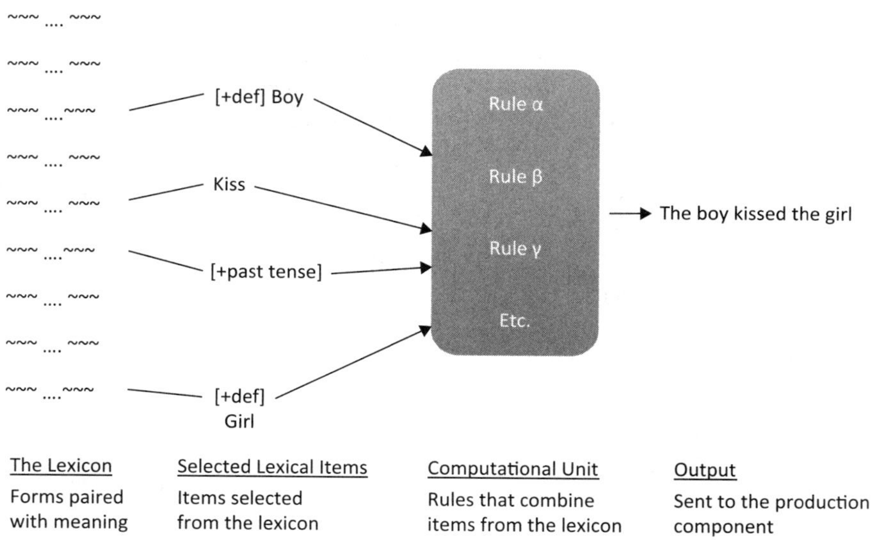

Figure 2.3
Words are drawn from the lexicon and fed into the computational unit, which then applies various procedures to produce the output sentence.

2.1.3 Competence versus Performance

The distinction between competence and performance was established most famously in 1965 by Chomsky, and it is important for students of generative linguistics to understand it. The term **linguistic competence** refers to the idealized state of one's linguistic potential: it is the knowledge base that allows you to produce and understand any sentence of your language. A native speaker's *ability* reaches a threshold such that they possess the same level of knowledge as others in the language community, so in a sense all native speakers have the same capacity for and knowledge of their language. However, this competence may not be expressed by individuals to the same degree, because the actual use of language in daily life involves something called **language performance**. Performance can be affected by things like fatigue or competing demands on cognitive processing. For example, you may be a perfectly articulate person, but if you are put on a ledge 100 feet off the ground and blindfolded, you will likely not be able to express yourself all too well. Likewise, if you get drunk, you likely slur your speech and you may even make subject-verb agreement errors because you can't keep track of who you are talking about. You might also make more slips of the tongue if you are simultaneously performing a cognitively demanding task. So while in principle we may possess the ability to understand and produce any sentence of our language (*competence*), we may not possess the ability to do that in real time to the same degree in all situations (*performance*).

Why is this important? Well, it means that we cannot take any piece of language produced by anyone as evidence against their competence. If we judged your language ability only by the time you were on that 100-foot-high ledge, we might think you are not a native speaker of *any language*! This is especially important when we study the language of children: the fact that they fail to perform language in an adultlike manner (they don't always produce well-formed sentences) does not necessarily mean that they lack competence in language (in the sense of lacking knowledge of the computational unit). They *might* be lacking competence, but it does not *necessarily* mean so.

There are many reasons why children may not perform to adult standards. Children do not possess vocabularies as large as those of adults. This includes nouns and verbs, but importantly it also includes various kinds of morphology, like prepositions and verb endings. Moreover, children have a much more limited working memory capacity than adults do, and they can't process information as quickly. This impacts how they can express themselves as well as how they comprehend language. And children tend to get distracted very easily, so they may produce ill-formed sentences in part because they change thoughts mid-sentence. We could go on, but the point is that there are numerous reasons why children might fail to perform like adults, but that's just what it is: a failure to perform. So it is possible that despite their errors, children actually have full (or fuller) competence in language. As researchers of language acquisition, we are fundamentally interested in uncovering the nature of that underlying linguistic competence.

2.1.4 Flavors of UG Approaches

When linguists say that children have innate knowledge of language, this claim can take different forms. It can mean that children are born expecting grammar to have a hierarchical structure and grammatical categories like verb, noun, and modifier (e.g., adjective) but that they take time to acquire the functional parts of language like tense marking and auxiliary verbs. Or it can mean that children are born with the full architecture of a syntax tree but fail to use it in an adultlike way because of limitations on their ability to process information (and their limited vocabulary). Some researchers take a stance in between these positions. For the sake of simplicity, we present here just the basic concepts. For more details, see the "Further Reading" list at the end of this chapter.

2.1.4.1 Continuity

One approach to UG is to say that the child's grammar is of the same kind as that of adults, and that children have access to the basics of UG from the very start of life. The procedures, principles, and rules are all in place, as is the overall architecture of language. This approach is called **continuity** (meaning that the child system is fully *continuous* with the adult system). One of the most striking advantages of continuity is that it makes it relatively easy to explain how children acquire their grammar. The basics of UG are already there, so all the child has to do is match the input with their internal knowledge of UG. In many ways, this is the classic generativist approach to language acquisition.

However, there are some challenges to this position, most notably how to account for the late acquisition of various structures. As we will see later in this book, certain syntactic properties appear to be acquired quite late in development, and this challenges a strong version of continuity. For example, if children have full knowledge of UG, why can't they comprehend the passive voice in English until after age 5? Why do they sometimes appear to think that *John hugged him* means the same thing as *John hugged himself*? Why do children have difficulty comprehending and producing certain complex constructions, such as object relative clauses? These questions pose challenges, and they have led linguists to propose some alternative explanations for why children's grammars are so systematic and constrained, yet they can differ from the adult grammar in some respects.

There are generally three avenues researchers take in dealing with this issue. The first is to weaken the degree of continuity. On this view, some but not all aspects of UG are present for the child from birth. This means that the task of acquiring a language is still assisted by UG, but not fully, so this gives rise to late acquisition of some structures and various patterns of errors. A second approach is to maintain a strong version of continuity but to explain apparent delays and systematic errors as being due to something outside of grammar, such as memory structures or other cognitive development.

A third approach is referred to as **maturation**. The idea here is that children are born with UG, but that some aspects of it are not available at first and instead develop at some point after birth. Just like other biological developments, the suggestion is that UG does not endow the child with all the tools of the computational unit until later in life, perhaps age 5. This accounts for the late acquisition of some structures in language, as well as errors in the early stages.

The maturation account appeals to some, but not to others. Detractors see maturation as a stipulation. Rather than explaining the developmental patterns, they argue, it simply pushes the unexplained delay in development into something mysterious like biological maturation. But supporters of this view say that maturation is a well-attested fact of human biology. Children begin to lose their baby teeth at around age 6 years—why does it happen at that age and not a few years earlier or a few years later? Puberty occurs between the ages of roughly 10 and 15 years, but the reason for why it happens at those particular ages is somewhat mysterious. Therefore, it is reasonable to assume that some aspects of the biological property of human language are likewise programmed to develop at some point after birth.

2.1.4.2 Principles and Parameters

Yet another way to reconcile the concept of UG with the fact that children don't speak like adults from the outset comes from the hugely influential theory of **principles and parameters** (Chomsky, 1981). In our discussion so far, we have assumed that UG consists of a series of biases specific to language. But the principles and parameters framework considers a more specific and very intriguing way of seeing UG.

We begin with an observation from linguistic typology—a discipline that looks at the similarities and differences across the world's languages. We know that while languages differ a great deal from each other, they don't vary in infinitely many ways. Rather, languages come in a limited number of types, while other types seem to not exist or to be vanishingly rare. For example, consider a simple phenomenon: the word order of the subject, verb, and object. English is a subject-verb-object (SVO) language in that the basic word order is such that the subject precedes the verb, and the object follows the verb. Other languages differ from this order. Japanese, for example, prefers the order SOV, while Arabic prefers VSO. But when you consider that there are six possible permutations (SVO, SOV, VSO, VOS, OSV, OVS), we find that the huge majority of known languages prefer one of three word orders: SVO, SOV, and VSO. The other three word orders just seem to be rare. In fact, the OVS order is so rare that it was long thought to be unattested—perhaps impossible in human language. We now know that it does occur (e.g., in the Native American language Hixkaryana, spoken in Brazil), but out of almost seven thousand human languages spoken today, it is found in only a handful. Similar typological facts can be found in numerous other phenomena.

Ideally, UG would provide a way to account for this kind of general property while also accounting for the Logical Problem of Language Acquisition (LPLA). One way that this might be done was proposed by Chomsky (1981) in the Principles and Parameters framework. The idea is that UG consists of a series of principles that are invariant across languages. We could think of these as the biases that we have discussed so far. For example, structure dependence might be a principle of language, in this technical sense. But according to this framework, some principles might contain two or more options (referred to as parameters).

Taking the word order example, it may be that there is an invariant **principle** that states that subjects must precede objects. This accounts for why the OSV, OVS, and VOS word orders are so rare in the world's languages. Moreover, there are two additional **parameters** at play here. One parameter says that in any given language, the subject may either precede the verb phrase or follow it, and the second parameter says that within the verb phrase (VP), the object may precede the verb or follow it.

(6) a. Principle: Subjects precede Objects
    b. Subject-VP Parameter: Subjects may precede or follow VPs
    c. Object-Verb Parameter: Objects may precede or follow verbs

Notice that even though we are using terms like *precede* and *follow*, which look like descriptions of linear arrangements, we are talking about structurally defined concepts (subject, object), which can themselves contain (in principle) an infinite number of words; therefore, these definitions are structure dependent. Principle 6a rules out OSV,

OVS, and VOS word orders, so children who are acquiring a language know from the outset that such word orders are not likely. This has the desirable effect of reducing the child's hypothesis space, easing the LPLA.[5] Additionally, children know that they will have to set each of the two parameters (6b–c), so they listen for evidence that will help them get this done.

If children set 6b as "subjects precede VPs," then the language will be either SVO or SOV, and if children set 6b as "subjects follow VPs," then the language will be VSO. The distinction between SVO and SOV comes from 6c: if children set the parameter as "objects precede verbs," then the language will be SOV, and if children set 6c as "objects follow verbs," the language will be SVO.

Another example of a principle and related parameter has to do with wh-movement. All languages have the ability to ask wh-questions: questions that ask *who* or *what* did something or was affected by an action and *when*, *where*, *why*, or *how* something happened. These are different from yes-no questions in that they are answered with a whole phrase rather than a simple *yes* or *no*. So, a principle of UG might be that language allows wh-questions. Where languages differ, however, is in the position that the wh-word occupies in the question. In English, Spanish, German, and many other languages, the wh-word moves to the beginning of the question, even if it is interpreted in some other part of the sentence (*What did John buy?* = 'John bought [what]?'). In Mandarin, Cantonese, Japanese, Swahili, and many other languages, the wh-word does not move anywhere. It remains, in the question form, in the part of the sentence where it is interpreted, as in the following Japanese example. Note that Japanese word order is SOV, so the statement form in 7b shows the object preceding the verb, just like the wh-word *nani* 'what' does.

(7) a.	Hanako-ga	nani-o	tabeta	no?
	Hanako-subject	what-object	eat-past	Q
	"What did Hanako eat?"			
b.	Hanako-ga	sushi-o	tabeta	
	Hanako-subject	sushi-object	eat-past	
	"Hanako ate sushi."			

The **wh-movement parameter** within UG states that languages come in two kinds: those that move the wh-word in questions and those that do not move it. Importantly, children know this and set about trying to figure out which language they have been born into.

In the Principles and Parameters framework, then, UG provides children with a series of these kinds of (ideally binary) parametric choices along the crucial points of difference that describe the languages of the world. It's almost as if the child has a switchboard of choices (provided by UG) that they must set according to the input they hear, and acquiring language amounts to setting the parameters (switchboard choices). Crucially, this means that the child does not need to learn these choices from induction, since the options are given to the child. However, it also means that there is the potential for children to make the wrong choice. It may be that children select the wrong option at first and later reset it. Child errors, therefore, are seen not as an indication of ignorance of language but rather as evidence of knowledge of the underlying UG system but not yet knowing which language they have been born into.

So here we have a framework that manages to (i) address the LPLA by providing children with the relevant principles through UG, thereby addressing the problem of induction; (ii) account for why children might not speak like adults from the outset (since children must set each parameter on the basis of their input); and (iii) capitalize on typological facts about the world's languages that otherwise would be unrelated facts about language. Because of this, the Principles and Parameters framework has been hugely influential over the years and remains an important part of our thinking.

We turn now to the first of our alternative (non-UG-based) approaches to how children learn language. The following section introduces an approach referred to as **statistical tracking**. While this is not generally seen as a direct competitor to the UG-based approach, it places greater emphasis on the influence of input on learning and so in that sense is distinct from the UG approach.

2.2 Statistical Tracking

Children have an amazing ability to track patterns in the world around them. They notice cause and effect—the cup falls from the table and Mom picks it up. They notice patterns of all kinds of events and behaviors, and there are many patterns in language that could potentially be tracked. Do children track these patterns in language? And if so, does this help them learn language?

Let's look at some examples of statistical patterns one might encounter in language. One type of pattern relates to something called **phonotactics**. Phonotactic constraints are constraints on which individual sounds are allowed to occur next to each other within a syllable or word. (A syllable is a unit of sound or sounds, usually consisting of a vowel and one or more consonants adjacent to it; for example, [ba] is a syllable with a consonant and a vowel. See chapter 3, sections 3.3.2–3.3.3.) Phonotactic constraints can vary by language. For example, in English we can have words that start with [tr] or [st] or [bl] (*tree*, *stick*, *black*)—these are considered to be phonotactically "legal" sequences at word beginning—but not *[tl] or *[ts] or *[rd]—these are phonotactically "illegal" at the start of English words. Because the first three clusters are phonotactically legal sequences at the beginning of a word or syllable, we can make up new words that have those patterns: *troob*, *stame*, *blorg*. But we cannot make up new words of English like **tloob*, **tsame* or **rdorg*. Some of these phonotactically illegal sequences are actually possible in other languages. For example, Cherokee has [tl], as in *tlvdatsi* [tlə̃datsi], 'lion'; German has [ts], as in *Zeit* [tsaɪt], 'time'; and Russian has [rd], as in рдеть [rdetʲ], 'to glow'. Phonotactically legal sequences are more likely to occur next to each other within a syllable or word than phonotactically illegal sequences are, even though those "illegal" sequences will be encountered by learners across syllable and word boundaries (Saffran et al., 1996; Storkel, 2001). For example, English learners will hear sequences such as *tha**t l**og, bu**t s**oon, he**r d**og.*

As we'll see in chapter 3, since the speech stream is continuous, there is an interesting question about how children begin to segment that stream of sound into smaller units like words. One idea about how children might begin to identify word boundaries is by noticing the relative probabilities with which a given sound follows another sound. If [t] is followed by [r] with a *high* probability, for example, then there's a good chance they form part of the same syllable, and so no word boundary occurs between [t] and [r] in a [tr] sequence. This is referred to as a **transitional probability**: the probability of one segment following another. Similarly, if [t] is followed by [l] with a *low* probability, then it is more likely that a word boundary occurs between [t] and [l] in a [tl] sequence. In this way, a learner might be able to predict word boundaries simply by tracking the probability of such combinations. The learner simply postulates word boundaries where low transitional probability is detected and ignores those places where high transitional probability is detected.

A fascinating demonstration of infants' ability to track transitional probabilities comes from a study by Saffran, Aslin, and Newport (1996). The researchers played for 8-month-old babies a two-minute stream of synthesized speech containing unbroken sequences of syllables. The syllables were arranged so that they formed four distinct groupings that could be called "words." Syllables were consonant-vowel (CV) sequences such as [pa], [ti], [go], [la], and [do], and the four "words" were

(8) a. pabiku
    b. tibudo
    c. golatu
    d. daropi

Babies heard unbroken sequences of these "words," so that there were no pauses between them or other prosodic information. For example:

(9) pabikutibudogolatudaropitibudopabiku ...

After the two-minute training phase (in which the babies simply listened to this stream of auditory input), the babies were presented with three-syllable test stimuli, some of which were the "words" presented in the training phase and some of which were "part-words," sequences of three of the syllables heard in training but not necessarily all together in one sequence.

Figure 2.4
Illustration of transitional probabilities within vs. between words in an artificial "language."

(10) a. word: golatu
b. part-word: tudaro

The sequence in 10b contains the final syllable of one of the "words" from the training set (final syllable of *golatu*), together with the first two syllables of another "word" from the training set (*daropi*). Crucially, the babies had heard all of these syllables adjacent to one another in the training phase. However, because the "words" always cohered as units, the transitional probability between syllables within a "word" was very high: in fact, the syllable [pa] was *always* followed by the syllable [bi], so the transitional probability between [pa] and [bi] was 1. But a "word" in this set could have been followed by any other "word," so the syllable [tu] could have been followed by [pa], [ti], [go], or [da] (the first syllable of any of the "words"). Therefore, the transitional probability between [tu] (the final syllable of one word) and any of these syllables was only 1 out of 4, or .25. This is illustrated in figure 2.4.

The result of the experiment was that the babies indicated surprise (as measured by visual fixation) when presented with the part-word test stimuli—the infants didn't expect to hear these sequences. The researchers concluded that 8-month-olds were able to identify plausible word units within which transitional probability between syllables was high, as distinct from syllable sequences with lower transitional probabilities.

This result raises the question of whether statistical tracking could be used for learning more abstract and complex aspects of grammar, such as sentence structure. Marcus et al. (1999) conducted an experiment similar to that of Saffran et al. (1996), but it differed in an important respect: the actual syllables used for training and for testing were *different*, which means that the transitional probabilities for all of the syllables in the testing phase were 0 (because they had not been encountered before). Marcus and his colleagues exposed 7-month-olds to sequences of syllables with either an ABA pattern or an ABB pattern. For example:

(11) a. ga ti ga (ABA)
b. ga ti ti (ABB)

Following a 2-minute training phase (exposure to a stream of these syllables in artificial speech), the babies were presented with sequences of new syllables that either *matched* or *did not match* the pattern they had heard during training (see table 2.1).

Table 2.1
Training and test items used by Marcus et al. (1999)

Training	Testing: Match	Testing: Mismatch
ga ti ga (ABA)	wo fe wo	wo fe fe
ga ti ti (ABB)	wo fe fe	wo fe wo

Source: Marcus et al. (1999).

The babies listened significantly longer to the mismatched patterns, indicating they were surprised by these sequences (see appendix B). So even though all of the syllables in the testing phase were brand-new and the babies couldn't use transitional probabilities to judge which test sequence was familiar, they could still tell the difference between the familiar and unfamiliar patterns. This result suggests that even 7-month-olds are capable of establishing an abstract representation of a pattern based on minimal exposure.

This is an important result because it shows that children's ability to track statistics in their input is not limited to the transitional probabilities between syllables. Rather, this shows that children (and adults, actually) can also track the frequencies of abstract symbolic structures like ABA and ABB. This is very important because grammar involves abstract symbolic structures. The fact that babies are attuned to this kind of abstract symbolic pattern shows us that children have a bias to look for such phenomena in their input.

This research program has been extended to investigate a variety of interesting questions about whether statistical tracking might be useful in acquiring syntactic patterns (Takahashi and Lidz, 2008). In fact, statistical tracking has proven to be a powerful tool in linguists' experimental arsenal for discovering what and how babies learn. Moreover, although statistical learning places an emphasis on how much babies can learn from input patterns, the existence of statistical learning procedures is not at odds with theoretical approaches that advocate innate linguistic knowledge. Together, statistical information and innate grammatical knowledge are used to acquire the target grammar. For example, Yang (2002) has proposed a learning algorithm by which learners use statistical properties of language input to decide between competing grammars that are specified by UG. Researchers disagree about how much knowledge about language needs to be innate in order to use statistical tracking to learn the specifics of one's target language (e.g., whether babies can learn that sentences are hierarchical structures simply by tracking statistical patterns of language, or whether statistical patterns are useful only if you start out expecting language to consist of such structures). In this book we adopt the latter approach, but we recognize that many important questions remain open about the interplay between innate knowledge and statistical learning.

We turn now to the constructivist approach. This approach provides an intriguing way to think about language acquisition, one that does not rely on any innate knowledge of language whatsoever. For researchers who favor a smaller role for innate knowledge, or only domain-general innate knowledge, this is an appealing approach indeed. However, as we will see at the end of the next section, this approach faces serious challenges in addressing the key issues that UG does: the problem of induction and the poverty of the stimulus.

2.3 Modern Constructivist Approaches

Put simply, constructivists adopt the view that a combination of (i) rich input and (ii) domain-general learning mechanisms suffice to account for both the LPLA and the DPLA (see chapter 1, section 1.1). Several questions might arise from this statement. For example, what exactly is a domain-general learning mechanism? How does constructivism work? After answering these questions we'll flesh out the difference between constructivism and UG.

2.3.1 What Is a Domain-General Mechanism?

The term **domain** refers to a cognitive function that is operationally distinct from other functions and depends on principles and procedures that are specific to that domain. For example, vision is a domain of cognition and can be considered separately from other domains of cognition, like number (the ability to count and represent numeracy) or memory. At least some of the principles that govern vision seem to apply only to vision and no other function in cognition. Vision scientists must master a particular set of facts that are unique to vision as well as understand the internal architecture of the function of vision. Because vision works in a way that is unlike any other function of the mind, it is considered a *domain* of cognition. Other domains of cognition might include music, social cognition, mathematics, and object perception. Importantly for our purposes, language is considered a domain of cognition by most researchers.

Think of a domain (such as vision) like a component, or a module, in a large machine. In this metaphor, the mind is the whole machine, and each domain of cognition is a smaller module of the machine. Each module has its own internal architecture and a particular function for which it is specialized. The different modules can interact with one another, but each module does its own work: the vision module only concerns visual perception, and the language module only concerns language, but the two can share information so that we can, for example, talk about what we see. Or imagine a car engine, which has several smaller modules, like a carburetor, ignition system, cooling system, and braking system. Each has its own job, and some modules feed into other modules, forming a large, complicated machine.

This view of the mind is often referred to as the **modular view** of mental architecture (Fodor, 1983). On this modular view, each domain (or module) has learning mechanisms that are unique to that domain. Such learning mechanisms are considered *domain-specific* learning mechanisms. Thus, a language-specific learning mechanism is one that applies within the domain of language only. It takes as its input some linguistic material (e.g., a sentence), and its output is knowledge of some aspect of that linguistic material (e.g., its structure and meaning). Crucially, this learning mechanism has no application outside of that domain, or else it would be considered a *domain-general* learning mechanism. A domain-general learning mechanism (for our purposes) is one that helps learn language as well as some other function(s).

A classic example of a domain-general learning mechanism is **analogy**. Analogy is a type of associative learning. In associative learning, the learner draws an association between one event and another. For example, a rat learns to associate a flash of light with an electric shock, so that even if no shock is present, the flash of light will still trigger whatever avoidance behavior the rat would engage in in anticipation of the shock. This is an example of a negative association, but positive associations can also be learned, as when a pigeon learns to associate the action of pecking at a certain key in order to get a food reward.

The specific type of associative learning we'll be mainly concerned with is analogy. Analogy is a very powerful process that takes certain salient properties (a feature or a relation) of one thing and then applies those properties to other things like it. It is a crucial tool in the workbag of the constructivist because, as we will see below, constructivism views language learning as essentially a process of drawing analogies between different grammatical expressions and associating these forms with a particular meaning. The constructivist workbag contains other tools too, all of which apply across domain boundaries. For example, the ability to organize your environment into discrete categories is domain general and applies to language as well as other domains (e.g., vision). The ability to track the frequency of certain things in your environment occurs in various domains of cognition (vision, olfaction), including language. In the rest of this section, we pay particular attention to analogy, a detailed example of which is presented below.

2.3.2 The Constructivist View of Language: Form-Meaning Pairings

At its heart, constructivism views language (and not just language acquisition) as a series of **form-meaning pairings**. Consider the most classic case of a form-meaning pairing: a word.

A common way to understand how words are represented in the mind is that there is a sequence of sounds (the form) that is paired with a meaning. Knowing this form-meaning pairing is part of what it means to be a speaker of a language. In the mind of each member of a language community, there is a list of form-meaning pairs, together forming the lexicon (the mental dictionary). When a listener hears a particular form, they look up that form in their lexicon, find the meaning it is paired with, and they know what the speaker intended. As a speaker, the reverse happens.

The form-meaning pairings in table 2.2 leave out a lot of other things that are typically thought to be part of a semantic representation (the word's meaning), like the difference between denotation and connotation as well as grammatical information like the fact that a verb like *poke* is transitive (takes both a subject and an object), its subject is typically

Table 2.2
Examples of form-meaning pairings

Form		Meaning
[but] (boot)	....	A sturdy item of footwear that covers more than the foot
[bet] (bait)	....	Food used to entice fish or other animals
[pok] (poke)	....	To jab or prod

Note: The meanings are not stored in the mind as sentences like this. This is just to exemplify the general structure of the lexicon.

either animate or a sharp or pointy object, and so on. Nonetheless, this illustrates the basic view of how a form-meaning pairing model works.

Such pairings must be learned on the basis of input. This is uncontroversial: constructivists and UG-based researchers alike all agree on this. Because the forms for these meanings vary between languages, children must learn the vocabulary of their language by hearing the language spoken around them; it makes no sense for particular word meanings to be determined by innate knowledge (though see Fodor, 1975, for the claim that the meaning portion of this pairing must indeed be innate). We will return in chapter 5 to the topic of how word meanings are learned, but for the moment what is significant is that the following statement forms the heart of the constructivist approach to language acquisition generally: learning depends on the input and on drawing analogies across expressions, something that requires no language-specific innate knowledge.

2.3.3 How Constructivism Works

Why is this approach referred to as the constructivist approach? Adherents to this approach argue that knowledge of language is *constructed*: it is built, piecemeal, one step at a time, from small pieces into (gradually) larger and larger pieces, rather than emerging the way an instinctual ability might emerge. And as the process progresses, domain-general mechanisms are used (such as analogy, categorization, statistical tracking) to grow the body of knowledge. Let's take an early and influential approach, referred to as the Verb Island approach (Tomasello, 1992). We simplify this model into its core principles, which we organize into a five-stage process.

Stage 1: The first stage begins with the fact that children have a tremendous talent for recognizing patterns (a domain-general skill). A child listens to language and initially hears a dizzying variety of forms. At first, it is a "blooming, buzzing confusion," to quote the classic psychologist William James (1890). But gradually, over time, the child notices some frequently occurring forms, for example, their own name, the words for *mother*/*father*, common verbs like *go*, *do*, *want*, and *eat*. This is the first step in breaking into the linguistic system—the wedges, so to speak, around which the rest of language acquisition proceeds. This may well be aided by the ability to track statistics—see section 2.2 for more on this.

Stage 2: Once the child has identified some commonly occurring forms, the next step is to form **fully specified schemata** (**schemata** is a plural form; the singular is **schema**). *Schema* simply refers to an organized pattern, but it is a technical term in this field. Children begin to notice patterns around these commonly occurring forms. For example, the child having already identified the word *eat* in Stage 1 might notice that *eat* occurs with predictable words after it, like *eat more* (as in, *Do you want to eat more peas*?), *eat it* (as in, *Are you going to eat it now?*), and *eat some* (as in, *I'm going to eat some cookies*). The child therefore notices that *eat* is like an island—things vary around it, but it remains constant (to some degree). This makes this island a pivotal element in the language (in fact, another term for an island is a **pivot**), and the child uses it to organize the rest of their learning.

Importantly, the child creates **associations** between the island and the items that co-occur with it. The child creates a list of associated pairings (e.g., *eat+more*, *eat+it*, *eat+some*), which contain no syntax per se but are just units of language. Such units are often referred to in the literature as **chunks**, denoting the fact that they are unanalyzed pieces of language that may themselves be paired with meaning, on par with single lexical items or idioms. Each of these chunks is referred to as a *fully specified schema* because it consists of specific words or phrases rather than the grammatical categories those words or phrases belong to (e.g., *it* instead of *pronoun*). The next stage is when abstraction begins.

Stage 3: Now is when the magic happens. This is when the child realizes that all the things that *eat* combines with can be put into a class of their own. At this point, the child creates what's referred to as a **partially specified schema**, in the form of something like [eat+Z], where Z is a variable or an abstract category. Forming an abstract category is a more efficient way to organize things in memory, since it only requires learning one (abstract) thing and having some way to switch specific items in and out of that abstract slot.

The members of the category Z gain their membership in this class because they have been linked through analogy. Because analogy is such a powerful mechanism, there is no way to really predict what goes into Z—it all depends on the experience of the child. If one child happens to hear lots of examples of *eat some soup* and *eat more cereal*, that child might well initially restrict the variable Z to mass nouns only.[6] But a child that hears more utterances like *eat the apple* and *eat another banana* might restrict the variable Z to countable nouns. Stage 3 is the most important stage in this process because it runs squarely into the problem of induction. We will discuss this shortly, but for now, keep this stage in mind.

Stage 4: The next stage with *eat* is when this same process of creating abstract categories continues to other parts of the sentence, for example, creating a subject category in the same way an object category was created or an indirect object category for ditransitive verbs. This results in a *semi-abstract schema* that looks something like this: [X+eat+Z]. This is still only semi-abstract in that the schema is still based on a specified lexical verb (*eat*).

Stage 5: The final stage is when the child creates an abstract category for the verb itself. That is, the child notices that several complex schemata that have been created are themselves similar to other complex schemata that have been created in parallel. So the *eat* schema looks similar to other schemata, like the one for *drink*. The child then forms a **fully abstract schema**, which is something like $[X_{subject} + Y_{verb} + Z_{object}]$. This fully abstract schema might be called the **transitive sentence schema**. Similar processes are happening in parallel that result in many other abstract schemata, e.g., **intransitive sentence schema** and **ditransitive sentence schema**. When all of these abstract schemata are combined together, they form the totality of an abstract grammar. These stages are summarized in table 2.3.

In this way, the abstract nature of sentences is *constructed* in a piecemeal fashion, starting with the verb and working outward. This is all based on associative learning, including analogy and categorization (all domain-general mechanisms), and in that sense, this approach successfully provides a framework for how children learn (some aspects of) language without depending on any innate linguistic knowledge.

It's important to note that the end state that constructivists assume (that is, the state of language in the adult mind) is significantly different from what UG researchers assume. The latter assume that language consists of a large lexicon coupled with a powerful computational unit. Together, these allow us to produce infinite numbers of novel sentences and to be infinitely creative, all while maintaining a structure that is recognizable to other speakers of our language. Researchers who adhere to the constructivist view, on the other hand, assume that the adult language consists of a very large set of form-meaning pairings, some fully abstract, some partially abstract, and some specific. Each theory makes significantly different predictions for all aspects of acquisition and psycholinguistics more broadly, but this is far beyond the scope of this book.

Returning to the constructivist approach, one problem, alluded to in the discussion of Stage 3, is that analogy is a very powerful tool. It actually does far more than we need it to. As suggested above, it may be that some children hear lots of mass nouns after the verb *eat*, and they will incorrectly postulate a schema that restricts the object to mass nouns. We need a way to restrain analogy from making these kinds of mistakes, or else the model will predict that language acquisition is far messier than it actually is. Let us consider two mechanisms that constructivists have developed that restrain this all-powerful analogy tool.

Table 2.3
Five-stage process to abstract schema formation

Stage 1	Stage 2	Stage 3	Stage 4	Stage 5
Verb Island	Fully specified schema	Partially specified schema	Semi-abstract schema	Fully abstract schema
eat	eat + more eat + some eat + it	eat + Z_{obj}	X_{subj} + eat + Z_{obj}	$X_{subj} + Y_{verb} + Z_{obj}$

These two mechanisms are **entrenchment** and **preemption**. When a child creates a generalization, it may well be the wrong generalization. Analogy will do that: it is very powerful. A learner needs a way to rid themselves of spurious generalizations. What do we mean by a spurious generalization? Well, the mass noun example above is one example, though one in which the association is too narrow (mass nouns are a subset of all nouns). Let's consider another spurious generalization that results in an association that is too broad. Imagine a child in Stage 3 hears the word *eat* in various contexts and postulates that *eat* means 'to consume anything by mouth.' This is an overly broad meaning since this could include things like *eat milk*, *eat water*, and perhaps *eat bubble gum*. The child must retreat from this incorrect generalization somehow. Similarly, the child might make the wrong generalization about the subject of the verb. The child needs some way to limit these incorrect generalizations, but not limit *correct* generalizations. Correct generalizations need to be safeguarded in some way, and this is what the first mechanism does.

Entrenchment is one way to safeguard the correct generalizations. The basic idea is that once a child has formed a generalization, the more confirmatory evidence they receive, the stronger that generalization becomes. On the one hand, if that initial generalization is correct, then entrenchment (based on the frequency of confirmatory evidence in the input) allows that generalization to become so strong that it will withstand evidence that is potentially disconfirming. For example, if the child develops a stricter generalization about *eat* such that the child thinks it means *to consume solid or semisolid edible objects* (as opposed to *consume anything by mouth*), then with each instance of *eat* that involves consuming such objects, that generalization is strengthened. It becomes so strengthened that when the child hears *eat some soup*, although that is more liquid than solid, the child's generalization will withstand the odd pairing, and *soup* will simply be considered an exception to the generalization.

Similarly, if the child develops a generalization that the subject of *eat* is animate and the object is inanimate, and this becomes strongly entrenched, then when the child hears something like *the blanket ate up Teddy* (perhaps when a parent is being playful with the child), the child won't disregard the [+animate subject] part of their generalization. Instead, the child will look for some other way to accommodate that potentially troublesome example (like *blanket* is metaphorically animate). This is because the correct properties of *eat* are so strongly entrenched that odd uses of *eat* won't derail the whole system.

A second process that works in concert with entrenchment is preemption. Here, if a child forms a generalization, this makes the postulation of a competing generalization less likely. That is, one generalization *preempts* other generalizations, unless evidence for an alternative generalization is very strong. If the child initially postulates the correct generalization, then the correct one preempts other competing (incorrect) generalizations. For example, if the child correctly postulates that *eat* takes only solid or semisolid direct objects, then this will preempt a possible incorrect generalization when the child hears *eat your soup*, for example. Because the correct generalization is so strongly entrenched, the possibility of the child entertaining the generalization of *eat* meaning to *consume anything by mouth* will be preempted by the strongly entrenched (correct) hypothesis.

One final point is to be made about the constructivist approach. Recall that we said that this approach posits that language is learned through a combination of (i) rich input and (ii) domain-general learning mechanisms. We have seen what is meant by domain-general mechanisms, but we have not yet discussed what is meant by rich input. We claimed in section 2.1.1 that the language input to children is *impoverished*: children are exposed to input that is fundamentally insufficient for generalizing abstract rules and structures. But constructivists argue that the input to children is actually very rich when one considers different kinds of information beyond just what children hear. Children recruit many kinds of information to help them with the language they hear. For example, we know that children attend very closely to eye gaze. When the parent looks at an object, the child's gaze follows. When the parent then looks at the face of a sibling, the child's gaze follows. So eye gaze is a powerful attention getter, and this helps the child decode their environment. This process of the child's attention being drawn by that of the adult is often referred to as **joint attention**, and it is thought to make the input to children richer because more can be interpreted.

Together, then, analogy, entrenchment, preemption, and rich input (through mechanisms like joint attention) work to produce an appealing framework for understanding language development. The constructivist approach lays out an intuitive, simple, and powerful process for language learning that depends on nothing but skills known to occur in other domains of cognition.

2.4 How Does Constructivism Differ from the UG-Based Approach?

Let's recap the main differences between constructivism and the UG-based approach. Constructivism is a theory built on the idea that, armed with powerful domain-general learning mechanisms, the input is rich enough for children to acquire language. This approach is founded on the core belief that language is an emergent phenomenon that arises through a composite of numerous other higher-level cognitive functions. Together, they give the illusion of an independent domain.

The UG-based approach, on the other hand, is founded on the idea that language is not a mix of other cognitive functions. Instead, language is an autonomous cognitive domain and, as such, has dedicated knowledge and learning mechanisms associated with it. Thus, the claim is that children are born with a disposition that favors certain structural analyses of incoming language data (i.e., the speech children hear around them). This disposition, which we may refer to as Universal Grammar, is essentially a series of biases that allow the child to make sense of hugely ambiguous linguistic data in a uniform and efficient manner. But crucially, the UG-based approach assumes that all of these biases are specific to the domain of language and cannot be derived from other domains at all.

Why do we, in this book, adopt the UG approach and not the constructivist approach? The short answer is that we view human language as not just form-meaning pairings but as a system of abstract hierarchical structures whose internal structure can't be directly inferred from hearing (linear) sentences. Such a system, almost by definition, cannot be learned only from the kind of language input children receive (which consists of sentence strings, not the hidden structures that generate them).

For a longer explanation, let's look at our yes-no question example once more and see how the constructivist approach handles the data.

Remember that the issue is that a child could analyze the rule to form simple yes-no questions either using a rule that depends on structure (the correct rule) or one that depends on linear order (the incorrect rule). Without any evidence that distinguishes between these two hypotheses (questions that involve one auxiliary that has moved over another auxiliary), at least some children (if they are not biased to use structure-dependent rules) should produce errors with multiple auxiliary questions. And remember that Crain and Nakayama (1987) found children never produced such errors.

So how does the constructivist approach deal with this issue? One way is to refute the poverty of the stimulus. For example, Pullum and Scholz (2002) conducted a corpus analysis of several child corpora and found some examples of multiple auxiliary questions. They argued that because the relevant evidence was not *completely* nonexistent, there is no poverty of the stimulus problem. In other words, they argued that children can learn to invert the correct auxiliary by attending to these few examples. Other researchers (e.g., Legate and Yang, 2002) have argued that the rate of occurrence of these multiple-auxiliary questions (0.06% of the questions in the input) is simply too small to lead children to learn the correct rule.

The problem of extremely low-frequency input becomes clearer when one considers the mechanism of entrenchment. Recall that entrenchment states that when the child hears a particular pattern that fits with an already-established analysis of the data, that analysis of the data becomes stronger. So if children are indeed unbiased learners, and if they initially select a hypothesis based on linear order, then they are in serious trouble in terms of getting to the correct analysis of yes-no questions: they will receive thousands of confirming pieces of evidence for the wrong hypothesis—99.94% of input will confirm the linear order hypothesis. As seen in figure 2.5, that means they will encounter 1,666 single-auxiliary yes-no questions before they encounter a question with two verbal elements, with the second fronted over the first. Those 1,666 single-auxiliary questions will do nothing but further entrench the already existing analysis of the data. Therefore, this approach predicts that children should exhibit massive evidence of linearity in their question formation.

But this kind of misanalysis is not observed in any studies of child language. How do we account for that? In our view, the simplest solution is that children never even consider the ambiguity. If it were the case that when they encounter their first yes-no question, chil-

Initial hypothesis: Linear order	→	1,666 single-auxiliary questions	→	Strong entrenchment of linear order hypothesis	→	1 double-auxiliary wh-question	→	Linear order preempts structural dependence

Figure 2.5
If children are unbiased learners, and if the initial hypothesis is based on linear order, subsequent data should result in strong entrenchment of that hypothesis, such that any recalcitrant data will be preempted.

dren think, 'Ah, to form a yes-no question, all I need to do is front the *main auxiliary*' (a structure-dependent rule), then this whole problem goes away. But notice that this would be an innate bias involving domain-specific knowledge, something the child *expects* about language prior to actually encountering language experience. This is just one example that poses an intractable problem for constructivist approaches to language acquisition. The field of generative language acquisition focuses on the many other problems like this, and on providing explanations for how children overcome those problems.

2.5 Summary

We reviewed three theories for how children might learn language: the UG-based approach and the two input-based approaches: statistical tracking and constructivism. The latter two rely on the input and domain-general learning mechanisms (tracking statistical patterns, drawing analogies), while the former relies on innate knowledge of abstract properties of human language (such as structure dependence).

The UG-based approach assumes that the child is hardwired with domain-specific knowledge about language that allows them to solve the Logical Problem of Language Acquisition, overcome the poverty of the stimulus, and thereby induce the fine-grained details of their own language. The beauty of this approach is that it says that the problem of induction is reduced to the child trying to decide between a small set of pre-given hypotheses. As such, the process of language acquisition can be seen more as a process of *confirmation*. There is significant room for variation in what any particular language will look like, but the child has a very broad blueprint for what the language they are about to acquire *could* look like, and they are spared the expense of considering a bunch of options for what it *could not* look like.

A child may not know that the language into which they are born has a rule for yes-no question formation in which the main auxiliary is moved to the front, but they know that whatever rule it is, it *won't* make reference to linear order. A child may not know that their language requires that wh-question words move to the front of the sentence, but they will know that such a rule is a perfectly good way to form wh-questions and is consistent with how human language works. They will also know that leaving the wh-word in its original position is another good strategy that human language permits. And they will know not to consider outlandish rules for wh-question formation—for example, a rule that says in order to form a wh-question, reverse the order of words in the sentence. That is a logically possible rule, but it simply is not one that any human language follows.

UG-based researchers disagree about the degree to which UG is present at birth. Some think that all principles of UG are present from birth and that development of language is essentially development of aspects of cognition outside of the computational unit. Others within the UG camp think that the computational unit itself develops over time, perhaps through maturation.

The input-based approaches do very well in explaining how children acquire abundantly available patterns in language, but when it comes to abstract properties or infrequent patterns, they face many challenges. In this book, we adopt the UG-based approach to language acquisition because we are concerned with how children acquire the ability to produce and comprehend the totality of their language, including the rare and abstract parts of it.

2.6 Further Reading

Ambridge, Ben, and Elena V. M. Lieven 2011. *Child Language Acquisition: Contrasting Theoretical Approaches*. Cambridge: Cambridge University Press.

Elman, Jeffrey, Elizabeth Bates, Mark Johnson, Annette Karmiloff-Smith, Domenico Parisi, and Kim Plunkett. 1996. *Rethinking Innateness: A Connectionist Perspective on Development*. Cambridge, MA: MIT Press/Bradford Books.

Yang, Charles. 2016. *The Price of Linguistic Productivity: How Children Learn to Break the Rules of Language*. Cambridge, MA: MIT Press.

2.7 Exercises

1. We discussed yes-no questions as an example of a construction whose acquisition requires formulating a structure-dependent rule. Another example sometimes used is the *wanna* contraction. That is, the words *want* and *to* can sometimes be contracted to make *wanna*, as in (i), but not always, as in (ii):

(i) Who do you wanna invite?
(ii) *Who do you wanna invite Jim? (= Who do you want to invite Jim?)

While not all English speakers agree that (ii) is ungrammatical, most find it worse than (i). The explanation for why (ii) sounds worse than (i) is that the statement forms of (i) and (ii) have different structures:

(iii) I want to invite Jim.
(iv) I want Sue to invite Jim.

In (iii) there is nothing standing between *want* and *to*, but in (iv) there is. A trace of that word between *want* and *to* in (iv) remains there when it's turned into a question, and apparently you can't contract *want* and *to* into *wanna* when there's a trace in the way.

Now for the question: How could you tell if children *know* this rule implicitly? That is, what kinds of sentences could you elicit from children in order to find out if they know where *wanna* can be contracted and where it can't?

For a further challenge, explain why this construction provides another example of the poverty of the stimulus.

2. Constructivists take the position that the input is *rich*, while proponents of UG take the position that the input is *poor*. Spell out some of the ways in which the input can be said to be rich. Then spell out ways in which the input is poor. How do these two positions lead investigators to ask different questions about how language is acquired?

3. Consider how a child might construct knowledge of the verb *try*. Think about the kinds of ways children might hear the verb *try*, and create a list of members of the X and Z categories for the semi-abstract schema for *try*, in which the schema looks like [X + *try* + Z]. You may listen for examples of how *try* is used in the speech around you, or you can draw from the following list of utterances spoken by a mother. To take a hypothetical example, the child might hear Mom say, "I tried to buy some cookies, but they didn't have any," in which case X would be *I* and Z would be *to*. The child might hear, "Shall we try some peas?", in which case X would be *we* and Z would be either *some* or *some peas*. Create two lists, one for X and one for Z, with ten members each.

*MOT: He's trying his new words out.

*MOT: Go in the street trying to find the moon?

*MOT: (.) Try it now.

*MOT: Trying to get the ball in the basket.

*MOT: Try him up there.

*MOT: Where are you trying to send your rocket?

*MOT: Why don't you turn it around and try?

*MOT: Try to put them in the oatmeal box.

*MOT: Close it and try again.

*MOT: I'm trying to put it together but I don't know how.

Now do the same for the verb *want*. Again, you can listen for and record uses of *want* by speakers in your environment, or you can draw from the following list of a mother's utterances:

*MOT: Want to?

*MOT: Do you want to see what I have?

*MOT: You want Mommy to sit on the floor?

*MOT: What do you want Mommy to do?

*MOT: Do you want Mommy to stand up?

*MOT: What do you want that for?

*MOT: The dog doesn't want any paper does he?

*MOT: Do you want to write on here?

*MOT: Adam (.) want to close the box?

*MOT: Oh (.) he wants to shake hands.

4. Looking at the two lists in Exercise 3, do you think the child would go on to the next stage and assume that *want* and *try* are verbs of the same class? (Do you know if they actually do belong to the same class?)

2.8 References

Chomsky, Noam. 1957. *Syntactic Structures*. The Hague: Mouton.

Chomsky, Noam. 1965. *Aspects of the Theory of Syntax*. Cambridge, MA: MIT Press.

Chomsky, Noam. 1971. *Problem of Knowledge and Freedom*. London: Fontana.

Chomsky, Noam. 1981. *Lectures on Government and Binding: The Pisa Lectures*. New York: Mouton de Gruyter.

Crain, Stephen. 1991. Language acquisition in the absence of experience. *Behavioral and Brain Sciences* 14(4): 597–612.

Crain, Stephen, and Mineharu Nakayama. 1987. Structure dependence in grammar formation. *Language* 63: 522–543.

Fodor, Jerry. 1975. *The Language of Thought*. Sussex: Harvester Press.

Fodor, Jerry. 1983. *The Modularity of Mind*. Cambridge, MA: MIT Press.

Gallistel, C. R. 1990. *The Organization of Learning*. Cambridge, MA: MIT Press.

Gentner, Dedre, and Jose Medina. 1998. Similarity and the development of rules. *Cognition* 65: 263–297.

James, William. 1890. *The Principles of Psychology*. New York: Holt.

Legate, Julie Anne, and Charles Yang. 2002. Morphosyntactic learning and the development of tense. *Language Acquisition* 14: 315–344.

MacWhinney, Brian. 2000. *The Child Language Data Exchange System*. Mahwah, NJ: Lawrence Erlbaum Associates.

Marcus, Gary F., Sugumaran Vijayan, S. Bandi Rao, and Peter M. Vishton. 1999. Rule learning by seven-month-old infants. *Science* 283: 77–80.

O'Grady, William. 2005. *Syntactic Carpentry: An Emergentist Approach to Syntax*. Mahwah, NJ: Lawrence Erlbaum Associates.

Piattelli-Palmarini, Massimo (ed.). 1980. *Language and Learning: The Debate between Jean Piaget and Noam Chomsky*. Cambridge, MA: Harvard University Press.

Pinker, Steven. 1994. *The Language Instinct*. New York: Harper Perennial Modern Classics.

Pullum, Geoffrey, and Barbara Scholz. 2002. Empirical assessments of stimulus poverty arguments. *Linguistic Review* 19: 9–50.

Saffran, Jenny, Richard Aslin, and Elissa Newport. 1996. Statistical learning by 8-month-old infants. *Science* 274: 1926–1928.

Storkel, Holly. 2001. Learning new words: Phonotactic probability in language development. *Journal of Speech, Language and Hearing Research* 44: 1321–1337.

Takahashi, Eri, and Jeffrey Lidz. 2008. Beyond statistical learning in syntax. In Ana Gavarró and Maria João Freitas (eds.), *Proceedings of the 2007 Generative Approaches to Language Acquisition*, pp. 444–454. Cambridge: Cambridge Scholars Publishing.

Tomasello, Michael. 1992. *First Verbs: A Case Study of Early Grammatical Development*. Cambridge: Cambridge University Press.

Yang, Charles. 2002. *Knowledge and Learning in Natural Language*. New York: Oxford University Press.

Yang, Charles. 2004. Universal Grammar, statistics, or both? *Trends in Cognitive Sciences* 8: 451–456.

II Module 2: Building a Sound System

3 Early Speech Perception

For all children with typical hearing ability who are born into a family using spoken language, sound is the medium through which children encounter language. So we will begin with the sound system in exploring children's discovery of their language system.

Exposure to language begins even before a baby takes its first breath of air. Through the amniotic fluid of the womb, a fetus can perceive environmental sounds, including linguistic sounds, as early as 7 months of gestation (DeCasper and Spence, 1986; Querleu et al., 1988). The loudest and clearest sounds, naturally, will be those coming from the fetus's mother. The quality of linguistic sound, however, is quite different when perceived through fluid rather than through air, and the acoustic signal reaching the fetus, whether from the mother or from other people, is compromised. The way in which the signal is compromised has to do with which acoustic frequencies are preserved and which are lost. Higher frequencies are filtered out, so the prenatal infant perceives only the lower frequencies in the acoustic signal. Research suggests that frequencies below 400 Hz reach the fetus without disruption, but frequencies above that level are less available in utero, with availability decreasing as frequency goes up (Peters et al., 1993; Griffiths et al., 1994). Some sounds between 500 Hz and 1000 Hz will reach the fetus but in degraded form, and frequencies above 1000 Hz will suffer a greater reduction in quality. This is important because the higher frequencies (above 1000 Hz) contain the most acoustic information about actual segments, or "phones" (as opposed to information about tone, pitch, stress, or other properties of the sound). For example, high-frequency information distinguishes different vowel sounds or consonants that have different places of articulation (e.g., [b] vs. [d], [s] vs. [f]).

Figures 3.1 and 3.2 show waveforms and spectrograms (depictions of sound waves and their component frequencies) of a human voice producing a series of words in English. Figure 3.1 shows the full range of frequencies that were produced by the speaker, while figure 3.2 shows only the lowest 1000 Hz.

Consider for a moment the difference in how speech sounds to you when you are speaking with someone in person compared with over the phone. Telephone connections preserve a large portion of the acoustic speech signal, and we generally have no trouble

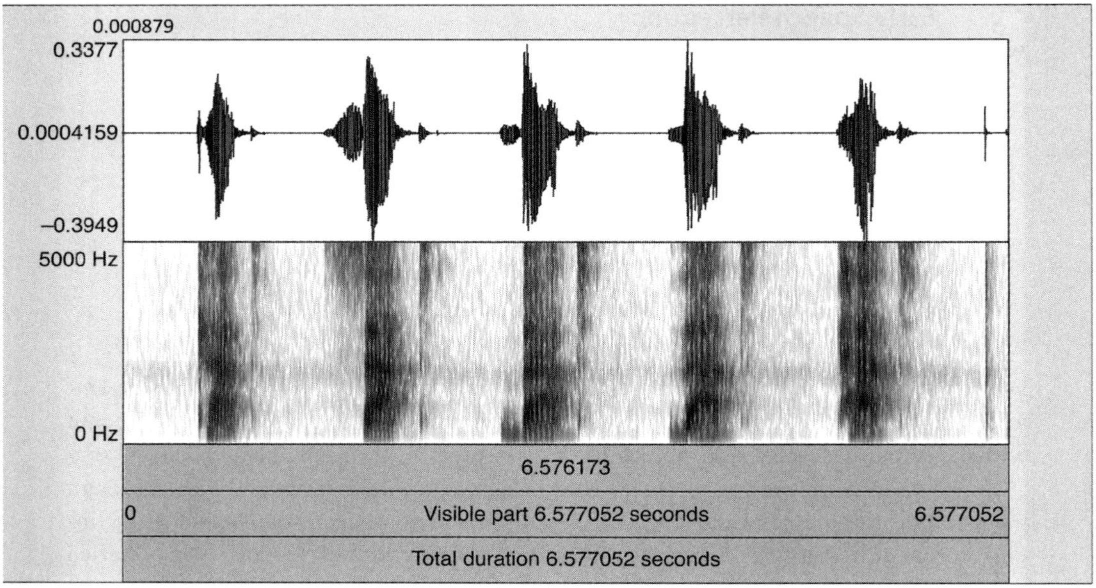

Figure 3.1
A waveform (upper half) and spectrogram (lower half) of the English words *cat*, *sat*, *bat*, *that*, *pat*. The spectrogram shows the component frequencies, with peaks of intensity shown by dark bands. The waveform shows the amplitudes (loudness) of the sound waves (a bigger wave means a larger amplitude, which means a louder sound).

understanding one another using this medium. However, we often have to provide extra cues and information when we are using certain sounds out of the context of a word or sentence. For example, if you have to spell your name for someone, you might need to say, "S as in Sally," to distinguish it from "F as in Fred." The high-frequency sound waves (not to mention visual cues) that make this distinction easy in face-to-face conversation are not preserved over the phone. This is because the phone lines are able to convey only a subset of the frequencies found in normal human speech. While humans can produce sounds up to about 14 kHz (14,000 Hz), and the human ear can perceive frequencies up to about 20 kHz, the telephone line transmits frequencies up to only about 3.4 kHz.

If the telephone connection is poor, we lose even more acoustic information. You might mistake the word *met* for *net* if contextual cues to the target word are missing or insufficient. Yet even with a poor signal, you will be able to tell whether the speaker's voice is rising or falling in pitch. This prosodic information is what is contained in the lower frequencies of the acoustic signal, and it is this information that is most robustly available to the infant in utero. The fetus, therefore, has access to some prosodic information in utero, but far from everything, and so after birth there is a lot of learning left to do.

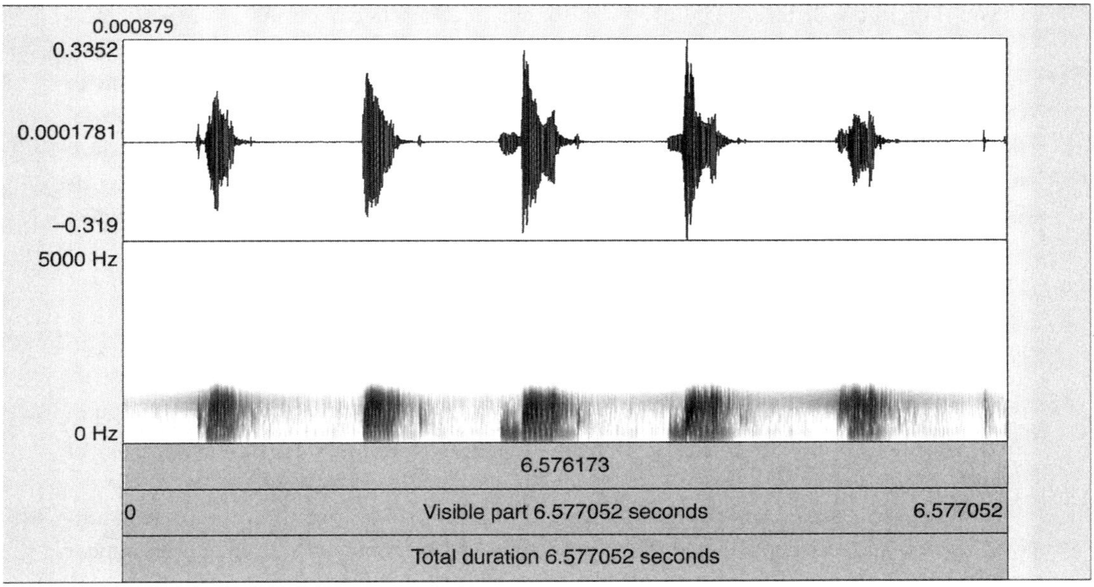

Figure 3.2
This waveform and spectrogram depict the same recording shown in figure 3.1, but with all frequencies above 1000 Hz removed and replaced with white space. Notice how much of the information in the spectrogram is missing.

In this chapter we will focus on how the infant's discovery of language proceeds from the moment of birth. We will mainly focus on babies' **perception** of speech, reserving discussion of their abilities in speech and language production until the next chapter.

Sidebar 3.1: Filtered Sounds

Listen to the sound files A.wav and B.wav. File A.wav has sound filtered above 500 Hz, and file B.wav has sound filtered above 1000 Hz. Before listening to C.wav (the full signal), how much of the sentence do you understand? Was it easier to understand B.wav than A.wav? Which sounds in particular are missing in the filtered versions? Which sounds and aspects of the sentence are preserved?

3.1 Speech Sound Discrimination

One of the most striking aspects of speech sounds is that we hear many speech sounds (especially consonants) as categories of sounds rather than as continua of sounds. That is, a tiny phonetic difference between two sounds will make those sounds "sound different" to you if that little difference puts the two sounds in different categories, while a big phonetic difference between two sounds may not cause the sounds to "sound different" if they still belong to the same perceptual category. To give a quick illustration of what we mean by sound "categories," say the word *neat* as you normally would. Then say it with the vowel drawn out, as if you are really emphasizing the word: *neeeeaaaat*. It is still the same word; it means the same thing. Now say the word *need*. This is a different word, ending in a different consonant. But the vowel is also *acoustically* different from the vowel in *neat* when you say it as you normally would. You may not be aware of it, but when you say *need* at your normal speaking rate, the vowel is slightly longer than when you say *neat*. However, while lengthening or shortening the vowel causes a *phonetic* difference in the sound (it is acoustically different), changing the length of the vowel doesn't make it a different vowel, at least in English.[1] Another way to say this is that the longer and shorter versions of the [i] vowel ([iː] and [i]) are not different categories of sound. If you changed the [i] vowel to an [ɪ] vowel (*knit*), this small change in phonetic features produces a different *category* of sound—if you are an English speaker, you hear it as a different vowel.[2]

Some of these sound categories take several months to acquire, and we will see examples of them and learn how they develop in section 3.3. But some of these categories appear to be universal and available from birth. Let's look at such an example.

When a consonant is "released" (the articulators restricting airflow come apart) and the next sound is a vowel, at some point the vocal cords begin vibrating. A speech sound characterized by vibration in this way is said to be **voiced**. If you touch your fingers to the front of your throat (where your voice box is located) while you say the sound "aaaaaa," you will feel a vibration: this is voicing. If the vocal cords are already vibrating when the consonant is released (that is, voicing begins *before* the consonant is released), you will hear the consonant as being voiced too. Consider the sounds [z] and [s], which are almost identical sounds, except the former is voiced and the latter is voiceless. Try saying "zzzzzaaaaaa" and hold the [z] sound. You should feel the vibration throughout. If the vocal cords start vibrating only after the consonant is released, you will hear the consonant as voiceless. Try saying "sssssaaaaaa." Now you should not feel any vibration until you get to the [a] sound.

Some languages, including English, have a category of stop consonants that are pronounced with voicing beginning *after* the release of the stop. In this case, you have already begun articulating the vowel, but there is a brief moment in which there is no voicing, and this results in a little puff of air known as **aspiration**. If you hold your hand in front of your mouth while you say the word "pat," you should feel a little puff of air as you release the [p] sound (if you are a native speaker of English). In contrast, if you say the word "bat," you should not feel any puff of air after the [b] consonant. That little puff of air is the result of something called **voice onset time** (VOT): the period of time between the release of the stop and the onset of voicing. VOT is essentially a short period of silence

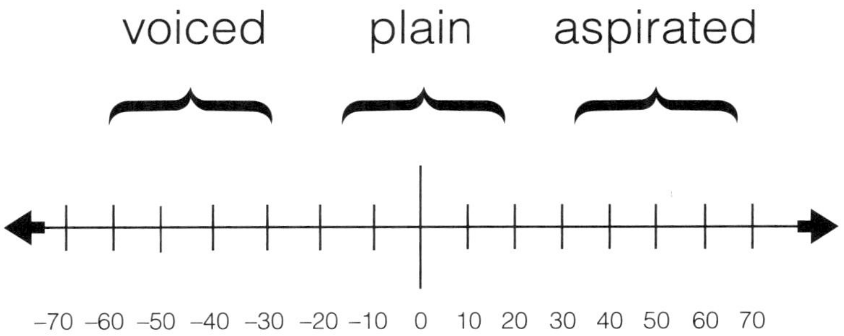

Figure 3.3
The VOT timeline marked off in 10 ms increments. The point marked 0 indicates the moment of release of the consonant. The three potential categories that languages may distinguish in terms of VOT are labeled with their phonetic voicing values of voiced, plain, or aspirated.

between the release of the stop and the vowel, and this gives rise to the perception of voicelessness. A VOT of more than 25 ms (milliseconds; that is, voicing begins at least 25 ms *after* the stop is released) yields aspiration, while a VOT closer to 0 does not. A VOT of 0 means that voicing begins *simultaneously* with the release of the stop, and it yields what we'll call a plain form of the consonant (neither aspirated nor voiced). A negative VOT—meaning that the voicing begins *before* the stop is released—means the consonant will be voiced as well as the vowel.

We can draw a timeline extending from before the release of the stop until after that release, indicated as 0, and mark off time in some interval. Let's use 10 ms intervals, as shown in figure 3.3.

Although the timeline is continuous and we could theoretically divide it into as many different categories as we wanted to, experimentation with adult speakers of different languages has revealed that speakers produce only two or three categories of stop consonants with respect to VOT (Lisker and Abramson, 1964). A few languages, such as Thai, distinguish three categories: voiced (also sometimes called prevoiced), plain, and aspirated stops. Most languages distinguish two categories: roughly speaking, a voiced category and a voiceless category, though the location of the boundary may be different in different languages. For English speakers, the boundary between what is perceived as a voiced sound (you hear it as /b/) and a voiceless sound (you hear it as /p/) is +25 ms (see figure 3.4). For Spanish speakers, the boundary between voiced and voiceless sounds is much closer to the moment of release.

Early work in babies' perception abilities suggested that their ability to discriminate sounds was linked to the practice they had articulating these sounds. That is, a baby who can articulate [bababa] and [$p^hap^hap^ha$] will be able to discriminate [b] and [p^h] in perception as well. But a remarkable discovery in the 1970s led to the realization that infants can

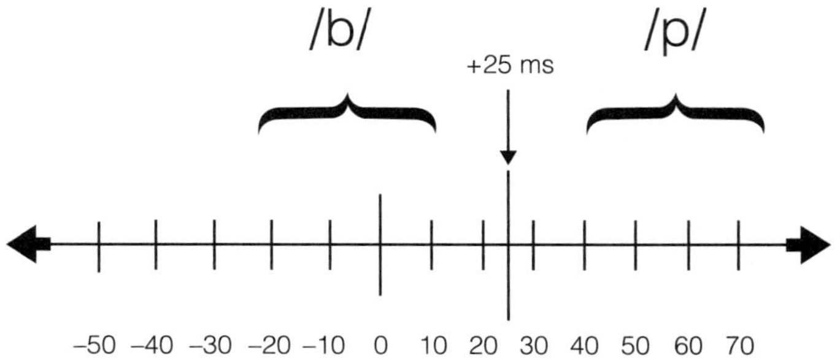

Figure 3.4
For English speakers, the +25 ms mark is the boundary between sounds perceived as /b/ and sounds perceived as /p/.

Sidebar 3.2: Square versus Angled Brackets

A note about notation: Sometimes we put square brackets around an International Phonetic Alphabet (IPA) symbol, like [a], and sometimes we put the symbol between slashes, like /a/. Why? Is it just a matter of personal preference?

No. A symbol between square brackets denotes the *phonetic representation* of the sound: how the sound is actually realized phonetically. For example, [p^h] is an aspirated bilabial stop. A symbol between slashes denotes the *phonemic representation* of the sound: how the sound is mentally categorized by speakers. So the bilabial consonants in the phonetic words [p^hɪl] (*pill*) and [spɪl] (*spill*) are both phonemically /p/.

distinguish these categories of sound long before they ever articulate them on their own. An experiment by Eimas et al. (1971) presented artificially generated consonant sounds to 1- and 4-month-old infants. These infants heard pairs of syllables in which the initial consonant fell at different points along the [ba] to [p^ha] continuum. There were three pairs of syllables in the experiment, each beginning with a bilabial stop. One pair (condition 1a in table 3.1) included stops with VOTs of 0 ms (voicing began exactly at the moment of the stop's release) and stops with –20 ms (voicing began 20 ms before the stop's release); a second pair (condition 1b in table 3.1) included stops with VOTs of +60 ms and +80 ms (so, both sounds had the voicing start long after the consonant's release); the third pair (condition 2 in table 3.1) had stops with VOTs of +20 ms and +40 ms (see figure 3.5). As you can see, the syllables were constructed so that the consonants in each pair differed in VOT by the same amount (there was a 20 ms difference in their VOTs), but across experimental conditions the pairs of sounds either came from the same category (/b/ in condition 1a; /p/ in condition 1b) or from different categories (/b/ vs. /p/ in condition 2). The researchers wanted to know whether these newborn babies would distinguish between the sounds in each pair with equal facility.

Table 3.1
Experimental stimuli in Eimas et al. (1971)

	Condition 1a: Same category	Condition 1b: Same category	Condition 2: Different category
Stimuli	/ba/—/ba/	/pa/—/pa/	/ba/—/pa/
VOT (ms)	−20, 0	+60, +80	+20, +40
Increase in sucking rate at change	not significant	not significant	significant increase

Source: Eimas et al. (1971).

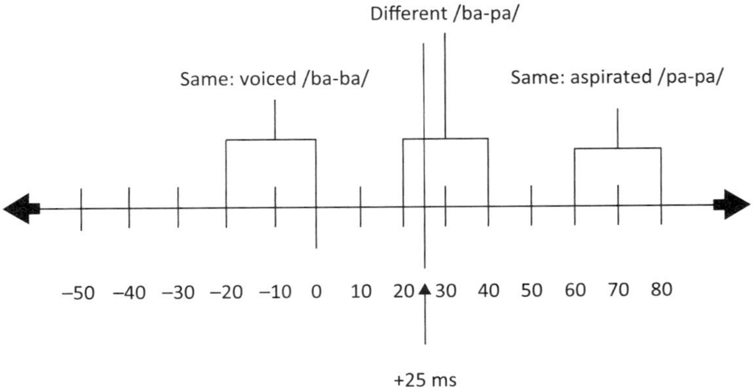

Figure 3.5
Illustration of the stimuli used by Eimas et al. (1971) to determine whether infants perceived stop consonants categorically.

How can you tell whether a 1-month-old or a 4-month-old perceives the difference between two sounds? Using an ingenious method, Eimas and his colleagues placed an electrode into a pacifier nipple and had the infants suck on the nipple. Sucking is a reflex, so until about 5 months of age it can be used to measure an infant's response to a stimulus. When babies perceive something new and different they get excited, and when babies get excited, they suck at a faster rate. When they get bored, their rate of sucking goes back down to their baseline rate. Using a technique called **high-amplitude sucking** (HAS), the researchers could tell whether the babies were perceiving something as new and different. They simply played one sound for the baby (e.g., [ba]) for a certain amount of time, until the baby's sucking rate decreased by a substantial amount and settled at a steady rate, indicating the baby had **habituated** to the first sound. At a predetermined time point, they played the second sound (either another version of /ba/ or a new sound, /pa/) and measured whether the sucking rate increased. If children were able to perceive the second sound as different from the first sound (the one they had habituated to), their suck rate was expected to rise. But if they did not hear the sounds as different, then their suck rate was expected to remain the same (see table 3.1). The boundary between the sounds we perceive as voiced and voiceless is +25 ms. Given this, which pair of sounds do you think you would hear as different sounds?

The babies heard the sounds exactly as you would: sounds that came from different *categories* were perceived as different sounds, while sounds that came from the same category were not, even though the physical difference between the sounds was the same in each case (20 ms). Thus, Eimas and his colleagues had discovered that with very little exposure to the ambient language, and with virtually no real "practice" uttering these sounds (see chapter 4), these tiny infants were hearing sounds categorically. This phenomenon is known as **categorical perception**. Moreover, infants' perception appeared to be structured like that of adults: the boundary between plain and aspirated stops was roughly +25 ms of VOT.

Infants' ability to distinguish voiced from voiceless consonants across the +25 ms VOT boundary appears to be universal and likely innate. We can see evidence for this from a study conducted by Streeter (1976), which looked at infants acquiring Kikuyu, a language whose category boundary between voiced and voiceless sounds is different from the English one; it is *before* the release of the consonant and so has a negative VOT value, unlike the positive (+25) value in English. In other words, both Kikuyu and English have two voicing categories (rather than three, like Thai), but the acoustic features of the boundary between the respective voiced and voiceless categories are different in the two languages (English's boundary is at +25 ms; Kikuyu's is a negative value). Nevertheless, Streeter found that infants exposed to Kikuyu speech were able to categorically distinguish the *English* voiced-versus-voiceless categories just as the American babies were. A similar result is found for infants acquiring Spanish, a language whose voiced-versus-voiceless category boundary is similar to the one found in Kikuyu. The fact that infants can perceive certain contrasts not found in their target language is important, because infants have had *no* exposure to these sounds and contrasts. It suggests that human infants are, in a sense, primed to detect certain categories of linguistic sound, even if not all of those categories will turn out to matter in their language.

There is a caveat, however: while babies appear to universally recognize the English-type boundary between voiced and voiceless consonants, it is less clear whether there is a universal recognition of boundaries found in other areas of the VOT spectrum. Studies have shown that American babies do perceive the voiced category found in languages like Kikuyu and Spanish. However, in order to differentiate the voiced versus plain sounds, an acoustic difference of 20 ms between the pairs of sounds is not sufficient. Instead, the sounds need to have a larger acoustic difference between them: closer to 70 or 80 ms (Eimas, 1975; Aslin et al., 1981).

3.2 Perceiving Phonemic Contrasts

In the preceding section we talked about infants' ability to perceive and distinguish one type of contrast: the contrast between voiced and voiceless stop consonants. Although having a distinction between voiced and voiceless consonants appears to be close to universal, languages differ in where they demarcate voiced from voiceless (e.g., Spanish versus English) and in whether they have three VOT categories or two (e.g., Thai versus English). Here we need to introduce the notion of a **phoneme**. A phoneme is an abstract representation of sound, and it is important for understanding how we identify and distinguish *words*, which we'll talk about more in the next chapter. For now, you can think of it as "how you hear a sound." For example, when you hear the words *pill* [pʰɪl] and *bill* [bɪl], you hear the first sound in each word as a different sound. That's not only because they are phonetically different (one has aspiration and the other doesn't) but also because they are different phonemes in your language. Exchanging one for the other actually changes the word (these words are called **minimal pairs**). When you hear the words *pill* [pʰɪl] and *spill* [spɪl], you hear the /p/ sounds as the same. But these sounds are just as different from each other phonetically as [p] and [b] are: the [pʰ] in *pill* has aspiration but the [p] in *spill* does not. If someone pronounced the word *spill* as [spʰɪl], it would sound like a weird pronunciation of that word, not a completely different word (the way [pʰɪl] and [bɪl] are different words). You hear these /p/ sounds as the same because their difference *does not matter* for your language. They are **allophones** (versions) of the same phoneme.

What we just said about the /p/ and /b/ sounds is true for English and some other languages, but not for every language: in Khmer and Hindi, for example, aspirated /pʰ/ and unaspirated /p/ are different phonemes, so that exchanging one for the other yields a different word (e.g., in Khmer, /pɔːŋ/ 'to wish' versus /pʰɔːŋ/ 'also'). Just as languages can differ from each other by which VOT categories are **phonemic**, they also differ in how they create phonemic categories along other dimensions. Here we'll look at examples of different phonemes according to **place** and **manner of articulation**. (See the chart of the International Phonetic Alphabet, or IPA, in appendix A for a review of places and manners of articulation.)

Tables 3.2 and 3.3 list some of the contrasts perceived by infants, the infants' ages, and the target language of those infants. Table 3.2 shows sound contrasts found in the target language and when infants perceive those contrasts; table 3.3 shows sound contrasts not found in the infant's own target language that can nevertheless be perceived at certain ages.

The examples in tables 3.2 and 3.3 show that up to the age of 6 months, infants are able to perceive categorically a wide range of sounds, including many found in other languages in addition to sounds found in their own language. This is a good thing, since a baby must be prepared to acquire whichever language he or she is presented with in the environment. But as we saw in earlier in this section, not all contrasts matter in every language. In particular, only phonemic contrasts matter for identifying *words* in a language, so in order to

Table 3.2
Target language contrasts perceived by English-acquiring infants by 6 months

Contrast	Language exhibiting contrast	Age of infants	Infants' target language	Source
ba-ga	English	5 months, 2 months	English	Moffit, 1971; Morse, 1972
bæ-dæ	English	2–3 months	English	Eimas, 1974
ba-wa	English	2 months	English	Eimas and Miller, 1980
ba-ma	English	2–4 months	English	Eimas and Miller, 1980
ra-la	English	2–3 months	English	Eimas, 1975
wa-ja	English	2 months	English	Jusczyk, Copan, and Thompson, 1978
ma-na	English	2–3 months	English	Eimas and Miller, 1980
fa-θa, va-ða	English	2 months	English	Levitt et al., 1988

Table 3.3
Nontarget language contrasts perceived by English-acquiring infants by 6 months

Contrast	Language exhibiting contrast	Age of infants	Infants' target language	Source
pa-pã	French	1–4 months	English	Trehub, 1976
ẓa-ʒa	Czech	1–4 months	English	Trehub, 1976
ta-ṭa	Hindi	6 months	English	Werker and Tees, 1984a
k'i-q'i	Nthlakampx (Salish)	6 months	English	Werker and Tees, 1984a
ejectives	Tigrinya (Ethiopian)	6 months	English	Best, 1991
ra-la	English	6 months	Japanese	Tsushima et al., 1994

begin learning words an infant should shift their perceptual focus from all sounds to those that are contrastive phonemically in their language. This happens during the second half of the first year of life.

As indicated in table 3.2, 6-month-olds acquiring English are quite good at perceiving nonnative contrasts, such as the Hindi dental /t/ versus retroflex /ṭ/ contrast. In fact, their performance at this age is similar to the rate of performance of native Hindi-speaking adults (that is, they are nearly perfect). But by 10 months of age their performance drops to about 50%, and by their first birthday they cannot make the distinction—their performance is now below 20% and nearly as poor as English-speaking adults. The same result bears out for English-acquiring babies hearing the Nthlakampx or Tigrinya ejective contrasts and for Japanese babies hearing the English /l/ versus /r/ contrast, a contrast not found in Japanese. In fact, all of the nonnative contrasts listed in table 3.3 are perceivable at 6 months of age, but not by 12 months of age for babies whose target language lacks those sounds. Thus, by the end of the first year of life, a baby's ability to discriminate speech sounds has dwindled from the full range of human speech sounds down to only those found in the baby's target language.

What explains the apparent loss of ability to discriminate sounds not found in the target language? One early idea was that since the nonnative sounds were not present in the babies' environment, the ability to perceive those sounds simply atrophied, the same way a muscle might weaken over time from lack of use. In other words, the neural structures used in the brain to respond to these sounds become inactive. This view is known as the **maintenance-loss view**, sometimes summarized as "use it or lose it" (Eimas, 1975; Strange and Jenkins, 1978; cited in Werker and Tees, 1984b).

A different explanation of the loss of this perceptual ability is that instead of completely losing the neural capacity to respond to certain sounds, our perceptual system is reorganized in such a way that sounds that are phonetically similar to native sounds are *assimilated* to those familiar sounds (retroflex [ʈ] is phonetically similar to alveolar or dental [t]), and sounds that are not phonetically similar to any native sounds are simply not heard *as* linguistic sounds. This view is known as **perceptual assimilation** (Best et al., 1988; Best and McRoberts, 2003; see also Werker and Tees, 1984b, for a slightly different account and Kuhl et al., 2008, for yet another approach). One piece of evidence supporting this view is that 1-year-olds are able to discriminate certain linguistic sounds that are completely unlike sounds found in their ambient language, such as the clicks found in some African languages (Best et al., 1988). Another type of support for the view that discrimination ability does not completely disappear is that older-child second-language learners can often become native-like in their perception of contrasts in the second language (Flege, 1993), and even adults can be trained to hear nonnative sounds as distinct categories if they are given sufficient exposure and training. The exposure period may need to be quite long: one study of English-speaking adults' ability to perceive the Hindi dental-versus-retroflex contrast found that five years' exposure to Hindi was needed; one year was not sufficient (Werker and Tees, 1984b).

Nevertheless, it is clear that by 1 year of age, babies are unable to discriminate many sound contrasts that are not phonemic in the language they are acquiring. Let's think for a moment about why the *loss* of this ability might be advantageous. Do all speakers articulate the same sound in exactly the same way? No. Is the same sound realized exactly the same way acoustically in different contexts? No. For example, the aspiration that occurs with voiceless stops at the beginning of a word in English (*pot* [$\mathrm{p^h}$at]) disappears when the stop is preceded by an [s]: *spot* [spat], not *[s$\mathrm{p^h}$at]. So when hearing the word *spot* or *spill*, English-speaking children need to perceive the [p] in these words as the same sound as the [$\mathrm{p^h}$] at the beginning of *pot* even though they are acoustically different. For another example, the sound [k] is articulated much farther forward in the mouth when it precedes the vowel [i] than when it precedes the vowel [u]. Say the words *key* and *coo* and feel where your tongue touches the roof of your mouth. Babies learning English need not be concerned with these kinds of minor variations (called **coarticulatory effects**) or with interspeaker differences. Part of the reason for this is that around their first birthday, babies are starting to acquire the **lexicon** of their language: they are starting to learn words. Learning what words mean presents its own set of problems and puzzles (see chapter 5), so for a baby to focus on solving those puzzles, it is advantageous to perceive the sound system in terms of a relatively small set of sound categories rather than perceiving all the myriad phonetic variations that sounds come in.

3.3 Finding Word Boundaries: Speech Segmentation

We just mentioned that around their first birthday babies are beginning to learn the words of their language and that perceiving linguistic sounds as categories (phonemes) is an essential component of this process. A related aspect of the challenge of word learning is that of locating the beginnings and ends of words within the continuous stream of speech. In the spectrograms in figures 3.1 and 3.2, it was easy to see the boundaries between the words because the sound recordings contained simply a string of individual words with pauses in between each. But normal everyday speech contains full sentences with the words run together so that there are no clean breaks and pauses. Consider the waveform and spectrogram in figure 3.6, which depicts the sound file you listened to in sidebar 3.1 (file C.wav).

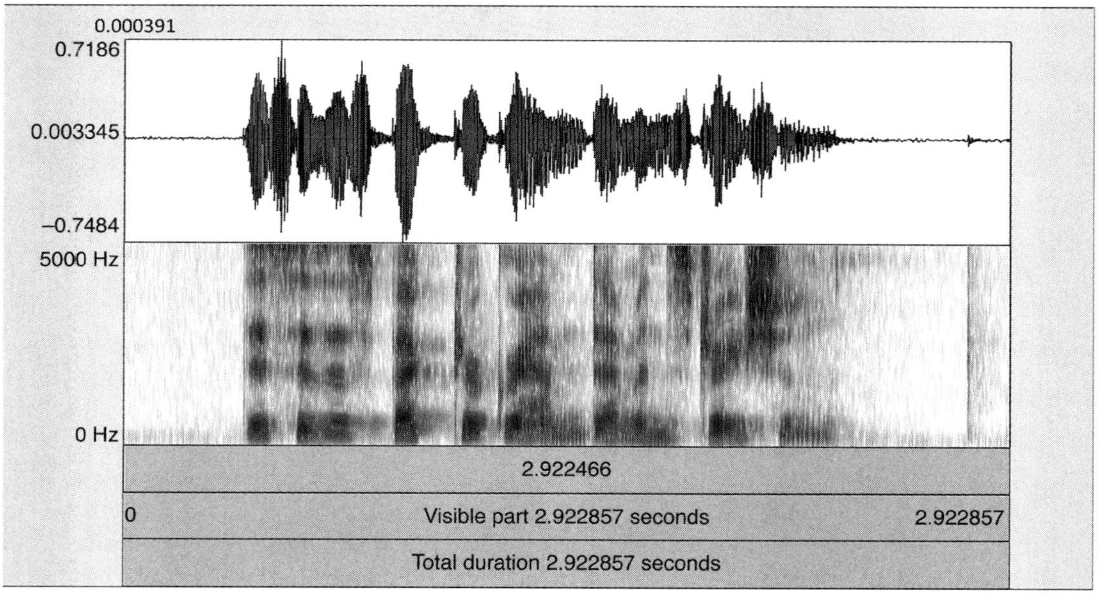

Figure 3.6
A spectrogram depicting the sentence *This is a spectrogram demonstration.*

In figure 3.6, it is much harder to "see" the breaks between individual words. In fact, the places where the amplitude of the waveform reduces to almost nothing are points where stop consonants are being articulated, not pauses between words.

The question for language acquisition is how an infant cracks the code of language and begins to hear language in individual units of words and morphemes rather than simply a continuous stream of undifferentiated sounds. A big part of the answer is found in infants' sensitivity to rhythmic properties of language. But before we see how and why babies can use the rhythm of their language to locate word boundaries, let's first consider some of the properties of the type of language directed to them, known as **infant-directed speech**.

3.3.1 Infant-Directed Speech

Many adults, both male and female, have a tendency to speak to babies and young children in a slightly different way than how they speak to older children, adolescents, and other adults. This mode of speech is sometimes mistakenly referred to as baby talk; in fact, adults do not alter their grammar to produce the kinds of grammatical errors or mispronunciations that babies and toddlers actually produce when they speak (we will learn more about these errors in chapters 4, 6, and 7). So how is infant-directed speech (IDS) different from adult-directed speech (ADS)?

The primary characteristics that distinguish IDS from ADS are as follows (Fernald, 1985; Cooper and Aslin, 1990; Kuhl et al., 1997; Soderstrom, 2007; Cristià, 2013):

(1) Properties of IDS
- a. higher pitch and wider pitch range
- b. exaggeration of vowel space
- c. slower rate of speech
- d. more frequent and longer pauses
- e. shorter continuous sequences
- f. more frequent stresses
- g. repetition

Sidebar 3.3: Finding Word Boundaries

Find a speaker of a language you do not know, and ask them to say a few sentences in their language, using sentences that are at least several words long. Now have them repeat those same sentences, but ask them to modify their speech in some way to help you find the word boundaries. What specific modifications would you request? Did their modified speech actually help you locate the breaks between words?

If you can't find a speaker of a language you don't know, listen to this audio file: audiufon.hum.uu.nl/sounds/kekchi_priester_lang.wav (if this doesn't work, go to the website http://audiufon.hum.uu.nl/data/onbekende_taal.html and click on the green circle with the ear under Kekchi). Imagine the speaker is there with you. What would you ask them to modify about their speech if you could ask them to repeat what they said in the recording?

These features are found in the IDS of speakers of languages ranging from English and French to Japanese and Mandarin Chinese. For example, Kuhl et al. (1997) found that English-, Russian-, and Swedish-speaking mothers produced vowels [a], [i] and [u] that were more widely separated from each other in the vowel space when producing IDS compared to ADS. Another example of the exaggeration of speech properties comes from a study of Mandarin-speaking mothers (Mandarin is a tonal language). This study found that these mothers produced exaggerated tones in speaking to their infants, but that these exaggerations did not distort the lexical tonal contours needed to acquire words in Mandarin (Liu et al., 2007).

While many of these features of IDS are found across languages, there are also language-particular ways in which adults modify their IDS. One study found that Japanese-speaking mothers produced larger distinctions among vowels in terms of *length* than English-speaking mothers (that is, there was a large difference between long and short vowels in their duration for the Japanese-speaking mothers), while English-speaking mothers produced larger distinctions among vowels in terms of qualities that relate to the *tenseness* (as opposed to laxness) of the vowel, compared to the Japanese-speaking mothers (Werker et al., 2007). Significantly, tenseness or laxness is a more relevant property for distinguishing vowels in English than length is, and length is a more relevant property for distinguishing vowels in Japanese than tenseness is. Thus, in both cases, mothers were exaggerating vowel features, but the features they were exaggerating depended on the particular language. In sum, some IDS features are found universally in adults who produce IDS (shorter sequences, wider pitch range), while other features may be language specific and relevant to the target language.

If you completed the task in sidebar 3.3, do any of the characteristics in 1 match what you asked (or would have asked) the speaker to employ in his or her speech to you? Undoubtedly, some or all of these characteristics can be helpful in breaking down the speech code into smaller chunks, thereby helping a language learner find the boundaries between sentences, phrases, and words. A study by Fernald (1985) found that 4-month-old babies showed a significant preference for listening to speech with the above characteristics over the less exaggerated and faster ADS. Another study, by Cooper and Aslin (1990), found that 1-month-old infants and even 2-*day*-old newborns also significantly preferred IDS. Recall from the beginning of the chapter that the type of low-pass filtered sound available

Sidebar 3.4: Differences in Varieties of Speech

We note in this section that adults produce exaggerated "point" vowels ([a], [i], and [u]) in IDS compared to ADS—that is, these vowels are farther apart from each other in the vowel space. You might have noticed that people often use a form of speech similar to IDS when speaking to their pets. Interestingly, while higher pitch and wider pitch range are properties of pet-directed speech, exaggeration of the vowel space is not (Burnham et al., 2002).

Sidebar 3.5: Infant-Directed Signing

Parents of babies who are deaf, who are themselves speakers of sign language, also exhibit alterations of their signs akin to the speech alterations found in spoken IDS. That is, when parents sign to their children, they tend to take more time to produce each sign, they increase the size of each sign, and they tend to repeat signs more than they do when signing to adults (Masataka, 2000; see also Lieberman, Hatrak, and Mayberry, 2014). Studies show that babies (both deaf and hearing) prefer infant-directed signing over adult-directed signing, even if they have no prior exposure to sign language (Masataka, 1996, 1998).

(i) In what ways might these modifications help babies who are deaf segment the "signing" stream?

(ii) What does this tell you about the nature of sign language? That is, does the existence of infant-directed signing support the notion that sign language is a natural human language, equivalent to spoken language? Why or why not?

(iii) Why do you think hearing babies might prefer infant-directed signing?

to fetuses in the womb omits much information about phonological *segments* (individual sounds) but maintains a lot of *prosodic* information, such as pitch. It may be that babies prefer IDS, at least in part, because it exaggerates many of the features of spoken language available to babies before birth. In fact, while IDS has been found to exaggerate prosodic patterns in language, there is some disagreement about how reliable IDS information is about individual phones or segments. For example, one study found that there was more variability in the acoustic values of vowels in IDS compared with the same vowels in ADS (Cristià and Seidl, 2014); other studies have found that IDS vowels were more distinct from one another than ADS vowels were (Kuhl et al., 1997; Adriaans and Swingley, 2017).

IDS, which exaggerates many of the salient prosodic features of language so as to provide extra clues to language learners about the units of speech, is widespread and preferred by babies from the earliest moments of life. Does this mean these helpful clues are *necessary* for speech segmentation?

The answer is not entirely clear. There is some evidence that IDS is *not* universal, suggesting that while its characteristics may provide some clues to cracking the language code, children may be able to solve the puzzle even if this help is not available. For instance, while many adults in many cultures employ some (or all) of these characteristics in their speech to infants, there is considerable cultural and individual variation in the degree to which IDS bears these features and whether adults employ them at all. A study of the Gusii tribe in Kenya found that with both 4- and 10-month-old infants, mothers were less likely than American mothers to respond to their baby's cries or other vocalizations by looking at or talking to their baby (instead, the Gusii mothers responded to their baby's cries by holding their baby) (Richman, Miller, and LeVine, 1992). Similarly, a study by Ochs and Schieffelin (1984) of the Kaluli people of Papua New Guinea found that mothers rarely spoke directly to their prelinguistic babies. In these cultures, the assumption is that it is pointless to speak to children before they can comprehend language, so mothers are unlikely to try to engage their prelinguistic infants in verbal conversation (importantly, these mothers attend to their babies' needs, but they do so more by touching and holding them than by directing words at them).

On the other hand, some researchers have pointed out that although Kaluli mothers may not use IDS with their infants, they may use this register in contexts in which their babies are present and they are speaking on behalf of their children (Cristià, 2013). Furthermore, Cristià (2013) notes that studies that report a low use or nonuse of IDS among particular populations within the United States (Heath, 1983) may not have properly controlled for the naturalness of the situations observed. So we don't actually know the extent to which properties of IDS are universal. See also Falk (2004) for suggestions about evolutionary precursors to IDS in early hominins.

3.3.2 The Importance of Prosody and Rhythm

While IDS may provide (many) babies with hyperarticulated cues that help them begin to segment the continuous stream of language around them, children will also need to segment natural ADS. How do children do this? Even in adult-to-adult speech there are a number of segmentation clues from **prosody**, which concerns aspects of speech such as **pitch**, **rhythm**, and **timing of pauses**. Pauses are an important source of information about where the boundaries between syntactic elements of a sentence are. Even though speakers do not necessarily pause between words, phrases, or sentences, there is evidence that by 7 months of age babies are sensitive to the likelihood that pauses will occur between, rather than within, clauses. A clause is a unit containing a subject and a predicate (i.e., roughly a sentence). One study by Hirsh-Pasek et al. (1987) found that 7-month-old infants showed a preference for IDS that contained artificial pauses *between* clauses but not for speech with artificial pauses *within* a clause. A later study by Jusczyk and his colleagues (1992) found that by 9 months of age babies also showed a preference for IDS containing artificial pauses at a *phrase* boundary but not for speech containing artificial pauses *within* a phrase. A phrase is a smaller chunk of speech than a clause, since a clause contains both a subject (itself a phrase—usually a noun phrase [NP]) and a predicate (itself a phrase—the verb phrase [VP]). To illustrate, 9-month-olds preferred to listen to sentences like 2a than sentences like 2b:

(2) Did [$_{\text{NP}}$ you] [$_{\text{VP}}$ spill [$_{\text{NP}}$ your cereal?]]
  a. Did you <pause> spill your cereal?
  b. Did you spill <pause> your cereal?

The subject in 2 is *you* (NP) and the predicate is *spill your cereal* (VP). In addition, 9-month-olds showed a preference for speech with a 1-second pause between the subject and predicate when the subject contained several words, as opposed to speech with a 1-second pause within that longer subject. That is, babies preferred to listen to sentences like 3a rather than 3b:

(3) [$_{NP}$ Many different kinds of animals] [$_{VP}$ live in the zoo.]
   a. Many different kinds of animals <pause> live in the zoo.
   b. Many different kinds <pause> of animals live in the zoo.

In sentence 3 the subject is *many different kinds of animals*, and the predicate is *live in the zoo*. Interestingly, 6-month-olds did not show any preference between 3a- and 3b-type pause patterns.

Taken together these results tell us that babies can use information about the locations of pauses to begin finding chunks of language, and as they grow from about 6 to 9 months they begin to break those chunks down into smaller and smaller units, from clause chunks to phrase chunks. This is one example of a process known as **prosodic bootstrapping**: babies can "pick themselves up by their bootstraps" by noticing patterns in the prosody of their language, such as the location of pauses in speech.

Another important ingredient in prosodic bootstrapping involves rhythmic structure. All human speech contains rhythms, and these rhythms are built out of units known as **syllables**. Speakers typically have good intuitions about identifying syllables, but let's make it more formal: a syllable is defined as a "peak of sonority" (Clements and Keyser, 1983), and it minimally contains a vowel (called the **nucleus** of the syllable) and possibly one or more consonants either before the vowel (called the **onset**) or after the vowel (called the **coda**; the nucleus and the coda together form what is called the **rime**). A representation of a syllable is given in figure 3.7, where C = *consonant* and V = *vowel*.

In some languages, like English, syllables can be stressed or unstressed. Consider the bisyllabic words in 4, where the word in 4a has stress on the first syllable, and the word in 4b has stress on the second syllable.

(4) a. pencil   PENcil   Strong-weak
   b. giraffe   giRAFFE   weak-Strong

In many languages, finding the syllable with primary stress is often a good way to locate a word boundary. For example, about 85% of lexical category words in English have primary stress on the first syllable (Cutler and Carter, 1987), so a strong (stressed) syllable is

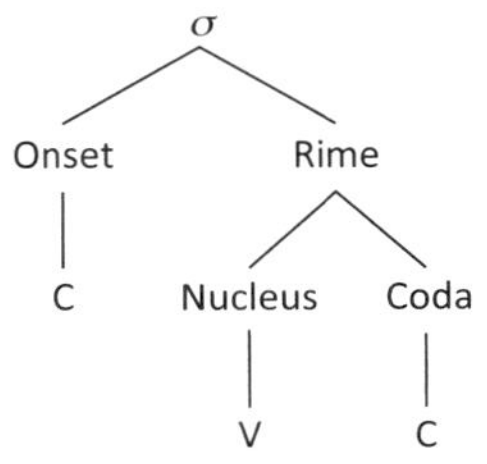

Figure 3.7
The basic structure of a syllable.

very likely to indicate the beginning of a word. Researchers have found that by about 7 to 9 months of age infants acquiring English are able to segment speech into bisyllabic units with stress on the first syllable (Jusczyk et al., 1999; Curtin et al., 2005). In French, stress tends to occur at the ends of words, and infants acquiring this language appear to segment speech into bisyllabic units with final stress, though the French-learning babies exhibit this segmentation ability a bit later, between 12 and 16 months (Nazzi et al., 2006).

Thus, we can see that very early in life, babies are able to use the rhythmic pattern of their target language to begin to find syllabic units within the speech stream. This, combined with their ability to use pauses to find clause and phrase boundaries, is the first step toward identifying the boundaries of those units we know intuitively as words.

3.3.3 Phonotactic Constraints

Another type of clue babies can use to find word boundaries comes from something called **phonotactics**. Phonotactics has to do with the ability of sounds to occur next to each other in a word or syllable. For example, in English the sound [z] is never followed immediately by the sound [t] in a syllable, but [s] and [t] frequently occur together, both at the beginnings (e.g., *store*) and ends (e.g., *roast*) of words. The sounds [z] and [t] can occur next to each other, but there has to be a word boundary between them, as in *has to* [hæz tu]. And some sequences of sounds can occur within a word, but they only co-occur very rarely, such as [sf] (e.g., *sphere*). Recall from the last chapter (section 2.2) that babies are really good at keeping track of statistical patterns. We saw in that section that babies can use the **transitional probabilities** across syllables to make guesses about where to put word boundaries: if two syllables have a very high transitional probability, they can probably co-occur within a word, but if two syllables have a very low transitional probability, they probably don't belong to the same word.

Babies can do this with an artificial language, but can they do it with naturalistic language input too? There is some experimental evidence that they can. We know that 9-month-old infants prefer to listen to nonword sequences containing possible and highly frequent phonotactic patterns than to impossible (in their language) and very infrequent phonotactic patterns (Jusczyk et al., 1993, 1994), suggesting that by 9 months of age babies are at least sensitive to phonotactic patterns. (Interestingly, 6-month-olds do not show any preference of this kind.)

A further study by Mattys and Jusczyk (2001) used the **headturn preference procedure** (see appendix B) to find out whether phonotactic cues could help babies recognize a word they had heard in a series of sentences. They tested this using the infrequent word *gaffe* and the nonword *tove*. During the familiarization phase of the experiment, 9-month-old babies listened to sentences in which *gaffe* (or *tove*) was surrounded either by "good" phonotactic cues—adjacent sounds that *could not* precede [g] or follow [f] within a word in English—or "poor" phonotactic cues—adjacent sounds that *could* precede [g] or follow [f] within English words. For example, the sequence "bean gaffe hold" contains good phonotactic cues because [ng] and [fh] are not possible within-word sequences in English.

(Note: Even though we write the letters *ng* together in a lot of English words, the nasal consonant is actually [ŋ]!) So, the surrounding environment for the word *gaffe* here can provide a kind of spotlight for identifying *gaffe* as a word. On the other hand, the sequence "fang gaffe tine" provides poor phonotactic clues for segmentation because [ŋg] and [ft] are possible sequences within an English word. Therefore, the environment for *gaffe* in this case does not provide any spotlight—the word just blends in with its surroundings.

After the familiarization phase, the researchers played for babies either the word they had heard in the familiarization sentences (*gaffe*, *tove*) or words they had not heard at all during familiarization (e.g., *pod*). What they found was that babies listened significantly longer to the word presented in the midst of good phonotactic cues than to words either heard with poor phonotactic cues or not heard at all. In fact, the babies listened just as long to the words surrounded by poor phonotactic cues as they did to words they hadn't even heard during familiarization, suggesting that the spotlight effect of phonotactics is very powerful indeed.

3.4 Summary

In this chapter we have seen how babies begin to decode the language around them. They are aided by an innate propensity to perceive certain types of linguistic sounds categorically, as we saw in the example of voiced and voiceless stop consonants (VOT). Categorical perception is crucial for decoding speech, both because it is necessary to disregard coarticulatory effects and interspeaker differences that are "unimportant" for interpreting language and because mature speakers perceive and represent linguistic sounds categorically. Essentially, being a speaker of a language means representing the sounds of the language as categories, so babies must learn to do this.

We also saw in this chapter that within the first 6 to 10 months of life, babies go from a universal perception of linguistic speech sounds to recognizing a narrower range of sounds that matter for distinguishing the words of that language. This reduction in perceptual ability allows babies to focus their perceptual energies on finding and learning words, the next step in language development. And the ability to break up the speech stream into smaller units so they can in fact find those units called words comes from babies' attention to the rhythmic properties of language: they use the location of pauses and either stressed syllables or syllable boundaries to find where words and phrases begin and end.

In the next chapter, we will look at how babies' and toddlers' sound system develops on the production side of things, from babbling to pronouncing words.

3.5 Further Reading

Jusczyk, Peter. 1997. *The Discovery of Spoken Language*. Cambridge, MA: MIT Press.

Morgan, James, and Katherine Demuth. 2014. *Signal to Syntax: Bootstrapping from Speech to Grammar in Early Acquisition*. New York: Psychology Press.

3.6 Exercises

1. In this chapter we saw evidence that young babies (under 6 months of age) hear a variety of language sounds *categorically*. To some of these sounds, the babies have had very little exposure, and to others, they have had no exposure at all (the sounds are not found in their target language). Is the ability to perceive these sounds categorically an innate capacity that is special to human language? Consider the following additional finding: Some studies of nonhuman animals have revealed an ability to discriminate sounds categorically that is very similar to the ability seen in human infants. Chinchillas and macaque and rhesus monkeys can discriminate voiced from voiceless stops (Waters and Wilson, 1976; Kuhl and Miller, 1978; Kuhl, 1981; Kuhl and Padden, 1982); Japanese quail can be trained to discriminate certain contrasts based on place of articulation (Kluender, Diehl, and Killeen, 1987). Do these results mean that categorical perception is not specific to human language?

Now consider the following further details:

(i) The quail require thousands of training sets, while human infants do not.

(ii) Monkeys' ability to perceive the contrasts requires larger VOT differences and may depend on how they are trained.

(iii) Monkeys discriminate *both* cross-category sounds *and* within-category sounds.

Do these additional facts change the way you think about what's going on with human infants?

2. Infants acquiring Spanish can perceive the same voiced/voiceless distinction as infants acquiring American English. However, their target language, Spanish, has a slightly different contrast from that found in English. Namely, Spanish distinguishes between voiced and plain stops. Interestingly, Spanish-acquiring infants cannot perceive this contrast at the same age as they (and English-acquiring babies) can perceive the English contrast. What do you make of this?

3. Bilingual babies are presented with two sets of phonetic inventories. Bilingual babies are slightly delayed on some of the milestones of early language acquisition, such as first words and first multiword utterances (they catch up with monolinguals by the time they start school, though). Are bilingual babies also delayed in losing the ability to perceive nonnative contrasts? How would you find out? Design an experiment to test this question with regard to Hindi-English bilingual babies.

In thinking about your experiment, you'll need to state the following:

(i) What are the ages of the babies you would test?

(ii) What are the stimuli you would use (what kinds of sounds would be played for the babies)?

(iii) What methodology would you use? This means: How would stimuli be presented to babies? What kind of response on the part of babies would you record and measure?

(iv) What outcome would lead you to think that babies could still perceive nonnative contrasts? Conversely, what outcome would lead you to think that babies could no longer perceive nonnative contrasts?

4. We saw in section 3.4 that by the end of the first year of life, babies have stopped being able to discriminate sounds not found in their native language. There is an important exception to this generalization: at age 12 months, babies can still perceive contrasts involving sounds totally unlike any sound in their own language. For example, English-acquiring babies, at 12 months, can still perceive the difference between dental /|/ and lateral clicks /||/, sounds found in isiZulu (Best, McRoberts, and Sithole, 1988). Why do you think babies can still perceive these sounds? (Hint: *How* are babies perceiving these sounds?)

5. We noted that the majority of English words with two syllables have stress on the first syllable (*PENcil*) and that children acquiring English can use stressed syllables in order to locate word boundaries in this way. But English certainly contains many words with stress on the second syllable (*giRAFFE*). What segmentation errors might occur in the following sentences for the words in boldface?

(i) The **giraffe is** tall.

(ii) The **guitar is** pretty.

(iii) The **platoon is** large.

3.7 References

Adriaans, Frans, and Daniel Swingley. 2017. Prosodic exaggeration within infant-directed speech: Consequences for vowel learnability. *Journal of the Acoustical Society of America* 141: 3070–3078.

Aslin, Richard, David Pisoni, Beth Hennessy, and Alan Perey. 1981. Discrimination of voice onset time by human infants: New findings and implications for the effects of early experience. *Child Development* 52: 1135–1145.

Best, Catherine. 1991. The emergence of native-language phonological influences in infants: A perceptual assimilation model. *Haskins Laboratories Status Report on Speech Research* SR-107/108: 1–30.

Best, Catherine, and Gerald W. McRoberts. 2003. Infant perception of non-native consonant contrasts that adults assimilate in different ways. *Language and Speech* 46: 183–216.

Best, Catherine, Gerald W. McRoberts, and Nomathemba M. Sithole. 1988. Examination of perceptual reorganization for nonnative speech contrasts: Zulu click discrimination by English-speaking adults and infants. *Journal of Experimental Psychology: Human Perception and Performance* 14: 345–360.

Burnham, Denis, Christine Kitamura, and Uté Vollmer-Conna. 2002. What's new, pussycat? On talking to babies and animals. *Science* 296: 1435.

Clements, George N., and Samuel Jay Keyser. 1983. *CV Phonology: A Generative Theory of the Syllable*, pp. 1–191. Linguistic Inquiry Monographs 9. Cambridge, MA: MIT Press.

Cooper, Robin P., and Richard N. Aslin. 1990. Preference for infant-directed speech in the month after birth. *Child Development* 61: 1584–1595.

Cristià, Alejandrina. 2013. Input to language: The phonetics and perception of infant-directed speech. *Language and Linguistics Compass* 7: 157–170.

Cristià, Alejandrina, and Amanda Seidl. 2014. The hyperarticulation hypothesis of infant-directed speech. *Journal of Child Language* 41: 913–934.

Curtin, Suzanne, Toben Mintz, and Morten Christiansen. 2005. Stress changes the representational landscape: Evidence from word segmentation. *Cognition* 96: 233–262.

Cutler, Anne, and David M. Carter. 1987. The predominance of strong initial syllables in the English vocabulary. *Computer Speech & Language* 2(3–4): 133–142.

Cutler, Anne, Jacques Mehler, Dennis Norris, and Juan Segui. 1986. The syllable's differing role in the segmentation of French and English. *Journal of Memory and Language* 25: 385–400.

Cutler, Anne, Jacques Mehler, Dennis Norris, and Juan Segui. 1992. The monolingual nature of speech segmentation by bilinguals. *Cognitive Psychology* 24: 381–410.

DeCasper, Anthony J., and Melanie J. Spence. 1986. Prenatal maternal speech influences newborns' perception of speech sounds. *Infant Behavior and Development* 9: 133–150.

Eimas, Peter. 1975. Speech perception in early infancy. In Leslie B. Cohen and Philip Salapatek (eds.), *Infant Perception: From Sensation to Cognition*, vol. 2, pp. 193–231. New York: Academic Press.

Eimas, Peter, and Joanne L. Miller. 1980. Discrimination of the information for manner of articulation. *Infant Behavior and Development* 3: 367–375.

Eimas, Peter, Einar Siqueland, Peter Jusczyk, and James Vigorito. 1971. Speech perception in infants. *Science* 171: 303–306.

Falk, Dean. 2004. Prelinguistic evolution in early hominins: Whence motherese? *Behavioral and Brain Sciences* 27: 491–541.

Fernald, Anne. 1985. Four-month-old infants prefer to listen to motherese. *Infant Behavior and Development* 8: 181–195.

Flege, James Emil. 1993. Production and perception of a novel, second-language phonetic contrast. *Journal of the Acoustical Society of America* 93(3): 1589–1608.

Grabe, Esther, and Ee Ling Low. 2002. Acoustic correlates of rhythm class. In Carlos Gussenhoven and Natasha Warner (eds.), *Papers in Laboratory Phonology*, vol. 7, pp. 515–546. Berlin: Mouton de Gruyter.

Griffiths, Scott, W. S. Brown Jr., Kenneth J. Gerhardt, Robert M. Abrams, and Richard J. Morris. 1994. Perception of speech sounds recorded within the uterus of a pregnant sheep. *Journal of the Acoustical Society of America* 96: 2055–2063.

Heath, Shirley B. 1983. *Ways with Words: Language, Life and Work in Communities and Classrooms*. New York: Cambridge University Press.

Hirsh-Pasek, Kathy, Deborah G. Kemler Nelson, Peter W. Jusczyk, Kimberly Wright Cassidy, Benjamin Druss, and Lori Kennedy. 1987. Clauses are perceptual units for young infants. *Cognition* 26: 269–286.

Jusczyk, Peter, Heather Copan, and Elizabeth Thompson. 1978. Perception by two-month-olds of glide contrasts in multisyllabic utterances. *Perception and Psychophysics* 24: 515–520.

Jusczyk, Peter, Angela D. Friederici, Jeanine M. I. Wessels, Vigdis Y. Svenkerud, and Ann Marie Jusczyk. 1993. Infants' sensitivity to the sound patterns of native language words. *Journal of Memory and Language* 32: 402–420.

Jusczyk, Peter, Kathy Hirsh-Pasek, Deborah Kemler Nelson, Lori Kennedy, Amanda Woodward, and Julie Piwoz. 1992. Perception of acoustic correlates of major phrasal units by young infants. *Cognitive Psychology* 24: 252–293.

Jusczyk, Peter, Derek M. Houston, and Mary Newsome. 1999. The beginning of word segmentation in English-learning infants. *Cognitive Psychology* 39: 159–207.

Jusczyk, Peter, Paul A. Luce, and Jan Charles-Luce. 1994. Infants' sensitivity to phonotactic patterns in the native language. *Journal of Memory and Language* 33: 630–645.

Kitamura, C., C. Thanavishuth, D. Burnham, and S. Luksaneeyanawin. 2001. Universality and specificity in infant-directed speech: Pitch modifications as a function of infant age and sex in a tonal and non-tonal language. *Infant Behavior and Development* 24: 372–392.

Kuhl, Patricia, Jean Andruski, Inna Chistovich, Ludmilla Chistovich, Elena Kozhovnikova, Viktoria Ryskina, Elvira Stolyarova, Ulla Sundberg, and Francisco Lacerda. 1997. Cross-language analysis of phonetic units in language addressed to infants. *Science* 277: 684–686.

Kuhl, Patricia, Barbara Conboy, Sharon Coffey-Corina, Denise Padden, Maritza Rivera-Gaxiola, and Tobey Nelson. 2008. Phonetic learning as a pathway to language: New data and native language magnet theory expanded (NLM-e). *Philosophical Transactions: Biological Sciences* 363: 979–1000.

Levitt, Andrea, Peter Jusczyk, Janice Murray, and Guy Carden. 1988. The perception of place of articulation contrasts in voiced and voiceless fricatives by two-month-old infants. *Journal of Experimental Psychology: Human Perception and Performance* 14: 361–368.

Lieberman, Amy, Marla Hatrak, and Rachel I. Mayberry. 2014. Learning to look for language: Development of joint attention in young deaf children. *Language Learning and Development* 10: 19–35.

Lisker, Leigh, and Arthur S. Abramson. 1964. A cross-language study of voicing in initial stops: Acoustical measurements. *Word* 20: 384–422.

Liu, Huei-Mei, Feng-Ming Tsao, and Patricia Kuhl. 2007. Acoustic analysis of lexical tone in Mandarin infant-directed speech. *Developmental Psychology* 43(4): 912–917.

Masataka, Nobuo. 1996. Perception of motherese in a signed language by 6-month-old deaf infants. *Developmental Psychology* 32(5): 874–879.

Masataka, Nobuo. 1998. Perception of motherese in Japanese sign language by 6-month-old hearing infants. *Developmental Psychology* 34(2): 241–246.

Masataka, Nobuo. 2000. The role of modality and input in the earliest stage of language acquisition: Studies of Japanese Sign Language. In Charlene Chamberlain, Jill P. Morford, and Rachel I. Mayberry (eds.), *Language Acquisition by Eye*, pp. 3–24. Mahwah, NJ: Lawrence Erlbaum Associates.

Mattys, Sven L., and Peter W. Jusczyk. 2001. Phonotactic cues for segmentation of fluent speech by infants. *Cognition* 78: 91–121.

Mehler, Jacques, Jean Yves Dommergues, Uli Frauenfelder, and Juan Segui. 1981. The syllable's role in speech segmentation. *Journal of Verbal Learning and Verbal Behavior* 20: 298–305.

Mehler, Jacques, Peter Jusczyk, Ghislaine Lambertz, Nilofar Halsted, Josiane Bertoncini, and Claudine Amiel-Tison. 1988. A precursor of language acquisition in young infants. *Cognition* 29: 143–178.

Nazzi, Thierry, Josiane Bertoncini, and Jacques Mehler. 1998. Language discrimination by newborns: Toward an understanding of the role of rhythm. *Journal of Experimental Psychology: Human Perception and Performance* 24: 756–766.

Nazzi, Thierry, Galina Iakimova, Josiane Bertoncini, Séverine Fredoníe, and Carmela Alcantara. 2006. Early segmentation of fluent speech by infants acquiring French: Emerging evidence for cross-linguistic differences. *Journal of Memory and Language* 54: 283–299.

Ochs, Elinor, and Bambi Schieffelin. 1984. Language acquisition and socialization: Three developmental stories and their implications. In Richard Shweder and Robert A. LeVine (eds.), *Culture Theory*, pp. 276–322. New York: Cambridge University Press.

Peters, Aemil J. M., Kenneth J. Gerhardt, Robert M. Abrams, and Jeffrey A. Longmate. 1993. Three-dimensional intraabdominal sound pressures in sheep produced by airborne stimuli. *American Journal of Obstetrics and Gynecology* 169: 1304–1315.

Querleu, D., X. Renard, F. Versyp, L. Paris-Delrue, and P. Vervoort. 1988. La transmission intra-amniotique des voix humaines. *Review of French Gynecology and Obstetrics* 83: 43–50.

Richman, Amy L., Patrice M. Miller, and Robert A. LeVine. 1992. Cultural and educational variations in maternal responsiveness. *Developmental Psychology* 28: 614–621.

Soderstrom, Melanie. 2007. Beyond babytalk: Re-evaluating the nature and content of speech input to preverbal infants. *Developmental Review* 27: 501–532.

Strange, Winifred, and James J. Jenkins. 1978. The role of linguistic experience on the perception of speech. In Richard D. Walk and Herbert L. Pick (eds.), *Perception and Experience*, pp. 125–169. New York: Plenum.

Streeter, Lynn. A. 1976. Language perception of 2-month-old infants shows effects of both innate mechanisms and experience. *Nature* 259: 39–41.

Trehub, Sandra. E. 1973. Infants' sensitivity to vowel and tonal contrasts. *Developmental Psychology* 9: 91–96.

Trehub, Sandra. E. 1976. The discrimination of foreign speech contrasts by adults and infants. *Child Development* 47: 466–472.

Tsushima, Teruaki, Osamu Takizawa, Midori Sasaki, Satoshi Siraki, Kanae Nishi, Morio Kohno, Paula Menyuk, and Catherine Best. 1994. Discrimination of English /r-l/ and /w-y/ by Japanese infants at 6–12 months: Language specific developmental changes in speech perception abilities. Paper presented at the Third International Conference on Spoken Language Processing, Yokohama, Japan.

Werker, Janet F., Ferran Pons, Christiane Dietrich, Sachiyo Kajikawa, Laurel Fais, and Shigeaki Amano. 2007. Infant-directed speech supports phonetic category learning in English and Japanese. *Cognition* 103: 147–162.

Werker, Janet F., and Richard C. Tees. 1984a. Cross-language speech perception: Evidence for perceptual reorganization during the first year of life. *Infant Behavior and Development* 7: 49–63.

Werker, Janet F., and Richard C. Tees. 1984b. Phonemic and phonetic factors in adult cross-language speech perception. *Journal of the Acoustical Society of America* 75: 1866–1878.

4 Speech Production and Phonological Development

In the last chapter we saw that as babies progress toward their first birthday they gradually unlearn the ability to distinguish all the possible human speech sounds and instead zero in on the ones that matter for their language—the phonemes. Even though the child's linguistic knowledge is still growing during this phase, in another sense it is a period of pruning, or cutting back. Linguist Roman Jakobson described this transition by saying that "the phonetic richness of the babbling period … gives way to a phonological limitation." That is, while very young infants can both produce and perceive a wide range of speech sounds, even those not found in the language they will acquire, by the time they begin to learn their first words, their sound system has shrunk. In some cases, in terms of early word production, this means that children fail to produce certain sounds in their language that they may have used in babbling just 6 months earlier. For example, sounds articulated at the back of the vocal tract, such as velar sounds ([k], [g]) can be found in early vocalizations but are rare in children's first words (Stoel-Gammon, 1989).

In this chapter we'll look at how this happens: how children's phonological inventory takes shape, the kinds of phonological patterns and errors we see, and how we can explain those patterns and errors by writing phonological rules or by ordering universal constraints.

4.1 When Are Vocalizations Part of Language?

Just as we considered in the previous chapter the question of when language *perception* starts, here we can ask when language *production* begins. When a newborn produces all manner of cries and coos and vocalizations, are these part of language? Probably not. So when is a baby's sound production really a linguistic sound? It turns out we can distinguish two kinds of vocalizations within the first year of life. The very earliest vocalizations are unstructured and include a wide array of types of sounds: gurgles, coos, clicks, and so forth. The infant's vocal tract within the first 3 months of life is actually quite different from an adult's: as shown in figure 4.1, the infant's larynx is high up in the throat, which allows the baby to drink and breathe at the same time and shortens the pharyngeal cavity (the space between the larynx and the epiglottis). The infant's tongue is proportionally

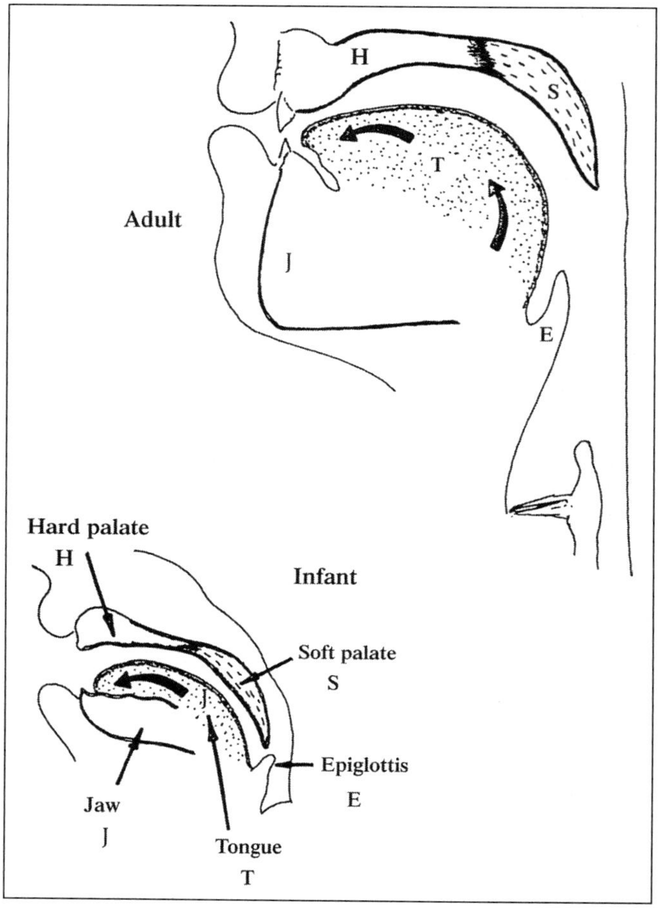

Figure 4.1
Illustration of adult and infant vocal tracts. (From Kent and Miolo 1995. Reproduced with permission.)

larger in the mouth than an older child's or an adult's. Moreover, infants' fine motor control abilities, necessary for phonetic articulation, are not quite developed until later (around 7 to 10 months). The vocalizations produced at this early stage are sometimes called simply "phonation" (Oller, 1978, 1980) or "vegetative sounds" (Stark, 1980).

At around 3 to 4 months of age, the vocal tract reconfigures: the larynx drops down into the throat, lengthening the pharyngeal cavity. If you listen to a newborn baby and a 4-month-old baby cry, you'll notice that they sound quite different from each other. The newborn's cry has an almost animal-like quality to it, while the slightly older infant's cry will sound more like a prototypical baby's.

Sidebar 4.1: What Infant Speech Sounds Like

> Listen to the sound files OM2-24-08 and OM3-30-08 to hear what an infant sounds like at 6 weeks and 3 months of age.

Shortly after this time, around 4 to 6 months of age, babies' vocalizations change in an important way: they become structured units of sound we recognize as **syllables**. As we saw in chapter 3, a syllable is a unit of sound centered on a "peak of sonority" (Clements and Keyser, 1983). What is a peak of sonority? Usually it's a vowel (V), and it can have one or more "obstructions" of sound, or consonants (C), on either side of it. In one study of the types of syllables babies used in babbling, the breakdown of syllable types was the following, which accounted for 98% of these children's vocalizations at about 1 year of age (Kent and Bauer, 1985).

These vocalizations are known as **canonical babbling**. They are language-like in a couple of ways: First, they contain syllables. Syllables are what make speech sound "speechy," as opposed to moans, groans, cries, giggles, whistles, and other nonlinguistic sounds humans can produce. Second, the *kinds* of syllables we find in canonical babbling are significant: looking at table 4.1 we can see that although babies at this stage produce a lot of vowels alone (60% of their vocalizations are just V syllables), when children do include a consonant it almost invariably *precedes* the vowel. Thirty-four percent of their productions comprise a CV syllable, its reduplicated form (CVCV), and a VCV sequence. A mere 4% of their productions involve a syllable ending in a consonant. Looking across the adult languages of the world, CV is the most common kind of syllable. The fact that babies' canonical babbles share this property with adult language suggests that these productions are not just random but instead are part of the child's budding language system.

There are some other links between canonical babbling and language. One is that although babies may babble with sounds not found in their target language (e.g., French-acquiring babies may produce [ha] or [hə], even though French does not have the phone [h]), as they progress toward the end of their first year, their babbles come to sound more and more like the target language (de Boysson-Bardies and Vihman, 1991). One study revealed that adults could, to a significant degree, determine the target language of an 8-month-old baby simply on the basis of the baby's babbles (de Boysson-Bardies, Sagart, and Durand, 1984). In fact, it can be difficult to tell the difference between a babble and a word. When a child acquiring English says [mama], is this simply a reduplicated CVCV babble, or is it the word *mama*? We normally assume that it is a word only if it has a fixed meaning associated with it (the child uses [mama] to refer to their mother or another person), but it can be surprisingly hard to tell sometimes if a baby is using a "word" with a fixed meaning. One child used the form [mama] to mean any human.[1] Another link between babbling and language is that deaf babies who are exposed to sign language in infancy

Table 4.1
Predominant vocalizations of 1-year-olds

Syllable structure	Frequency
V	60%
CV	19%
CVCV	8%
VCV	7%
VC	2%
CVC	2%

Sidebar 4.2: Changes in Infant Speech

Listen to the sound files OM5-10-08, OM5-29-08, and OM6-29-08. The ages of the baby in these files are 4 months, 4½ months, and 5½ months, respectively. What shift do you notice between the first and second sound files?

begin to babble with their hands around the same time that hearing babies produce canonical babbles (Pettito, 1991; see chapter 8, section 8.2).

4.2 Building a Sound System

4.2.1 What Is a Phoneme?

Once a child starts producing words (as opposed to babbling, as discussed above), we can say that the child is developing an inventory of **phonemes**. Linguistic sounds are also called **segments** or **phones**, but the word *phoneme* has a special meaning: it is a sound that exists in the speaker's mental representation. We introduced this concept in chapter 3, but let's review it briefly. For an example of two different phonemes, think of two words you know that differ only by a single sound, such as *bat* and *pat* (linguists call such pairs of words **minimal pairs**). If you are an English speaker, you recognize these words as being different because the sounds /b/ and /p/ are represented in your mind as different sounds—they are different *phonemes*, and we say that they *contrast* with one another. (Recall that linguists write symbols for phonemes between two forward slashes, //, to differentiate the mental representation from the phonetic segment, which is written between square brackets [].)

Now consider two other words that you recognize: *pat* and *spat*. As we mentioned in section 3.2, the two /p/ sounds in these words are actually different phones. The /p/ sound at the beginning of *pat* is aspirated (there's a tiny puff of air that escapes your lips when you say it), while the /p/ sound in *spat* is unaspirated (no tiny puff of air, or at least a very much tinier one). If you put your hand in front of your mouth when you say these two words, you should feel the difference. We write that puff of air with a superscript 'h' symbol, so writing these words phonetically we have [p^hæt] and [spæt]. The sounds [p^h] and [p] are actually different phones, or segments (recall from chapter 3 that their difference has to do with **voice onset time**, or VOT), but if you are an English speaker they are the same *phoneme*—in your mind they are represented as being the same sound (i.e., they are non-contrastive for you).

4.2.2 Early Phoneme Inventory

The shift from perceiving sounds as *phones* to perceiving sounds as *phonemes* is a critical step in the development of language. Having a mental representation of a sound means that you can recognize it even when it is spoken in different words, by different speakers, and with different phonetic properties (like in the aspirated versus unaspirated /p/ example). This is just one example of a process we'll see repeatedly in many domains within language acquisition: the process of **generalization**. It's an interesting question to ponder how children arrive at these abstract mental representations of sound. We don't know exactly how this happens, but we can look at what children's early phoneme inventories look like and consider what might be possible influences on those inventories.

It turns out that across languages we see a remarkable amount of similarity in children's early phoneme inventories. Jakobson was one of the first modern linguists to notice that children acquiring various languages all start out with /a/ as their earliest vowel and a labial stop (a sound made with the lips, such as /p/, /b/, or /m/) as their earliest consonant. Soon children begin to develop other consonants that contrast with their first one along one phonetic dimension. For example, children will develop /m/ and /b/ or /p/, which contrast in the [+/- nasal] feature but share the [+ labial] feature, or /t/, /d/, or /n/, which contrast with /p/, /b/, and /m/ in their place of articulation (/p/ is labial; /t/ is coronal—made with the tip or blade of the tongue). Sounds produced much later include velars and palatals (place of articulation), and the early-produced oral stops and nasals are later joined by fricatives and liquids (manner of articulation).

Jakobson claimed that there was a **universal** order of acquisition of phonemes, and he argued that children adhered to this order because it was a fundamental property of human language that certain classes of sounds were more basic than others, and these sounds had to be acquired first. More recent research has revealed that his claim about the universal order was too strong, but nevertheless there are robust tendencies across children and across languages. For instance, de Boysson-Bardies and Vihman (1991) tallied the places and manners of articulation of consonants produced by babies acquiring English, French, Japanese, and Swedish in these babies' babbles and first words. In terms of manner of articulation, all of the babies produced many more stops than fricatives, nasals, or liquids. For place of articulation, the English- and French-acquiring babies produced more labials than dentals

Sidebar 4.3: Markedness

The idea of **markedness** plays a central role in much of generative linguistics. Markedness is the idea that some features of language are extremely common or robust across the world's languages, while other features are more unusual. Features or properties that are common are **unmarked**, while features or properties that are rare are **marked**. In phonology, we can say that certain classes of sounds are unmarked (stops and coronals—sounds articulated with the tip or blade of the tongue) while others are marked (pharyngeals, clicks) and that certain syllable shapes are unmarked (CV) while others are marked (CCCVCCC). As we'll see in future chapters, markedness surfaces in other parts of linguistics as well. In syntax, for example, certain word orders or sentence constructions can be considered unmarked (SVO and SOV word orders) while others are marked (OVS). Oftentimes unmarked forms are acquired earlier than marked forms, though this is not always the case (e.g., the "verb second" rule in the syntax of some languages is a marked construction, but it is acquired early by children).

(sounds made by touching the tongue to the teeth), while the Japanese- and Swedish-acquiring babies produced more dentals than labials; but all babies produced velars (articulated at the soft palate, or velum) at the lowest rate.

Interestingly, children's order of acquiring phonemes (in production) is not only widespread across child languages; it also shares some patterns that we find in adult language, in that children's earliest sound features tend to be the most common across adult languages. For example, all known languages of the world have stops, but not all languages have fricatives (Maddieson and Precoda, 1990).[2] Likewise, coronal and labial sounds are crosslinguistically extremely common (all consonant inventories have at least one coronal sound, and [m], a labial consonant, is the most widely attested phoneme, found in 95% of the world's languages [Hyman, 2008; Moran, McCloy, and Wright, 2014]), but palatal sounds are rarer. While it is true that some languages have no labial stops, such as the Iroquois languages, which may have [m] and a labio-velar glide [w] but not [b] or [p], this situation is extremely rare crosslinguistically.

While Jakobson claimed that children's phonological inventories adhere to a universal hierarchy of sounds, subsequent research has uncovered some variation in how children construct their sound systems. Some of these differences appear to be due to individual variation, such as an English-learning child who uniformly replaces the alveolar /s/ with the alveopalatal [ʃ], even though other speakers of the same language do not do this, or a child who turns word-final /g/ or /k/ into a glottal stop ([ʔ]). Another deviation from Jakobson's universal order is found more generally in English and other languages; namely, the palatal glide /j/ is produced quite early. Other differences appear to be related to the particular target language. For example, one comparison between toddlers acquiring English and toddlers acquiring Japanese found that while English-acquiring toddlers tend to front /ʃ/ to [s] (this is known as "fronting," a type of substitution process; see section 4.3.1), Japanese toddlers instead tend to "back" /s/ to [ʃ] (phonetically, the post-alveolar fricative in Japanese is actually [ɕ], but it sounds to English speakers like [ʃ], possibly due to perceptual assimilation [Li, Edwards, and Beckman, 2009]). A study of Finnish-speaking children's development found that the [d] was produced late in development, while [r] was produced earlier (Itkonen, 1977, cited in Dunbar and Idsardi, 2016).

What is it about individual languages that yields these differences? An important study by Pye, Ingram, and List (1987) looked at the phoneme inventories of children acquiring K'iche', a Mayan language, and asked whether the frequency with which particular sounds occurred in the target language might be related to the order in which they were acquired: more frequently occurring sounds might be acquired earlier. These researchers discovered some interesting differences between the early inventories of the K'iche'-speaking children as compared to English-speaking children at the same age. Restricting their study to word-initial consonants, they found that K'iche' children began to produce the affricate [tʃ], glottal stop [ʔ], and [l] earlier than English-speaking children, while English-speaking children began producing [s] and [n] earlier than the K'iche'-speaking children. Pye, Ingram, and List suggested that these crosslinguistic differences could be due to the relative frequency with which certain sounds appear in the target language. For example, the affricate /tʃ/ is more frequent in K'iche' than in English, and it is also produced earlier by K'iche'-speaking children than English-speaking children; conversely, /s/ is more frequent in English speech than in K'iche' speech, and it is produced earlier by English-speaking children.

On the other hand, we know that frequency of occurrence in the input cannot be the sole (or even the main) explanation for why children acquire sounds in the order they do. In the de Boysson-Bardies and Vihman (1991) study described above in this section, the English target words contained more dentals than labials, but the English-acquiring babies still produced more labial sounds. Moreover, according to one ranking of the frequencies of phonemes in spoken English, some of the most frequent consonants include [r], [l], and [ð], all of which are produced fairly late, while some of the earliest-acquired sounds, such as [p] and [b], are relatively less frequent.[3]

To summarize, children's development of their productive sound inventory exhibits certain orderings, such as the widespread acquisition of front sounds (labials and dentals) and stops before back sounds and non-stops (fricatives, affricates, liquids). To the extent that these orderings are found crosslinguistically, we can see children's development as adhering to universal markedness scales. However, variation is also to be found within and across languages.

4.3 Common Phonological Processes

When listening to small children speak their first words and sentences, it is easy to observe that they do not sound exactly like adults—they alter the phonological form of many words. These alterations are found from the time children begin producing words around age 1 year into the beginning of their third year. Many of them resolve by age 2;6 or 3. The following are examples of mispronunciations or phonological “errors”:

(1) *cup* [tʌp]
water [wawa]
banana [nænə]

Children’s phonological errors are sometimes cute or funny, but what’s interesting about them is that they are *highly regular*. That is, we see very consistent patterns across children acquiring the same or different languages, and we also see evidence that children apply their alterations in an across-the-board way. For example, if a child pronounces /k/ as [t] at the beginning of one word (e.g., *cup* as [tʌp]), they are very likely to pronounce /k/ as [t] at the beginning of other words, too (e.g., *coffee* as [tafi]).

Due to the strong regularity of children’s mispronunciations, we can identify a number of patterns that we call **phonological processes** (Ingram, 1979). There are three main categories of phonological processes: substitutions, assimilations, and syllabic processes. Within each category we can then identify a number of specific subtypes.

4.3.1 Substitutions

Children may **substitute** one sound for another, where the two sounds involved usually share several features but differ in either the place or manner of articulation. One very common type of substitution is **stopping**. Stopping involves substituting a stop for a fricative or other non-stop sound, such as a liquid. Another very common substitution is **fronting**, in which a more back sound, such as a velar or palatal consonant, is replaced with a more front sound, such as an alveolar consonant. Examples are given in table 4.2. Fronting and stopping are extremely common processes across many child languages.

Another kind of substitution process we observe is **gliding**. Gliding involves a liquid (/l/ or /r/) being replaced with a glide ([j] or [w]), as in *yellow* being pronounced [jɛjoʊ] or *room* being pronounced [wuːm]. The rhotic liquids (the ‘r’ sounds) can vary quite a bit from language to language (some are alveolar sounds like Spanish or Estonian /r/; some are central approximants like English /ɹ/, and some are uvular trills or fricatives like French /ʁ/). With respect to children’s phonological alterations, sometimes these phonetically distinct sounds nevertheless function similarly in the phonological system, as we see in table 4.2. However, some differences can also be found across languages: the alveolar trill /r/ is often replaced with the lateral liquid [l], or in some cases the stop [d] (Tessier, 2016).

Finally, sometimes a vowel is substituted for a consonant, often a liquid or glide (these classes of sounds are more similar to vowels—they are more *sonorant*—than certain other consonants, like stops or fricatives). This kind of substitution is called **vocalization**. Note that in the examples in table 4.2 (*bottle* pronounced [babu] and *apple* pronounced [apo]), it is a velarized /l/ (phonetically, [ɫ]) that is being vocalized.

Table 4.2
Examples of children's substitution phonological processes

Phonological process	Target word	Child's pronunciation	Sounds involved
Stopping	sea /si/	[ti]	/s/ → [t]
	that /ðæt/	[dæt]	/ð/ → [d]
	fleur 'flower' /fløχ/ (French)	[pø]	/f/ → [p]
Fronting	shop /ʃap/	[sap]	/ʃ/ → [s]
	Katharina /kataʁina/ (German)	[tataʁina]	/k/ → [t]
	cassé 'broken' /kase/ (French)	[tase]	/k/ → [t]
Gliding	yellow /jɛloʊ/	[jɛjoʊ]	/l/ → [j]
	room /ɹu:m/	[wu:m]	/r/ → [w]
	raha 'money' /raha/ (Estonian)	[jaha]	/r/ → [j]
	robe 'dress' /ʁɔb/ (Québec French)	[wɔb]	/ʁ/ → [w]
Vocalization	bottle /batəl/	[babu]	/l/ ([ɫ]) → [u]
	apple /æpəl/	[apo]	/l/ ([ɫ]) → [o]

Table 4.3
Examples of children's assimilation phonological processes

Phonological process	Target word	Child's pronunciation	Sounds involved
Consonant assimilation (place)	duck /dʌk/	[gʌk]	/d/ → [k]
	tape /teɪp/	[beɪp]	/t/ → [b]
Consonant assimilation (manner)	mouton 'sheep' /mutõ/ (French)	[potõ]	/m/ → [p]
Voicing	pig /pɪg/	[bɪk]	/p/ → [b] and /g/ → [k]
	bed /bɛd/	[bɛt]	/d/ → [t]

4.3.2 Assimilations

Sometimes children alter a feature of a sound to make it more similar to a neighboring sound; this is called **assimilation**. Some examples are given in table 4.3. Here we'll discuss **consonant assimilation** (or consonant harmony), which can involve place assimilation (e.g., *duck* pronounced [gʌk]) or manner assimilation (e.g., the French word *mouton* [mutõ] 'sheep' pronounced as [potõ]). In this latter case, the first consonant takes on the voicing feature ([-voice]) of the second consonant; when [m] loses voicing, it also loses its nasality and becomes [p].

Another process that is considered an assimilatory process involves **voicing**. However, the voicing process in child language is not straightforwardly assimilatory the way consonant and vowel harmony are. Instead, children generally voice word-initial consonants (or produce them as voiceless unaspirated consonants if they are stops) and devoice word-final consonants. For example, *pig* can be pronounced [bɪk] and *bed* would be pronounced [bɛt].

4.3.3 Syllabic Processes

A third way that children modify the phonological shape of words is by changing their syllable structure. First, let's review basic syllable structure. As we saw in chapter 3 (section 3.3.2), we can represent the syllable, symbolized with the Greek letter σ (sigma), as a hierarchical structure.

(2)

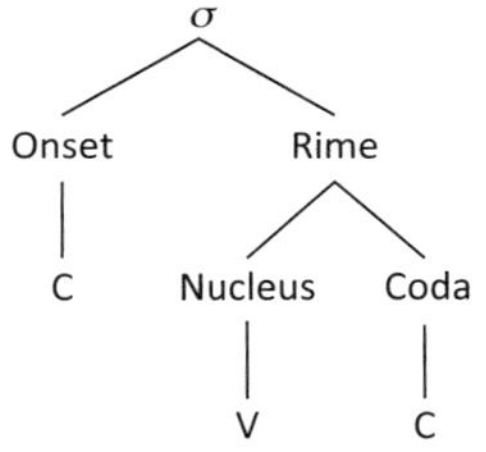

In many adult languages the onset and/or the coda can contain multiple segments. English allows this, as we see in the example in 3, which gives the syllable structure for the word *streets*.

(3)

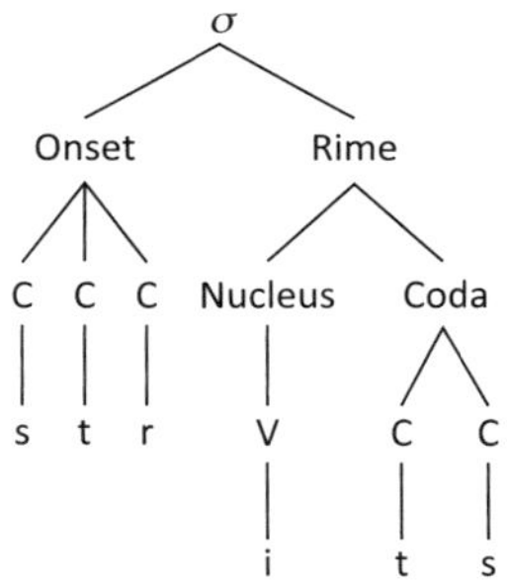

One very common syllabic process is **consonant cluster reduction** (CCR). CCR can happen either in onset or coda position, and it involves a complex onset or coda (more than one consonant) being reduced to a single segment. It is important to remember that sounds such as [θ] as in *think*, [ð] as in *that*, and [ʃ] as in *ship* are single phonemes, not clusters.

Sidebar 4.4: Phonological Processes

Which phonological process or processes apply when a child pronounces *that* as [dæt] or *ship* as [sɪp]?

Table 4.4
Examples of children's syllabic phonological processes

Phonological process	Target word	Child's pronunciation	Sound change involved
Consonant cluster reduction (onset)	clown /klaʊn/	[kaʊn]	/kl/ → [k]
	prune /pɹun/	[pun]	/pr/ → [p]
Consonant cluster reduction (coda)	box /baks/	[bak]	/ks/ → [k]
	desk /dɛsk/	[gɛk]	/sk/ → [k]
Coda deletion	egg yolk /ɛg jok/	[ɛ jok]	/g/ → ∅
	more /mɔɹ/	[mɔ]	/ɹ/ → ∅
Weak syllable deletion	banana /bənænə/	[nænə]	/bə/ → ∅
	Rapunzel /ɹəpʌnzəl/	[pãzow]	/ɹə/ → ∅
Reduplication	water /watɚ/	[wawa]	
	raisins /ɹezɪnz/	[ɹiɹi]	
	vache 'cow' /vaʃ/ (French)	[vava]	

So when a child pronounces *that* as [dæt] or *ship* as [sɪp], these are not examples of cluster reduction.

Sometimes a coda is deleted altogether, which is called **coda deletion**. Onsets are not deleted as frequently as codas, though this does sometimes occur. One child pronounced *lollipop* as [alɛbap], deleting the word-initial consonant /l/.

Two other syllabic processes can be identified: **weak syllable deletion** and **reduplication**. Weak syllable deletion involves a multisyllabic word, in which an unstressed syllable is deleted. For example, *banana* might be pronounced [nænə]. Sometimes a feature or segment of the deleted syllable is retained, so that *banana* could also be pronounced [bænə]. Reduplication involves the doubling of a syllable. Examples of all these syllabic processes are given in table 4.4.

Most of these phonological processes are found widely, both across children acquiring the same language and across children acquiring different languages. But does this mean that all children the world over acquire their phonological systems in exactly the same way? Just as we saw that there is some variation across children, and across languages, in the order in which children acquire their phoneme inventories, there is some individual and crosslinguistic variation in the phonological processes children apply in their speech (see table 4.5). For example, according to Ingram (1979), French-speaking children are more likely to simply delete a later liquid /l/ than replace it with a glide, as in *lapin* [lapɛ̃] 'rabbit' pronounced [apɛ̃]. In addition, certain substitutions are found in some languages but not others. Children acquiring English tend to replace the interdental fricative [ð] with [d] (this is **stopping**, as in *that* pronounced as [dæt]), but Greek-speaking children instead replace [ð] with [l], and children acquiring Spanish tend to replace it with either [l] or [r]. Another difference we find in substitution patterns is that while English-speaking children

Table 4.5
Crosslinguistic substitution patterns

Language	Target sound	Substituted sound
English	ð	d
	r	w
	l	j
Greek	ð	l
Spanish	ð	l, r
K'iche'	r	l

often substitute [r] with [w] and [l] with [j], children acquiring K'iche' substitute [r] with [l].

4.3.4 Covert Contrasts

It is important to bear in mind that children's phonological "errors" may not always be true errors. That is, children may produce two sounds in such a way that adult listeners cannot perceive a difference between them, but an acoustical analysis reveals that the two sounds, which sound identical to adults, are actually phonetically distinct. When such phonetic differences are statistically reliable, even if they are not perceivable by listeners, they are known as **covert contrasts**. Many studies of children's pronunciations rely on human transcribers, who by definition cannot accurately represent covert contrasts, but in some studies a finer-grained acoustical analysis has been done.

For example, Li et al.'s (2009) study of English-speaking and Japanese-speaking toddlers (see section 4.2.2) found that some of the English-speaking children who appeared to neutralize /s/ and /ʃ/ (pronouncing both sounds as [s]) actually showed a statistically significant difference between the two sounds at a phonetic level. Similarly, some of the Japanese-speaking children who appeared to neutralize both of their fricatives (pronouncing both /s/ and /ʃ/ as [ʃ]) exhibited a measurable acoustical difference between their two sounds, even if this difference was not audible to human listeners.

These findings suggest that at least some of children's apparent mispronunciations may be more adultlike than they appear. The notion that children's representations of sounds are actually quite adultlike will be further supported in the sections below.

4.4 Accounting for Patterns: Phonological Rules

We've just seen that some of children's apparent phonological errors may not be as far off the mark as we might at first think. However, in many cases when a child mispronounces a word, they truly omit or substitute a sound, resulting in a non-adultlike or non-target-like pronunciation. Putting aside the possibility of covert contrasts for a moment, we must try to explain children's mispronunciations. Are children unable to hear the target (adult-

Sidebar 4.5: Imitating Mom

> Listen to the sound file LM12.1.12pepper.WMA. How does the child respond to her mother's imitations of her word *pepper*? (Try to ignore the older sibling's playful productions.)

like) form or to tell the difference between their own pronunciation and the target one? Are children simply incapable of articulating certain sounds? There is a third possibility: children represent a word's phonological form correctly (i.e., the adult way) and may be capable of articulating the adult sound, but they alter it through the application of phonological *rules*. In this sense, children's errors arise not from problems with hearing or articulation but rather from their phonological grammar (Smith, 1973).

A couple of kinds of evidence support this view. One is that children often *reject* adults' attempts at reproducing the child's own pronunciation. In the following exchange between linguist Neil Smith and his son, Amahl, notice that Amahl does not perceive his father's [sɪp] as corresponding to [ʃɪp], even though [sɪp] is Amahl's own pronunciation of /ʃɪp/ (Smith, 1973).

(4) Father: What's a [sɪp]?
Amahl: When you drink.
Father: What else does [sɪp] mean?
Amahl: (puzzled, suggests *zip*)
Father: No; it goes in the water.
Amahl: A boat.
Father: Say it.
Amahl: No. I can only say [sɪp].

Amahl may have an unusual degree of awareness of his own pronunciations (compared to children whose parents are not linguists), but crucially, he does not hear his father's production of [sɪp] as equivalent to his own pronunciation of /ʃɪp/, even though phonetically they are the same.

That children's phonological "errors" do not arise (at least, not exclusively) from articulatory problems is evidenced by the fact that sometimes a particular phonetic form that the child *appears* unable to produce is produced for another lexical target. At age 2;1 Neil Smith's son, Amahl, produced the following words:

Word	Target	Child's form
puddle	/pʌdəl/	[pʌgəl]
puzzle	/pʌzəl/	[pʌdəl]
thick	/θɪk/	[fɪk]
sick	/sɪk/	[θɪk]

'e pronunciations, could you argue that Amahl's nontarget pronunciations of *thick* were due to an inability to pronounce the segments [d] and [θ] in the -get environments? What problem would such an argument encounter?

If children's phonological "errors" are not the result of problems hearing the way adults pronounce words or the result of problems articulating those sounds, then what accounts for children's mispronunciations? Many linguists believe that children actually store a mental representation of sounds the same way adults do. That is, their **phonemic representation** matches the adult's phonemic representation (so, even if a child pronounces *sick* as [θɪk], their own stored form is /sɪk/). But children apply a series of phonological **rules** that convert their underlying phonemic form into a phonetic form that is slightly different. So, what are those rules? First, let's review what a phonological rule looks like. Here is its basic form:

(5) /phonemic form/ → [phonetic form] / (preceding environment) ____ (following environment)

The blank corresponds to the sound that is getting changed by the rule. Depending on the rule, we may need to specify just the preceding environment, just the following environment, or both. For example, consider the following two lists of English words and one child's pronunciation of these words (from Smith, 1973).

(6) a.		b.
stamp [dɛp]		window [wɪnuː]
bump [bʌp]		handle [ɛŋu]
drink [gɪk]		finger [wɪŋə]
tent [dɛt]		angry [ɛŋiː]
uncle [ʌgu]		hand [ɛn]
empty [ɛbiː]		band [bɛn]
thank you [gɛguː]		

Consider first the words in 6a. Several different processes may be going on within each word in this set, but notice that in all of the words there is a nasal consonant getting deleted. If we had to write a rule for the *whole* set, we could write a "nasal deletion" rule—that is, a rule that specifies the environment in which a nasal sound is deleted. What is the environment in which it is getting deleted? If we look at the preceding environment (the sound preceding the nasal), we don't see anything interesting: in every case the preceding sound is a vowel, but these vowels do not form any kind of natural class. That is, they are not all high vowels or all back vowels, for example. In some words, there is a consonant preceding that vowel (as in *tent*), but not in all cases (as in *empty*), so that is not helpful for our rule either. Now consider the environment following the nasal in each word. Do these sounds have anything in common? Yes! They are all stops. Do they have anything else in common? Yes! They are voiceless stops. Does it matter that they are voiceless? We don't know yet, without looking at more data.

Now consider the words in 6b. These words contain nasal consonants too, but this time the nasals are not getting deleted. Rather, something else is getting deleted. What is it? Stops! And where are the stops being deleted? After nasals! But we just saw a set of words in which a nasal followed by a stop resulted in the nasal being deleted. Why isn't that happening here?

If you noticed that the stops that get deleted are all **voiced**, you are very observant. Now we can see that in the 6a list, when a nasal is followed by a voiceless stop, the nasal is deleted. But in the 6b list, when a nasal is followed by a voiced stop, the (nonnasal) stop is deleted. We can now write our rules as in 7.

(7) a. $\underset{\begin{bmatrix}+\text{nas}\\+\text{stop}\end{bmatrix}}{\text{C}} \rightarrow 0 \ / \ \_\_\_\_ \ \underset{\begin{bmatrix}-\text{voi}\\+\text{stop}\end{bmatrix}}{\text{C}}$ b. $\underset{[+\text{voi}]}{\text{C}} \rightarrow 0 \ / \ \underset{[+\text{nas}]}{\text{C}} \ \_\_\_\_$

In some cases, it is not the immediately preceding or following phoneme that conditions the rule but rather the position in the word or syllable in which the sound in question appears. For example, if a child produces the following pronunciations,

(8) a. *sun* [dʌn]
b. *scissors* [dɪdə]
c. *soon* [du:n]

we could write a rule stating that /s/ becomes [d] at the beginning of a word (we use the '#' symbol to mean a word boundary).

(9) /s/ → [d] / # ____

This rule simply says that the phoneme /s/ becomes the phonetic form [d] when it occurs right after a word boundary (i.e., at the beginning of a word).

Sometimes linguists come up with a rule that covers some data, only to discover other data that requires making the rule more precise. Suppose we now discover that the same child who produced the words in 8 also produces these forms:

(10) a. *sing* [gɪŋ]
b. *sock* [gɔk]

Here we see an /s/ at the beginning of the word that does not become [d]; rather, it becomes [g]. What could account for this difference? Think about the types of consonants that occur at the ends of these words, compared to the words in 8. What is their place of articulation? What place of articulation does the /s/ take on in the phonetic form?

By thinking about sounds in terms of their **features**, we can recognize that in the words in 8, /s/ is becoming an alveolar stop ([d]) when the next consonant in the word is alveolar, but in 10 it becomes a velar stop ([g]) when the next consonant in the word is velar. Thus, we can recognize what's happening here as a combination of stopping (the fricative /s/ becomes a stop) and place assimilation (it assimilates to the place of articulation of the next consonant; we notate the intervening vowel with V but leave out the vowel's place features, since it doesn't seem to matter which vowel intervenes). We can either write two rules, like in 11a, or we can use the convention shown in 11b, where the 'α' symbol simply means "whichever place variable" the following consonant (C) has.

(11) a. $/s/ \rightarrow [d]\ /\ \underline{\quad}\ V\ \underset{[+alv]}{C}$

$/s/ \rightarrow [g]\ /\ \underline{\quad}\ V\ \underset{[+vel]}{C}$

b. $/s/ \rightarrow \begin{bmatrix} +stop \\ \alpha place \end{bmatrix}\ /\ \underline{\quad}\ V\ C\alpha$

An important property of rule-based phonology is that rules must be **ordered** in a particular way. That is, if one rule alters the conditioning environment for another rule, ordering the rules the wrong way will result in one of the rules failing to apply. Consider the following two rules, one of which we saw above in 7.

(12) a. $\underset{\begin{bmatrix} +voi \\ +stop \end{bmatrix}}{C} \rightarrow 0\ /\ \underset{[+nas]}{C}\ \underline{\quad}$

b. $\underset{[-son]}{C} \rightarrow [-voi]\ /\ \underline{\quad}\ \#$

The first rule says "a voiced stop consonant is deleted after a nasal consonant" (this accounts for why *hand* was pronounced [ɛn], for example). The second rule says "an obstruent (fricative, affricate, or stop) becomes voiceless at the end of a word." This rule accounts for words like *bed* being pronounced as [bɛt]. Given Amahl's pronunciation of the words *hand* as [ɛn] and *mend* as [mɛn], how do these two rules need to be ordered with respect to each other?

4.5 Accounting for Patterns: Constraints

Phonological rules allow us a means of explaining how children could store sounds with an adultlike representation yet pronounce words so differently from adults. Rules also help us to detect patterns in children's pronunciations; for example, when a child applies a given rule to a sound in one word, the same rule will apply when the same sound appears in another word, provided the conditioning environment is the same. Finally, rules are helpful because they show the striking similarities between a child's mental grammar and an adult's: children are not employing a radically different mechanism to produce language than adults use. Instead, the same sorts of rules apply (deletion rules, insertion rules, assimilation rules, and so on), and they apply to the same classes of sounds (nasals, stops, fricatives, high vowels, and so on) as in adult language.

But one thing that is slightly odd about conceiving of children's phonology this way is that it implies that a child's mental grammar contains a lot of extra stuff that is then lost as the child matures and develops into an adult speaker. That is, the rules that apply to child's grammar to yield their pronunciations do not apply to the adult language. The same *kinds* of rules found in child phonology are found in various adult languages (assimilation rules, deletion rules, and so on), but the whole reason we needed to propose these particular rules in section 4.3 was to explain why children sound *different* from adults. Therefore, these particular rules can't apply in the child's particular target language.

There is another way of approaching phonology that doesn't require us to impute young children with extra knowledge that they have to unlearn. Phonologists studying adult sound systems using a rule-based system began to notice that there were crosslinguistic similarities that were not adequately captured by these rules. They began to notice that languages around the world employed the same *kinds* of rules (deletion rules, assimilation rules, and so forth) but not the same exact rules. Phonologists wondered if there was a way to capture these global similarities in the kinds of things human language sound systems were trying to achieve.

The framework that developed out of this effort is known as **Optimality Theory** (Prince and Smolensky, 1993). Optimality Theory proposes a set of **universal constraints** on how words can be phonologically structured. The idea is that all languages share these universal constraints, but different languages rank the constraints differently. An example of a constraint is *CodaVoice, which means "a coda cannot be voiced" (the '*' symbol means "not allowed," just like in syntax). This constraint captures the fact that some languages require final consonants to be voiceless. For example, in German this constraint is ranked quite highly, meaning it applies robustly in the language: German obstruents (stops and fricatives) are always voiceless when they occur at the end of a word (e.g., *Rad* 'wheel' is pronounced [ʁat]), but they remain voiced before a vowel (*Räder* 'wheels' is [ʁedɐ], not *[ʁetɐ]). But in other languages, like English, this constraint is not highly ranked, which simply means it doesn't have much effect. English freely allows consonants to be voiced in coda position (e.g., *bed* [bɛd], *please* [pliz]). (We can sometimes see minor effects of such constraints, however: even in English there is sometimes a tendency to devoice final consonants, especially voiced fricatives like /z/ [Smith, 1997].)

The idea behind Optimality Theory is that for each word in the lexicon there is a set of possible phonological forms the word could have. These forms (called **candidates**) compete with one another, and the candidate that violates the *lowest ranked* constraints (the least important ones for that language) is the optimal form; this is the form that is actually pronounced. Let's look at a concrete example. In English we have the word *bed*, which has a final consonant that is voiced. Imagine if this word were borrowed into a language that did not allow voiced codas—how would it be pronounced? In principle, language allows a number of options. The coda could be devoiced, yielding [bɛt], the coda could be deleted altogether ([bɛ]), or we could insert (epenthesize) a vowel so that the [d] is no longer the final sound in the word, yielding [bɛdɛ] or [bɛdə]. In this example, we say that /bɛd/ is the input form, or underlying form, and the options for pronunciation ([bɛt], [bɛ], [bɛdɛ]) are the possible pronunciations. We then rank the universal constraints *CodaVoice, in such a way that the winning candidate is the one d.

signed to reflect the kinds of patterns and preferences we actu- nguages. There are two main types of constraints. One type is **ints**. These constraints favor phonological forms that are very ow that the least marked syllable type is CV, so constraints that either sallow codas, weaken codas by devoicing them, or disallow complex onsets or codas are markedness constraints. We also know that languages prefer syllables with single onsets that exhibit the maximum contrast with a vowel—the least vowel-like consonant is a stop consonant. So constraints that disallow more vowel-like onsets (like liquids or glides) would also be markedness constraints.

We also need constraints that put the brakes on too much deletion or insertion, or else every word would just consist of V or CV syllables with no codas and no complex onsets. There are languages in which this is actually how words and syllables must be formed (such as Hawaiian), but many languages, like English, allow more complex syllable types. The second main type of constraint is **faithfulness constraints**, which favor candidates that are very similar to the underlying form of the word along some phonological dimension. So if the underlying form of the word (/bɛd/ in our example above) has a coda, a voiced coda, or a complex onset, a highly ranked faithfulness constraint would rule out forms that added or deleted sounds or changed features in order to make less marked syllables.

Let's propose some constraints to evaluate the different possible pronunciations of *bed* we saw earlier.

(13) Markedness constraint: *CodaVoice
Faithfulness constraints: IdentVoice, Max, Dep

The markedness constraint should be fairly transparent: *CodaVoice means "don't have a voiced coda." The faithfulness constraints require a little more explanation. IdentVoice means "each segment should have the same voicing feature as the corresponding underlying segment." In other words, don't change the voicing of any segments in the word. Max means "don't delete anything," and Dep means "don't insert anything."

While the process of doing rule-based phonology involves figuring out which rules apply in a particular case and how those rules must be ordered, doing phonology in Optimality Theory involves figuring out how we need to rank the constraints. Remember, all languages are assumed to have all the same constraints; they are just ranked differently (they have different degrees of importance) in different languages. In section 4.4 we wrote a rule that accounted for final devoicing; the rule in 12b is repeated here:

(12) b. C → [-voi] / ______ #
[-son]

How could we account for this same pattern using Optimality Theory (OT)?

To figure this out we set up a table (called a **tableau** in OT terminology) with the candidates going down the left side and the constraints going across the top. The constraint in the leftmost constraint column is the highest ranked, and the constraint in the rightmost column is the lowest ranked. The ordering of the candidates from top to bottom doesn't matter, and we indicate the winning candidate with the ☛ symbol. The constraint ranking in the tableau below reflects the phonology of adult English, and the winning candidate is the pronunciation English speakers use for the underlying form. For this small example let's just consider two candidates, one with a voiced coda and one with a voiceless coda.

"bed" /bɛd/	IDENTVOICE	*CODAVOICE
☛ bɛd		*
bɛt	*	

As noted above, this ranking reflects adult English phonology. We ended up with this result because we ranked the constraints so that the constraint that the winning form violates, the markedness constraint *CODAVOICE, is the lowest ranked. And the constraint that the other candidate violates, the faithfulness constraint IDENTVOICE, is ranked more highly.

But now suppose we want to model the way a child pronounces a word. We saw in section 4.2.2 that children sometimes alter the voicing of target sounds. In particular, they tend to voice onsets and devoice codas. So this time we want to model the fact that a child pronounces the word *bed* as [bɛt]. How can we rerank our constraints to make [bɛt] the optimal form?

"bed" /bɛd/	*CODAVOICE	IDENTVOICE
bɛd	*	
☛ bɛt		*

Notice what we did in order to get [bɛt] as the winning candidate: we made the markedness constraint (*CODAVOICE) more important (higher-ranked) and the faithfulness constraint (IDENTVOICE) less important (lower-ranked). In fact, we can make a fairly broad generalization about children's phonology: their errors tend to involve producing *less marked* forms than those found in the adult grammar. Going back to one of our motivations for explaining children's phonology using constraints, what is different about children versus adults is the relative ranking of these families of constraints. Children generally have their markedness constraints ranked *highly* in their grammar and faithfulness constraints ranked *lower*. Optimality Theory demonstrates that, unlike the rule-based approach, children do not have extra rules in their grammar they need to unlearn; rather, as they grow they must simply rerank their constraints so that the (relevant) markedness constraints end up lower and the faithfulness constraints end up higher. This is how children come to have an adultlike pronunciation of words and an adultlike phonological system (Tesar and Smolensky, 1998; Boersma and Hayes, 2001).

One advantage of explaining children's phonology using an OT approach is that it gives us an easy way of accounting for the variation we find across children in their pronunciations. While the processes and rankings we've talked about are quite widespread, we do find some individual differences, even across children acquiring the same language. In the *bed* example above, we had to rank the *CodaVoice constraint very high in order to get [bɛt] as the optimal form, since that is the way one child (the child we were trying to model) pronounced it. But another child acquiring English had such a strong dispreference for codas in general that he epenthesized a vowel at the end of every word that ended in a consonant. Thus, he produced forms like [baga] for *bag*, [buku] for *book*, and [dɔgɔ] for *dog*. So for this child, the Dep constraint ("don't insert anything") must be ranked extremely *low*, and we would need to introduce another markedness constraint, *Coda ("don't have a coda"), which would be ranked very *highly*, higher than his Dep constraint. In this child's grammar, then, it is better to insert a vowel in order to avoid having a coda than to remain faithful to the target form and allow a coda. The following tableau illustrates how we could rank our constraints to yield this child's pronunciation as the winning candidate.[4]

"dog" /dɔg/	*Coda	Dep
dɔg	*	
dɔk	*	
☛ dɔgɔ		*

4.6 Summary

In this chapter we have seen that children go through quite uniform stages of phonological development in speech production, from developing their phoneme inventories along similar trajectories to applying similar phonological processes to their early words. While there is some variation across languages (note differences in the early phoneme inventories of children acquiring K'iche' and children acquiring English) and variation across individuals acquiring the same language, variation is found to be of limited scope. Taking a cue from Grammont (1902, cited in Jakobson, 1971), "[the] child undoubtedly misses the mark, but he always deviates from it in the same fashion." These uniform deviations in children's phonological systems can be linked to principles of adult phonological grammar: broadly speaking, children tend to produce unmarked, or less marked, forms (stops, labials, CV syllables) and take longer to incorporate more marked forms into their language.

4.7 Further Reading

De Boysson-Bardies, Benedicte. 1999. *How Language Comes to Children: From Birth to Two Years.* Cambridge, MA: MIT Press.

Tessier, Anne-Michelle. 2016. *Phonological Acquisition: Child Language and Constraint-Based Grammar.* New York: Macmillan Education, Palgrave.

Vihman, Marilyn May. 1996. *Phonological Development: The Origins of Language in the Child.* Oxford: Blackwell.

4.8 Exercises

1. In the following table, put a check in the squares under the phonological process or processes that apply in each case. (CCR = consonant cluster reduction)

	Fronting	Gliding	Consonant assimilation	Reduplication	CCR	Coda deletion	Weak syllable deletion
cold [tod]							
not [na]							
yogurt [jogok]							
grass [dæs]							
tongue [gʌŋ]							
medicine [mɛsan]							
play [pwej]							
coffee [tafi]							
clean [kin]							
desk [gɛk]							

2. Consider the following lists of words. For each list, come up with a single phonological rule that captures what all of the words in that list have in common.

a. handle [ɛŋu]	b. ball [bɔː]	c. biscuit [bɪgɪk]	d. dark [gaːk]
pedal [bɛgu] beetle [biːgu] bottle [bɔgu]	bell [bɛ] trowel [dau] bolt [bɔːt] elbow [ɛbuː] milk [mɪk]	escape [geip] skin [gɪn] Smith [mɪt] spoon [buːn] scream [giːm] swing [wɪŋ]	drink [gɪk] leg [gɛk] ring [gɪŋ] singing [gɪŋɪŋ] snake [ŋeːk] stuck [gʌk] taxi [gɛgiː]

3. Adult Japanese has a rule that turns alveolar stop consonants into affricates when they occur before a high vowel ([i] or [ɯ]; [ɯ] is a high, back, unrounded vowel). Thus, /t/ becomes [tʃ] before [i], and it becomes [ts] before [ɯ]. No other stops are affected by this rule. Look at the following pronunciations of a child acquiring Japanese, and consider what *underlying representations* of sounds are pronounced as [t] (data cited in Fromkin, 2000, p. 669).

Target form	Child's form	Gloss (meaning)
[tama]	[tama]	ball
[terebi]	[terebi]	TV
[mikã]	[mitã]	orange
[nɛko]	[nɛto]	cat
[matʃi]	[matʃi]	city
[tsɯta]	[tsɯta]	ivy
[aki]	[ati]	fall
[kɯma]	[tɯma]	bear

Does this child have two different mental representations for /t/ and /k/ as adult Japanese speakers do? Or just a single underlying representation /t/? How can you tell?

4. Consider the following pronunciations by a child acquiring German.

Target form	Child's form	Gloss (meaning)
[dʀɛk]	[glɛk]	dirt
[andʀejas]	[æŋgle:əs]	Andreas (name)

Two sets of features are trading places in sounds in these words. What are those features and which sounds are changed? Instead of writing a rule, describe what processes lead to the child's pronunciation of these two words. (Hint: Think of both place and manner features.)

5. Consider the following pronunciations by a 2-year-old English-speaking child (Gnanadesikan, 2004):

(1) a. clean [kin] e. please [piz]
b. sleep [sip] f. friend [fɛn]
c. slip [sɪp] g. draw [da]
d. grow [go] h. cream [kim]

Focusing on the onsets, we can see that this child deletes one of the consonants in the cluster to yield a singleton onset. Which one does she delete and which does she preserve?

What manner feature is shared by the deleted consonants? What manner features do the preserved sounds have?

(i) Write a phonological rule to account for this child's process of onset consonant cluster reduction.

(ii) If you read section 4.5 on Optimality Theory, use the following constraints to construct a tableau to account for the data in exercise 1 above.

(a) *ComplexOnset: An onset contains only a single consonant

(b) Max: Don't delete any segments

4.9 References

Boersma, Paul, and Bruce Hayes. 2001. Empirical tests of the gradual learning algorithm. *Linguistic Inquiry* 32: 45–86.

Braunwald, Susan. 1978. Context, word and meaning: Toward a communicational analysis of lexical acquisition. In Andrew Lock (ed.), *Action, Gesture and Symbol: The Emergence of Language*, pp. 485–527. New York: Academic Press.

Butcher, Andrew. 2006. Australian Aboriginal languages: Consonant-salient phonologies and the "place-of-articulation imperative." In Jonathan Harrington and Marija Tabain (eds.), *Speech Production: Models, Phonetic Processes and Techniques*, pp. 187–210. New York: Psychology Press.

Clements, George N., and Samuel Jay Keyser. 1983. *CV Phonology: A Generative Theory of the Syllable*, 1–191. Linguistic Inquiry Monographs 9. Cambridge, MA: MIT Press.

De Boysson-Bardies, Benedicte, Laurent Sagart, and Catherine Durand. 1984. Discernable differences in the babbling of infants according to target language. *Journal of Child Language* 11: 1–15.

De Boysson-Bardies, Benedicte, and Marilyn Vihman. 1991. Adaptation to language: Evidence from babbling and first words in four languages. *Language* 67: 297–319.

Dunbar, Ewan, and William Idsardi. 2016. The acquisition of phonological inventories. In Jeffrey Lidz, William Snyder, and Joe Pater (eds.), *The Oxford Handbook of Developmental Linguistics*, pp. 7–26. Oxford: Oxford University Press.

Fromkin, Victoria (ed.). 2000. *Linguistics: An Introduction to Linguistic Theory*. Malden, MA: Blackwell Publishers.

Gnanadesikan, Amalia. 2004. Markedness and faithfulness constraints in child phonology. In René Kager, Joseph Pater, and Wim Zonneveld (eds.), *Fixing Priorities: Constraints in Phonological Acquisition*, pp. 73–108. Cambridge: Cambridge University Press.

Hyman, Larry. 2008. Universals in phonology. *The Linguistic Review* 25: 83–137.

Ingram, David. 1979. Phonological patterns in the speech of young children. In Paul Fletcher and Michael Garman (eds.), *Language Acquisition: Studies in First Language Development*, pp. 133–148. Cambridge: Cambridge University Press.

Itkonen, Terho. 1977. Huomioita lapsen äänteistön kehityksestä. *Virttäjä*, 279–308.

Jakobson, Roman. 1971. *Studies on Child Language and Aphasia*. The Hague: Mouton.

Kent, Ray, and Harold R. Bauer. 1985. Vocalizations of one-year-olds. *Journal of Child Language* 12: 491–526.

Kent, Ray, and Giuliana Miolo. 1995. Phonetic abilities in the first year of life. In Paul Fletcher and Brian MacWhinney (eds.), *The Handbook of Child Language*, pp. 303–334. Cambridge, MA: Blackwell Publishers.

Li, Fang-fang, Jan Edwards, and Mary E. Beckman. 2009. Contrast and covert contrast: The phonetic development of voiceless sibilant fricatives in English and Japanese toddlers. *Journal of Phonetics* 37: 111–124.

Macken, Marlys, and David Barton. 1980. The acquisition of the voicing contrast in English: A study of the voice onset time in word-initial stop consonants. *Journal of Child Language* 7: 41–74.

Maddieson, Ian, and Kristin Precoda. 1990. UPSID-PC: The UCLA Phonological Segment Inventory Database. http://www.linguistics.ucla.edu/facilities/sales/software.htm.

Moran, Steven, Daniel McCloy, and Richard Wright (eds.). 2014. PHOIBLE Online. http://www.phoible.org.

Munson, Benjamin, Jan Edwards, Mary E. Beckman, Abigail C. Cohn, Cécile Fougeron, and Marie K. Huffman. 2011. Phonological representations in language acquisition: Climbing the ladder of abstraction. *Handbook of Laboratory Phonology*, pp. 288–309.

Oller, D. Kimbrough. 1978. Infant vocalizations and the development of speech in infancy. *Allied Health and Behavioral Science* 1: 523–549.

Oller, D. Kimbrough. 1980. The emergence of the sounds of speech in infancy. In G. H. Yeni-Komshian, J. F. Kavanagh, and C. A. Ferguson (eds.), *Child Phonology*. New York: Academic Press.

Pettito, Laura-Ann. 1991. Babbling in the manual mode: Evidence for the ontogeny of language. *Science* 251: 1493–1496.

Prince, Alan, and Paul Smolensky. 1993. Optimality theory: Constraint interaction in Generative Grammar. Unpublished manuscript, Rutgers Center for Cognitive Science.

Pye, Clifton, David Ingram, and Helen List. 1987. A comparison of initial consonant acquisition in English and Quiche'. In Keith E. Nelson and Anne van Kleek (eds.), *Children's Language*, vol. 6, pp. 175–190. Hillsdale, NJ: Lawrence Erlbaum Associates.

Smith, Caroline. 1997. The devoicing of /z/ in American English: Effects of local and prosodic contexts. *Journal of Phonetics* 25: 471–500.

Smith, Neil V. 1973. *The Acquisition of Phonology: A Case Study*. Cambridge: Cambridge University Press.

Stark, Rachel E. 1980. Stages of speech development in the first year of life. In Grace H. Yeni-Komshian, James F. Kavanagh, and Charles A. Ferguson (eds.), *Child Phonology*, vol. 1: *Production*, pp. 73–92. New York: Academic Press.

Stoel-Gammon, Carol. 1989. Prespeech and early speech development of two late talkers. *First Language* 9: 207–223.

Tesar, Bruce, and Paul Smolensky. 1998. Learnability in Optimality Theory. *Linguistic Inquiry* 29: 229–268.

Tessier, Anne-Michelle. 2016. *Phonological Acquisition: Child Language and Constraint-Based Grammar*. New York: Macmillan Education, Palgrave.

III Module 3: Word Meaning and Word Structure

5 Word Learning

We spent a great deal of time talking about the problem of induction in chapter 2. We saw that there is a serious problem for learning through experience, and the most viable solution to that problem was an approach that involves innate linguistic biases, referred to as Universal Grammar. With that in mind, it is tempting to think that learning the meanings of words is the easy part of language acquisition. At first glance, it seems like syntax and phonology are the real puzzles because they involve abstract representations, and the thought of a parent explaining phonemic contrasts or syntactic structure to a child is absurd. How does a child learn an abstract system of rules—a computational unit? Word meanings, on the other hand, are sometimes modeled for children, even directly ("Look! That's a *bird!*"). So this part of the language-learning puzzle should be relatively easy. But that is far from the truth. In reality, the problems we discussed in chapter 2 about induction are equally applicable to the domain of lexical learning.

In this chapter, we begin with a description of the characteristics of early word learning—the actual empirical facts about how quickly children learn various kinds of words and some common mistakes children make in this process (underextension and overextension). Once we know what the broad facts are, we consider the logical puzzle of word learning—how children actually map a word (encountered as a string of sounds) onto a concept (a meaning)—much the way we thought about the logical puzzle of learning grammar in general in chapter 2. We then go over some of the major principles and constraints that have been proposed in the literature that show that children come to the game of word learning with inherent biases. These biases help them overcome the problems of word learning. We finish the chapter with a discussion of how syntactic structure can also aid word learning, especially for verbs.

5.1 Characteristics of Early Word Production

Children are simply voracious word learners. They munch through new words like candy, acquiring hundreds of words in the span of months. Think about when you learned a second language: How easy was it to learn the vocabulary? Remember the flash cards? The quizzes? The stammering trying to remember the right word? Well, kids don't do any of that. They do sometimes struggle to recall from memory a word they have learned, but they don't use explicit practice drills or conscious exercises in word repetition.

We know that, on average, children produce their first words sometime between age 10 months and 15 months. Some children do not produce their first word until closer to the second birthday, which is still considered to be within the normal range. Children who begin talking after the second birthday may be deemed late talkers and could be at risk for language and/or other cognitive delays, though certainly not all late talkers have language or cognitive disorders (see chapter 8, section 8.4, on language impairment). Moreover, we know that children comprehend words even earlier than they produce them. How and when do children amass their vocabulary?

Perhaps the single most influential study in this area is a study published by Bates et al. (1994), in which they tracked 1,803 infants in the beginning stages of language acquisition and assessed the words that they had acquired. They used a tool that has become a mainstay of acquisition research in recent decades: the MacArthur-Bates Communicative Development Inventory (MBCDI), which is essentially a list of words that young children are likely to know and which parents can use to report whether their children say or understand. More on this a little later, but for now, let's look at some numbers. Table 5.1 shows the average number of words that children produced at various ages, along with the range within the population of 1,803 children.

A few things pop out from this table. First, most children are using some words by the end of the first year of life. But this is not universal—there are children who are not producing any words by the first birthday, as seen in the range in parentheses. Second, there is an incredible amount of variation from child to child: at their first birthday, some children are producing no words at all, while others produce an amazing 52 words. This amount of variation is observed even at later ages, suggesting that *variation* in productive abilities might well be a hallmark of early lexical acquisition. One might be skeptical about these numbers: the MBCDI is essentially a parental report on their children's lexical abilities, and we all know parents who will gush about their kids. Perhaps some parents are overly

Table 5.1
Ages of children and the average (median) number of words produced, with range of words in parentheses, where provided

Mean age of children	Median number of words produced (range)
1;0	6 (0–52)
1;4	44 (6–347)
1;8	170 (3–544)
2;0	311 (57–534)
2;6	574 (208–675)

Source: Bates et al. (1994).

exuberant in interpreting their children's words while other parents are more conservative, leading to these large differences in the ranges. However, Bates et al. (1994) argue against this point, and argue that there are some genuinely precocious children. Moreover, in the years since the MBCDI was developed, research has shown that it is actually quite accurate in estimating children's vocabulary.

The third point about the data in table 5.1 is that the rate of development of productive abilities is quite impressive. Over the course of 18 months (from 1;0 to 2;6), children go from basically no words to more than 500 words in their productive lexicons. But how do they achieve this impressive feat? Do they learn these words at a regular pace? For example, do they learn one or two words a day over the course of that 18-month period? This seems not to be the case. It looks more like children go through what are called **vocabulary spurts**. That is, there are periods of time when a child acquires words very quickly, and then they go through periods when they acquire words more slowly.

5.1.1 The Vocabulary Spurt

After a child has acquired (roughly) the initial fifty words, there is often (though not always) a sudden rush of word learning. This rush is like a burst of energy, or an explosion, in which the child seems to take a big step up in their word production. Typically, the first such vocabulary burst (and many subsequent ones too) is a nominal explosion, meaning that many nouns are learned in a short amount of time. But later in development, when the child has several hundred words in their productive vocabulary, there may be an explosion of verbs. While not all children exhibit vocabulary bursts, most do. Firstborn children are more likely to exhibit this vocabulary burst than children of other birth orders (Goldfield and Reznick, 1990). Moreover, vocabulary bursts have been observed crosslinguistically, even in sign language (Petitto, 1992). Because of this, vocabulary bursts are considered a typical property of lexical acquisition.

5.1.2 Content of Early Vocabulary

The first fifty words typically consist of specific nouns (e.g., *mommy*) as well general common nouns like *ball*, *dog*, and *milk*. The general nouns are by far the largest group of words in early vocabulary, though there are other kinds of words too. Those fifty words also often include basic action words such as *go* and *look* as well as modifiers like *big*, *all gone*, and *outside*. There are often some social words like *no*, *want*, *please*, and *bye-bye* as well as a few grammatical function words like *what*.

Those first fifty words, therefore, are a mixed bag, though dominated by general and specific nouns. As a child acquires the next fifty words (increasing the vocabulary to 100 words), they tend to acquire more nouns than anything else, further increasing how noun-rich the child's vocabulary is. But after the first 100 words, children's vocabularies begin to take on verbs and adjectives at faster rates (this is also when children start to exhibit vocabulary bursts). Looking at table 5.1, this happens around age 1;6 on average.

ficantly later in development (around age 2;0 to 2;6), once the child has about 400 n their productive vocabulary, function words (like auxiliary verbs, agreement markers, certain modal verbs, the copula verb *be*, conjunctions, and articles) increase in number and frequency.

5.1.3 Early Uses of Nouns: Overextension and Underextension

What does it mean to *know* a word? One thing it means, of course, is how the word is pronounced (its phonological form). Another component is the word's lexical category: Is the word a verb? A noun? An adjective? A preposition? Knowing a word's lexical category means knowing how the word functions in a sentence. Probably what most people think of as knowing a word is knowing the word's **meaning**, which means linking the word to a **concept**. We will talk more in section 5.3 about what it means to link a word to a concept. In this section we'll look at some of the ways the mapping of a word to its meaning can go awry. In sections 5.2 and 5.3, then, we will talk about some explanations for *how* children perform this mapping process—that is, how they learn the meanings of words.

Sometimes children's early words are bound to the immediate context. That is, the child might use a general noun like *birdie* to refer to a stuffed animal bird, not to birds in general. This is known as **underextension**: when a word refers to a class that is smaller than the target class. Underextension is not extremely common, but it has been observed in early speech. Another example is a child who uses *dog* to refer to Labrador dogs only, not other kinds of dogs. This may be because the child first heard the word *dog* in the context of a Labrador and has not realized yet that the word *dog* refers to a larger class of animals than just Labradors. This is often idiosyncratic to the child. Fortunately, it is fairly straightforward to correct: eventually, the child who takes *dog* to mean only Labradors, or only the child's own dog, will hear the word *dog* being used to refer to the neighbor's poodle, or a Great Dane at the dog park, and the child will be forced to revise their initial hypothesis.

Opposed to underextension is a phenomenon referred to as **overextension**. Overextension occurs when a child takes a word's meaning to include a larger class than in the adult grammar. For example, the child might look up at the moon and say, "Ball!" This is presumed to be the child analogizing the shape of the moon onto the shape of a ball. Assuming the child does not yet have a word for *moon*, the child must use a related word in place of *moon*. It's almost like the child is saying, "Ball-like object!"

Overextension is usually based on some perceptual features, such as the following:

- shape (e.g., *ball* for *moon*)
- texture (e.g., *rabbit* referring to all furry things, like bedroom slippers or a shag carpet)
- color (e.g., *white* for anything that is light in color)
- natural kinds (e.g., *doggie* for any four-legged animal)
- function (e.g., *bus* for any moving vehicle)

Unlike underextension, overextension errors are very common and are found crosslinguistically. At first glance this type of error seems to present a more serious logical problem for the learner: Since children do not regularly receive (or make much use of) negative evidence (explicit correction), how will they retreat from their overly broad hypothesis? It is an open question to what degree explicit correction might be useful for labels (e.g., Mom might say, "That's not a bus; that's a car"), even if it is not useful for learning grammatical rules. On the other hand, overextension is only observed in production, not comprehension. We can tell this because if you have a child who refers to all four-legged animals as dogs, and you show that child a picture of a dog, a cat, a bear, and a horse and say, "Point to the dog," the child will in fact point to the dog and not to the other animals. Thus, overextension appears to be an error in production more than an error in the underlying concept, so the logical problem of retreating from an overextension error may not be as

Sidebar 5.1: Why Do Children Overextend?

There has been considerable debate as to why overextension errors occur. Here we mention five such proposals (see Hoek et al., 1986).

Incomplete meanings: Word meanings contain semantic features. A word like *puppy* might have features like [+animate, +four-legged, +mammal, +canine, +young, +cute]. But if your lexical entry for *puppy* is missing some features, like [+young] or [+canine], it could include other four-legged animals like cats, older dogs, and pigs, and this would appear as an overextension error to the adult. One problem with this explanation is that it could predict problems in comprehension, not just production.

Limited lexicon: It may be that children simply lack the vocabulary to talk about the things they want. If the child wants to talk about the moon but simply does not know the word for that bright, round object in the sky, the child will use the closest available word that might work. *Ball* might work, *plate* would work too—both attested examples of overextension. This squares well with the fact that children don't tolerate adults using overextended words with them—overextension is a phenomenon of production only, not comprehension.

Metaphoric extension: Related to the previous explanation, perhaps children are trying to use metaphor when they overextend. It's as if when the child says *ball* to refer to the moon, the child is trying to say, "The moon is like a ball." So it sounds like an overextension, but the child is being more sophisticated than the adult assumes.

Phonological complexity: The phonological complexity of the target word may be enough to sway children away from using it, so they might use related words instead. The slight inaccuracy in meaning is viewed as worth the phonological gain that the child enjoys. Hoek et al. (1986), for example, tested one child younger than age 2 on production and comprehension tasks involving nonsense words. They found that the child had no problem comprehending newly taught words that were phonologically similar or dissimilar to existing words in her lexicon, but in production she often replaced difficult-to-pronounce novel words with ones that more closely matched the phonology of words she regularly produced, suggesting that the phonological complexity of new words plays a role in lexical selection.

Lexical access error: We know that lexical access (the process of retrieving a lexical item from the lexicon in real time and using it in an utterance) is a taxing process for both children and adults. It's possible that child lexicons are even harder to access, so given this difficulty, it may be that the child reaches into the lexicon and retrieves the wrong word. This may seem odd, but we know that adults do the same thing once in a while. When we are under pressure, fatigued, or otherwise impaired, we often have trouble retrieving lexical items and sometimes retrieve the wrong words—words that are related to what we mean, but not quite the right ones. With children, such difficulty in lexical retrieval could well lead to the selection of words that are linked to the target word, but not quite the right word. *Dog* is semantically related to *cat*, so *dog* might be retrieved instead of *cat*.

serious as it first appears. Children do correct these errors spontaneously within a relatively short time.

5.1.4 Early Vocabulary Comprehension and Fast Mapping

The previous section provided some basic background on how children produce their first words. We turn now to some properties of children's receptive vocabularies, or words they understand. The first point of note is that receptive vocabularies are always larger than productive vocabularies. This is true for children at all ages, and is in fact true for adults as well. Think about the number of words that you hear or read and can understand but which you never produce. You know what an anemone is, right? How often have you used that word? Probably quite rarely, but you know what it is. And if you don't, look it up, and you will now know what it is but will probably use it very rarely unless you become an avid underwater sports enthusiast or a marine biologist. The point is that it is a property of human lexicons that receptive vocabularies are larger than productive vocabularies, and this is no different for young children.

When you ask parents, they often tell you that young infants who have not yet produced a single word understand some words. They know this because babies respond appropriately when they hear those words. For example, when the parent asks the child, "Where's your ball?" the child might point under the couch, suggesting that they understand at least *where* and *ball*. Larry Fenson and his colleagues investigated the receptive vocabularies in 100 children using the MBCDI and found that at age 10 months (when, remember, most children are not producing any words), they have a receptive vocabulary of between 11 and 154 words. Six months later, at age 1;4, that receptive vocabulary has skyrocketed to between 92 and 321 words (Fenson et al., 1994).

The receptive vocabulary is, like the productive vocabulary, dominated by nouns. However, there are more verbs in the receptive vocabulary than the productive vocabulary (Gentner, 1978). This may be due to the use of so-called *light verbs*, which are verbs used in an all-purpose manner, such as *do*, *make*, and *go*. These verbs can be used in a variety of patterns (e.g., as main verbs, as in "I did my homework already"; as support verbs, like auxiliary *do*, as in "I didn't eat yet"; as verb combinations, as in "Go bring me a clean plate please"), and children appear to make especially good use of them (Clark, 1978).

As we saw above, children acquire words very quickly. During a vocabulary spurt, they are sometimes acquiring one word per waking hour. That means learning needs to be *really fast* if children are going to learn that quantity of words in that amount of time. And, in fact, this has been found. Carey (1978) coined the term **fast mapping** to refer to the fact that children only require one exposure (or very few exposures) to a word in order to acquire some aspect of its meaning.

To test this idea, Carey and Bartlett (1978) taught twenty children aged 3;0–3;10 a novel color term. They used the word *chromium* (after checking to ensure the children did not already know that word), and they did so by asking one of the teachers to use the word within the context of a classroom activity. For example, if the teacher was setting up snacks, the teacher would turn to the child and say, "You see those two trays over there? Could you get me the chromium one? Not the red one, but the chromium one." Children saw two trays, one a red tray and one an olive-green tray. Most children tried to repeat the word *chromium*, and many were able to respond correctly. Moreover, of those that had responded correctly, when tested ten days later, some (though not all) were able to correctly identify "chromium" objects, showing that the one exposure ten days earlier had had a lasting impact on their lexicon. Children had acquired some aspect of meaning for the word *chromium*, despite just one or two exposures to that form.

Carey and Bartlett (in both their 1978 paper and in subsequent work together) go to great lengths to explain that fast mapping is not all that is needed to acquire a word. Rather, the initial creation of a lexical entry is *fast*, requiring minimal exposure. There is no bottleneck at the entry point of the lexicon, and this is a great feature for a learning system. However, subsequent learning of additional features of the meaning of individual words is a much longer, slower, and more laborious process (see, e.g., Carey, 2010).

But fast mapping is a very important part of the acquisition process. Without it, we would not see the steep trajectory of word learning that we see in most children. The average first grader has a productive vocabulary of 10,000 words and a receptive vocabulary of 15,000 words. Children would not get to this point if they did not have something like fast mapping to help them learn words.

5.2 The Problems of Word Learning, and the Limitations of Ostension

Children are remarkably fast at getting that first parse of a word's meaning, as we saw above. This makes it seem like word learning is easy. It looks like all the child has to do is be exposed to a word and, presto, that word is learned. "Children are like sponges"—we hear people say that all the time. And fast mapping gives this casual observation some teeth. However, as we mentioned above, fast mapping is only the first step of a long process, and it doesn't tell us why a child picks out the particular meaning they do as their first guess. Here we discuss some of the potential difficulties in fleshing out the full meaning of a word and why mere exposure to a word is not enough to ensure that the child acquires the (correct) meaning. As we will show, in the domain of lexical acquisition, there exists a serious problem of induction, just as we saw in chapter 2. And the solution to this problem of induction is, once again, innate principles or constraints that guide learning.

Before we demonstrate the principles that guide children in their word-learning adventure, let us take a moment to flesh out the reason why simply hearing a word is not sufficient to give children the correct meaning of that word. Here we need to introduce the term **ostension**. The meaning of *ostension* is "the process of showing or pointing out" or "instructing by exhibiting." In our context, then, learning by ostension is when the parent shows the child something and labels that thing with a word. Imagine a mother holding up a bottle to the child and saying, "Bottle. This is a bottle. Can you say bottle?" That's classic ostension. The philosopher John Locke, way back in 1690, suggested that children learn through ostension:

> If we observe how children learn language we will find that to make them understand what the names of simple ideas or substances stand for, people ordinarily show them the thing whereof they would have them have the idea; and then represent to them the name that stands for it, as 'white', 'sweet', 'milk', 'sugar', 'cat', 'dog'. (Locke, 1690/1964 Book 3.IX.9; cited in Gleitman, 1990)

The idea is that when you present the child with a word, the child hears the form, sees what the context is, and maps the word onto the concept gleaned from the environmental context and thereby learns the meaning of that word. Seems simple enough, right? This is indeed simple (and intuitive), but there are two major problems with this approach to learning the meanings of words. These two problems have led researchers to propose a number of principles that children use to overcome these problems. Let's discuss the problems before presenting the principles of learning.

Problem 1: The Mapping Problem The mapping problem contains two smaller subproblems, presented below. The essence of the problem is how children are able to map a particular given word onto the correct meaning. This will become clear once we have discussed the subproblems that add up to the mapping problem.

Problem 1A: The Gavagai Problem (Quine, 1960) The **Gavagai problem** (discussed by the philosopher W. V. O. Quine) is that when presented with a word in a context, there are an infinite number of logical hypotheses about the word's meaning that a child might consider. Imagine a parent and child are sitting on the couch and the family dog enters the room. The dog is a friendly golden retriever, tongue hanging out, breathing heavily, and it ambles over and settles at their feet. The parent says "Dog! That's a dog!" What is the child to conclude from this? A subset of plausible hypotheses are listed here:

Hypothesis 1: dog

Hypothesis 2: tongue

Hypothesis 3: eyes

Hypothesis 4: fur

Hypothesis 5: breathing

Hypothesis 6: breathing heavily

Hypothesis 7: friendly

Hypothesis 8: animal

Hypothesis 9: be careful

Hypothesis 10: teeth

Hypothesis 11: smelly

Hypothesis 12: four-legged animal

There is no reason why the child should immediately know that the word *dog* refers to the dog—all these other hypotheses are equally plausible and perfectly good hypotheses. (In addition, there is an infinite set of logically possible but *implausible* hypotheses about the word's meaning, e.g., 'dog before noon, but cat after noon'.)

Why is this called the Gavagai problem? The thought experiment that Quine laid out was the following: A nineteenth-century explorer travels to a faraway land, and when he arrives there, somehow he manages to get a local to take him back to the village. On the way back to the village, as they walk through tall grass and unfamiliar terrain, a rabbit jumps out of the grass and runs down the path, disappearing round the corner. When the explorer's guide sees the animal, he points at it and shouts, "Gavagai!" Quine asks: How is the explorer to know what this new word refers to? It could mean rabbit, but it could also mean a host of other things: fur, legs, eyes, ears, running, jumping, escaping, beautiful, ugly, tasty, "There goes lunch!", and so on. This is the same problem faced by a child when hearing a new word for the first time.

So the Gavagai problem says that for every word, there are multiple (in fact, infinite) potential meanings, and unless the child has some mechanism(s) to guide them through the very large hypothesis space, there is the potential for very poor word learning. But we generally don't see poor word learning—we do see some errors (like errors of under- and overextension), but not the wild kinds of hypotheses that the Gavagai problem might suggest.

Problem 1B: The Hot Stove Problem Imagine a child approaches a hot stove and reaches up to put their hand on the stove top. If a parent sees the child, what will the child quickly hear? Likely it will be something like "No!" or "Stop!" But they may also hear the following:

- "hot"
- "stove"
- "don't touch"
- "be careful"
- "bad"
- "fire"
- "ouch"
- "you'll burn yourself"

So the **hot stove problem** is the inverse of the Gavagai problem: every referent/action/concept may be labeled by many different words, because many different aspects of the situation could be talked about. That is, there is no necessary, one-to-one correspondence between events or situations in the world and a linguistic label for that event. Could the child assume that the word *hot* is the name of the hot object (the stove)? Or that the word *stove* refers to the sensation of heat? Those are perfectly plausible hypotheses, and without any guidance on this, children are predicted to make many such errors. Once again, such errors do sometimes occur, but not to the degree that these two problems might suggest.

Together, these two problems add up to what is referred to as the **mapping problem**. Given the inherent variability between what we experience in the world and how we talk about those experiences with language, it's not reasonable to think the child will know exactly which aspect of an object or situation is being labeled with a particular word without some inductive constraints.

Problem 2: The Categorization Problem In section 5.1.3 we brought up **concepts** and the fact that learning the meaning of a word involves linking that word to a particular concept. Because of how words are linked to meanings, the lexicon is often described as a list of form-meaning pairings. But since concepts are also linked to other concepts, the lexicon is actually much more complex than that. It is a highly structured matrix, where various form-meaning pairings (words) are connected to other related pairings because their concepts are connected. So words cluster in groups that are related by some feature (or set of features). For example, *Labrador* may cluster with *Doberman* because they are both breeds of dog; *dog* may cluster with *cat* because they are both domestic pets; while *cow* and *pig* might cluster because they are farm animals. Moreover, *dog, cat, cow,* and *pig* cluster because they are animals (as opposed to, e.g., *sandwich* and *burrito*, which are inanimate food items). Finally, *dog, cat, cow, pig, sandwich,* and *burrito* cluster at a different level because they are all tangible objects (as opposed to, e.g., *freedom* or *frighten*, an abstract noun and an action, respectively).

If learning a word means learning to map a sound sequence (e.g., [dɔg]) onto a particular meaning or concept ('dog'), and if that concept has features that are related to certain other concepts, how do children figure out exactly what the relevant features are and which other concepts are related?

This might seem like a trivial problem, but it actually isn't. Let's take a simple example: What defines the category *dog* such that all new examples of dogs will be correctly placed into that category? Go ahead, ask yourself: What exactly is a dog? Well, let's see ... what are the properties of a dog that are shared by all dogs? They have four legs, fur, wagging tails, cold noses. They bark, they like to sniff a lot, they pee and poop all over the place unless properly trained. These are prototypical properties of dogs, and one way to define concepts is in terms of **prototypes**. Let's say we filled out a list of prototypical dog proper-

Sidebar 5.2: The Structured Lexicon

We know from a whole wealth of psycholinguistic research that conceptual structure in the lexicon is real. For example, if you first hear the word *dog* and are then shown a picture of a cat and asked to name it, you will do so very quickly. The idea is that hearing the word *dog* activates that lexical item in the lexicon as well as the words that are connected to it (like *cat*). When you are then asked to name a cat, you can do so very quickly because the word *cat* is already semi-activated because of its relation to the previously mentioned word *dog*. If you first hear the word *tangerine* and are then shown a picture of a cat, you will still name the cat, but the time it takes for you to do so will be slightly longer. This shows that *tangerine* is not in the same matrix as *cat*. Such tasks allow researchers to show that the lexicon is highly structured and that this structure creates categories of related items (domestic pets, farm animals, and so on).

ties, and then along came a new breed of dog. Would it fit into that category? It would, assuming it possessed many of these prototypical properties, so that tells us we are on the right track.

But what if you encounter a dog that lacks some of these prototypical features? What if you encounter a dog that never barks? Is that still a dog? Most people would say yes, it is still a dog. What about a three-legged dog? Again, most people would agree that it's still a dog. Certainly, there are things that are not prototypical of their category: there are hairless cats, cats and dogs without tails, flightless birds, and flying mammals. Therefore, we use many different features as well as something ineffable called **essence** to judge category membership. There is a fascinating literature on children's judgment of essence and category membership (Carey, 1985; Keil, 1989; Gelman et al., 1994), and children are able to reason about category membership by late preschool, though their categorizations continue to shift and become more adultlike throughout primary school (Carey, 2010).

The **categorization problem**, therefore, refers to the difficulty in determining the correct set of features that regulate the inclusion of a referent into a particular class. The difficulty may be particularly acute for words for nontangible things, like mental states (*think*, *believe*, *guess*) and abstract qualities (*easy*, *hard*, *hidden*, *scary*), but as we have seen, categorization even for concrete words is not simple. We have argued that the difficulty posed by the categorization problem is not solved by ostension, and, moreover, individual form-meaning mappings cannot be innate as they vary from language to language: Japanese has Japanese words to be learned; French has French words to be learned. But just because the forms must be learned through input and the form-meaning mappings must be figured out, that does not mean there is no role for innate learning mechanisms. In the next section we discuss some of the innate principles that constrain children's hypotheses about the ways words map onto concepts and therefore help children solve logical problems of language acquisition in the domain of word learning.

5.3 Principles That Guide Word Learning

5.3.1 Principle 1: The Principle of Reference

In section 5.1.2, we mentioned that the earliest words in a child's vocabulary are typically nouns; more specifically, they are nouns that **refer** to objects and people in the world. Before a child can even learn what words mean, they must know that words actually can be linked to individual meanings—that is, that words (can) have the property of being **referential**.[1] Where does this knowledge come from? Imagine a child who does not realize that words are referential—that child would need to first learn that words carry meaning, and only then could they start learning the meanings of words. On the other hand, a child who expects words to be different from other sounds in the environment (words carry meaning, while whistles, beeps, thuds, and even human sounds like raspberries and burps don't) would have a significant leg up on the task of learning the meanings of words. The idea behind the principle of reference is that children are born expecting words to refer—that words will map onto meanings. They therefore immediately make use of the referential properties of words to help understand their environment.

This may seem like an obvious skill for children to possess, but we need to demonstrate it scientifically. Fortunately, we have some good evidence for this principle. The basic idea behind this evidence is that if children know that words refer, then they should attend more to objects in the presence of language than in the absence of language—a word should have the effect of directing children's attention to potential referents for that word.

Baldwin and Markman (1989) tested young infants aged 10–14 months by presenting them with an unfamiliar toy (e.g., a snorkel) in one of two conditions: a labeled condition and an unlabeled condition. In the labeled condition, children were presented with the unfamiliar toy along with a label for that toy (e.g., *See the snorkel? That's a snorkel!*) In the unlabeled condition, children saw the same toy, but the toy was presented without a label—the experimenter presented the toy to the child but did not label it in any way. The toys were purposely selected to be small, colorful, and interesting to babies. The experimenters then gave the infants a sixty-second play period after this introduction phase, and they measured how much of that sixty-second period the child played with the new toy. If the presence of language enhanced their interest in the object, then children should play for more of that sixty-second period when the toy was presented with a label. And this is indeed what they found. In the labeled condition, children played with the new toy significantly longer during that sixty-second play period compared to children in the unlabeled condition. The interpretation of this result is that children spending extra time attending to the new object is essentially them trying to figure out which features are crucial for the meaning of the new word and perhaps trying to categorize the new object.

In a subsequent experiment, Baldwin and Markman investigated how the effect of labeling on attention to objects compares to the effect of pointing (ostension) on attention. They tested children aged 10–14 months on two conditions: (i) a pointing and labeling condition and (ii) a pointing-only condition. In the former, children were presented with a pair of toys (e.g., a snorkel and a flipper) and the target object was pointed to while being labeled (e.g., *Look, a snorkel. See the snorkel?*). In the latter condition, the experimenter simply pointed silently at the target object.

The results showed no significant difference between the conditions during the immediately following play period. Pointing alone was so powerful that there was no additional benefit to labeling, suggesting that pointing alone was enough to reach the ceiling of children's attentional abilities. However, in a subsequent play period (that occurred later), the researchers observed that those objects that had been previously pointed to *and* labeled were attended to longer than those objects that had only been pointed to. So while ostension may be a powerful attention getter, the effect of linguistic labeling may have a more durative impact on children's interest in an object.

On the face of it, this might discredit our earlier point that children can't learn just from ostension, since in the initial observation period there was no difference between the pointing and labeling condition and the pointing-only condition. But the results of this study simply show that ostension and labeling increase attention to an object. We don't know which part or property of the object children were attending to, and the logical problems associated with ostension remain. Nonetheless, we know that labeling does increase attention, and this implies that the increased attention is due to the children's attempt to map the label onto some meaning associated with that object. This means that children must know that words are referential, or else they would not be trying to create those mappings.

Note that this principle applies to expressions that can actually refer, like nouns. There are many expressions in language that are nonreferential, such as verbs, adjectives, and quantifiers (e.g., *every*). Nonreferential expressions may be disadvantaged with respect to this principle, and we might predict that such expressions are acquired later than nouns. This prediction is confirmed by experimental studies that show that 14- and 18-month-olds easily map a new label onto an object but not a property of the object (Booth and Waxman, 2009). And recall from earlier in this chapter that children's first fifty words are dominated by referential nouns. However, there are other learning principles that account for the learning of nonreferential words, one of which is discussed in section 5.4.

5.3.2 Principle 2: The Whole Object Constraint

The principle of reference restricts a child to guessing that a word will refer to an object. But what *about* that object is the word labeling? When a child hears a new word, there are infinitely many possible meanings for that word, as we saw in section 5.2. The word *gavagai* could refer to the rabbit, the action of running, the rabbit's fur, the rabbit's ears, and so on. But we now know from many studies that children initially assume that a new word refers to the whole object, rather than its parts, location, or physical characteristics. In fact, even when color, texture, and other properties are manipulated to make these features more salient, children still tend to ignore such characteristics and assume that a new word refers to a whole object (e.g., Markman, 1989). This is a robust finding, and there is relatively little controversy surrounding the existence of this constraint. In fact, it is widely assumed that this constraint exists from the earliest testable ages, so children come equipped with this constraint from the onset of word learning.

What does it mean to say that a word refers to a "whole object"? One thing it means is that children take words to label discrete objects as opposed to substances. This was demonstrated in an experiment in which babies were shown a screen on which an object appeared, such as a ball, and another screen on which there was a video of flowing lava. The ball was rather boring, not moving, not doing anything. The lava, instead, was interesting: it was flowing, it had pretty colors, it was quite attractive. In the absence of any label (babies simply heard, "Look! Wow!"), babies preferred to look at the lava. But when a label was provided (they heard, "Look! A dax!") babies preferred to look at the ball, even though it was visually less interesting. Provision of a label encouraged looks to the object rather than the substance (Woodward et al., 1993).

Another thing that's meant by "whole object" is that for solid objects, children take a label to refer to the shape of the object and not its size, texture, or what it's made of. Landau, Smith, and Jones (1988) introduced children to an object that was like a square but with one side missing and taught children that it was a "dax." Then, they showed the children a series of objects with the same shape but different sizes (preserving proportions), different covering materials to make it differently textured, or modified shapes (e.g., one of the sides bent). For each new object, they asked the children if it was "also a dax." Children were willing to accept as "daxes" the new objects that had different sizes or textures but not the new objects with modified shapes.

The third thing that is meant by saying that a word labels a "whole object" is that children assume that the word does not label a part of the object. That is, when you hear *gavagai* in reference to a rabbit, you assume the speaker means the rabbit, not its nose or ears or tail. To see how this has been investigated we first need to introduce the third principle of word learning, mutual exclusivity.

5.3.3 Principle 3: The Principle of Mutual Exclusivity

So far, we've seen that children know that words refer to objects in the world and that a new word refers to a whole object rather than a substance or the object's size or texture. However, it can't be that these principles are inviolable—they must be soft constraints, or biases, that can be flexible and overcome in some way. As we've already mentioned, some words in language are not referential (e.g., *sincerity*, *seem*, *the*). Moreover, if the whole object constraint were inviolable, how would children ever learn the names of parts or traits of objects? How does the child ever learn the word *paw* or *tail* if every time they hear these words in the presence of an animal, they assume they refer to the whole object?

Obviously, this can't be the case. Researchers think that this is avoided because children have what we might call the **principle of mutual exclusivity**: each object has one and only one label (Markman, 1989). The idea here is simple: the whole object constraint applies to new words for new objects—that part is true. However, if the new word appears to refer to an object that the child already has a word for, then the child assumes that the new word refers to something other than the whole object (a part, a trait, and so on).

For example, assume the child already knows the word *dog*. Now the child hears *paw* in a context in which the parent is pointing to the dog and saying, "Look at that paw! It sure is big." By the whole object constraint, the child will initially think that *paw* refers to the whole object—the dog in this case. The context is, after all, consistent with this interpretation. However, because the child already has a word for this kind of animal, the child has a decision to make. Does the child entertain the possibility of synonyms for whole objects; that is, could *paw* just be another word for *dog*? The principle of mutual exclusivity says no, the child does not entertain the possibility of synonyms (initially, at least). Instead, the child rejects the whole-object meaning on the basis of mutual exclusivity. They instead consider other possible meanings, such as a part of the object (the correct meaning, here) or a trait/characteristic of the whole object, or perhaps that the word refers to a subtype of dog. Words that would fit each of these hypotheses would be *paw* (part), *cute* (characteristic), or *puppy* (subtype).

In a landmark study, Markman and Wachtel (1988) investigated how children extend labels in a familiar and an unfamiliar condition. In the former, 3-year-old children were presented with a picture of an object that they were already familiar with, such as a fish or a fire truck. They were then presented with a novel linguistic label, such as *dorsal fin* or *boom*, respectively. The child was asked if the new word referred to the whole object or a part of the object. In the unfamiliar condition, children were shown pictures of objects that they were not familiar with, such as a current detector or a model of a human lung. The unfamiliar labels were then presented in the same way (e.g., *detector* and *trachea*, respectively), and children were asked if the new word referred to the whole object or a part of the object (see table 5.2). The results show that children selected the whole object less than 30% of the time in the familiar condition but 57% of the time in the unfamiliar condition (a statistically significant difference). This shows that when children already have a label for a particular meaning, and when they hear a new word that potentially refers to the same

Table 5.2
Markman and Wachtel's (1988) experimental design

Familiar condition		Unfamiliar condition	
Object	*Novel label for part*	*Object*	*Novel label for part*
fish	dorsal fin	current detector	detector
fire truck	boom	lung	trachea

Source: Markman and Wachtel (1988).

referent, they tend to look for new meanings. Mutual exclusivity means that in the early stages of word learning, children avoid synonyms.

Thus, the principle of mutual exclusivity provides a way that children might overcome the whole object constraint and thereby learn property labels (adjectives) and substance terms as well as class inclusion terms (*dog* and *pet*). And this is a very sensible principle for a child to have. This makes the learning of words maximally efficient: children don't end up spending all their time learning object names only nor learning synonyms for already learned words.

Importantly, just as with the whole object constraint, the mutual exclusivity constraint will eventually have to be weakened. This is because adult language does allow multiple labels to apply to the same object: objects can be referred to at different levels of categorization (the same entity could be called *Fido*, a *dog*, or an *animal*), and synonyms do exist (e.g., *couch* vs. *sofa*, *apartment* vs. *flat*). Interestingly, there are very few true synonyms in language, perhaps because even adult language has a preference for applying mutual exclusivity to its lexicon.

5.4 Learning Verbs via Syntactic Bootstrapping

So far we have primarily considered how children learn nouns. They make up the bulk of children's earliest learned words; the objects they label are often tangible, physically present and available for tactile manipulation; and a great deal of research has taught us much about how this category of words is acquired. But word learning does not stop once nouns are acquired—we also learn words that label actions and events, properties and attributes, abstract concepts (which can be labeled with words of any category), and grammatical functions (determiners, conjunctions, and so forth). We argued above that the seemingly simple process of learning nouns is not really all that simple or straightforward. While that's true, the logical puzzle of learning the meaning of words in other categories is even more complex. Here we will focus on verbs, and we'll start by pointing out just a few of the problems that children face when trying to learn verb meanings.

Imagine a child is trying to learn what a verb means purely by hearing it spoken in the presence of some action the child witnesses. That is, the child hears a verb (*chase*) and tries to match that verb to an event (ideally a chasing event) happening in the real world. It seems straightforward, right? In reality there are a number of ways the learning process could go wrong. First, when the child hears an utterance that describes a scene, there are many ways to describe that same scene. For example, when viewing a scene of a dog chasing a cat, this scene could be described as *the dog is chasing the cat*, *the cat is fleeing* (or *escaping from*, or *running from*) *the dog*, *the dog and cat are running*, and so on. This is essentially what we described earlier as the mapping problem: When the child hears a description of that scene, how is the child to know which interpretation they should assume in order to assign a meaning to the verb in that sentence?

Sidebar 5.3: Two Additional Principles of Word Learning

There additional principles of word learning, two of which are the principle of natural categories and the taxonomic constraint.

The Principle of Natural Categories: When a child hears a new word, there are quite literally countless possible meanings that the child *could* consider. For example, if the child hears a nonsense word like *bleen*, the child could think it refers to a whole object, or a color, or an attribute. But what's to stop the child from considering a truly outlandish meaning? Maybe *bleen* means green before midday but blue after that, or green when sitting on a flat surface but blue when on a curved surface. These may seem like ridiculous meanings to consider, but why would children know these are ridiculous? The principle of natural categories says that children restrict the hypothesis space that they consider when learning a new word to natural categories only—categories that reflect the mental categories that they already have.

Much experimentation has been done investigating this, and there is good reason to believe that children do adhere to such a principle. For example, in a seminal study, Soja, Carey, and Spelke (1991) showed children (twenty-four 2-year olds) either an object (e.g., a wooden honey dipper) or a substance (e.g., a glob of hair gel) and said "This is my blicket." They then showed children two additional things. In the object condition, they showed children either a pile of wooden bits (same substance as target object) or a plastic honey dipper (same shape as target object, but different substance). Children were instructed to point to the blicket. In this condition, children picked the plastic honey dipper (same shape, different substance). Likewise, in the substance condition, children were presented with either small dots of hair gel (same substance as the target, but different shape) or a glob of face cream (different substance but same shape as the target), and here, children picked the dots of hair gel (same substance but difference shape). This shows that the categories that children entertain for objects and substances are constrained by what Soja et al. refer to as **natural categories**: when you hear a new word for an object, you construct a meaning around its shape, not its substance. And when you hear a new word for a fluid-like substance, you construct a meaning around its substance, not its shape. This is important because it helps children constrain the hypotheses for new words that they entertain and therefore helps explain how children solve the categorization problem.

The Taxonomic Constraint: The mental lexicon is organized as a complex web of connections. Words are stored with memory links to other words that represent things that occur in the world together. For example, the word *cow* is linked in our lexicon to words like *milk*, *grass*, *moo*, *beef*, and *horns*. The relation between *cow* and these words is referred to as **thematic**. But at the same time, the word *cow* is stored with links to words like *Bessie* (a particular cow), *heifer*, *farm animal*, *mammal*, and *animal*. This latter kind of connection refers to *cow* as it relates to super categories (mammal, animal, and so on) and subcategories (heifer, Bessie). The relationship between these words is referred to as **taxonomic** because the words form a taxonomic line from an individual (e.g., Bessie) to a maximally broad category (e.g., Bessie>heifer>cow>farm mammal>mammal>animal>living being>thing).

Markman and Hutchinson (1984) presented children (as young as 2 years old) with a target picture (e.g., of a poodle) along with two selectable pictures. These latter two pictures were of things that were linked to the target either via taxonomic relation (e.g., a German Shepherd, another kind of dog) or thematic relation (e.g., dog food). In the *unlabeled condition*, children were shown the target object and they heard, "See this? Can you find another one like this?" Children of all ages picked essentially at chance, selecting the thematically related object (i.e., dog food, in this example) on average 41% of the time. However, in the *labeled* condition, children were introduced to the target picture with a novel word, for example, "See this? This is a fep. Find another fep that is the same as this fep." In this condition, even the youngest children picked the taxonomically related object (i.e., German Shepherd, in this example). On average, children picked the taxonomically related object 83% of the time (the difference being statistically significant).

This shows that in the absence of language, children seem to think thematically, but when language is introduced, children prefer to think taxonomically. Again, this helps with the categorization problem since it narrows the possible hypotheses that children naturally entertain.

Second, several verbs come in reversible pairs, such as *buy-sell*, *take-give*, and *flee-chase*. In these cases both verbs could be used to describe the same action, but the two verbs in fact have opposite meanings. We just saw that a "chasing" scene could also be described as a "fleeing" scene. Similarly, when the child sees a person buy some fruit from another person and hears *he bought some fruit*, how does the child distinguish between the *buy* versus *sell* interpretation of that scene (or the *give* versus *take* interpretation)?

Third, many verbs express an intention, a belief, or the perspective of the speaker and do not refer to tangible, visible events (e.g., *think*, *feel*, *know*, *believe*). How does the child assign a meaning to these kinds of verbs if all they have is the experiential world to use as a reference? Relatedly, many of the most frequent verbs in English have multiple meanings. Consider the verb *get*. When you think of the meaning of this verb, you probably think of a meaning like "receive," as in *I got a present for my birthday*. But this verb can also be used in the following ways, each of which has a meaning very different from "receive":

(1) a. Get over here.
    b. Get your coat on.
    c. Get me some water.
    d. It got cold all of a sudden.
    e. The player got tackled by the opposing team.

Fourth, verbs are not reliably uttered at the exact time the specific action they describe is happening. Many verbs are used when there is no relevant event going on at all. For example, a child is told, "*Eat your peas, dear!*" precisely when she is *not* eating her peas, so there is no event to look for in order to understand the meaning of *eat*.

Lastly, and related to the previous point, many events occur without being labeled at all. For example, imagine a parent that gets up to go to the other room. The parent walks to the door, opens it, and closes it behind them. But the parent does not narrate this whole series of events with matching language: "I'm getting up now. I'm walking toward the door. I am approaching the door. I am reaching out with my hand to grab the doorknob. I am grasping the doorknob. I am turning the doorknob. I am opening the door. I am stepping through the door. I am closing the door behind me. I am going out of sight now." So many events in the world are unlabeled for the child.

Sidebar 5.4: Meanings of Most Common Verbs in English

After *be*, the most frequent verbs in English are *have*, *do*, *say*, *make*, *go*, and *take*. What do you think of as the *basic* meaning of each of these verbs? What are some of the other meanings or usages these verbs can have? Make a list of at least two sentences with different sentence frames (different syntactic structures) for each of these verbs. Compare your list with that of a classmate.

Another consideration is that events are conceptually more complex than objects. Many objects that children interact with early on are discrete, are countable, have definite edges, cohere and move through space as one solid thing, and, as we noted above, can be quite salient (especially the objects parents are likely to label). Events have none of these properties. They can have a beginning, a middle, and an endpoint (e.g., a throwing event), and a verb could refer to any or all parts of the event; they can be continuous or momentary; and verbs themselves may encode various properties of the event, such as the path of a motion event (*enter, descend*) or the manner of a motion event (*bounce, roll*). This is a lot to unpack, both conceptually and linguistically.

In sum, verb acquisition is significantly more complex than noun acquisition, so there is a tremendous logical problem in acquiring verb meanings. Fortunately, verb meanings are related to how many participants their actions or states entail, which in turn relates to how many noun phrases they "select" (a chasing event requires two participants—the chaser and the chasee—and these will correspond to a subject noun phrase and an object noun phrase, while a sleeping event requires only one participant—the sleeper, the subject), so the acquisition of verbs is much more tied in to syntax than is the acquisition of nouns (though nouns too interact with syntax, as we will see below).

Researchers have long wondered whether children might be able to use knowledge from one domain of language to help acquire properties of a different domain of language. This idea is referred to as **bootstrapping**—children use information and knowledge from one domain of language to *bootstrap* into another domain of language. We saw in chapter 3 that infants can use prosody (intonation and pauses) in speech to bootstrap into speech segmentation, and this was called **prosodic bootstrapping**. Here we will talk about **syntactic bootstrapping**—the process of using syntactic information to bootstrap into verb meanings. Gleitman (1990) proposed that if children could use prosodic bootstrapping to identify clause and phrase boundaries, and if they could learn some basic nouns, then children could build up very basic structures (see figure 5.1).

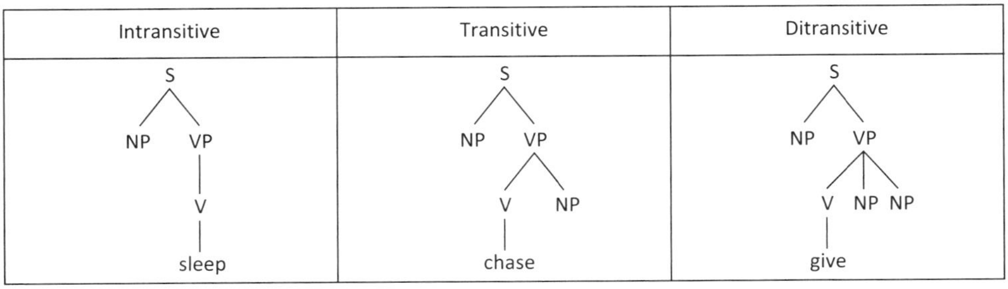

Figure 5.1
Basic structures for verbs and the "participants" in their events.

Imagine that a child knows that intransitive verbs (those with only a subject and no object, like *laugh*, *sleep*, or *run*) select only one participant (the sleeper, mapped to the subject position in the tree); transitive verbs (like *push* or *chase* or *tickle*) select two participants (a chaser, mapped to the subject, and a chasee, mapped to the object); and ditransitive verbs (like *give* or *tell*) select three participants (a giver, mapped to the subject, a thing given, mapped to the direct object, and a recipient, mapped to the indirect object). We refer to the noun phrases (NPs) that label these participants as **arguments**. An argument is a noun phrase or prepositional phrase that is required by the verb in order for the sentence to be well formed; an argument can be a subject, a direct object, or an indirect object.

The reason figuring out the basic structures shown in figure 5.1 can help children learn what the verbs mean is that verbs that select the same number of participants, or arguments, often share certain semantic features. For example, the relationship between the subject and the object of a transitive verb is typically some kind of causative or affected relationship—that is, the subject of a transitive verb *affects* the object in some way. For example, in *Jason pushed Paul*, Paul is affected by the action of Jason. Such an affectedness relationship does not exist with intransitive verbs. Instead, verbs that select only a single argument, like *sleep*, typically label actions or events you do by yourself. Verbs that select three arguments, like *give*, typically have a meaning that relates to transfer, either of objects (like *give*) or ideas (like *tell*). A verb that means something you do by yourself can't be used in a sentence with three arguments (**The girl slept the baby to her mother*). This is something that children's syntax gives them: the syntax of a sentence restricts what the main verb could mean. For this reason, we refer to this idea as the **syntactic bootstrapping hypothesis** (Gleitman, 1990; Gleitman et al., 2005).

Experimental evidence for syntactic bootstrapping abounds in the literature. One of the earliest studies of children's use of argument structure in limiting verb meanings was conducted by Naigles (1990), who showed 2-year-old children videos of two characters who were simultaneously engaged in two activities. For example, in the training phase, both characters were wheeling one of their arms, while one character forced the other into a squatting position. In the testing phase, the two activities were dissociated: one screen showed both characters wheeling their arms (but there was no forcing-to-squat activity), and the other screen showed one character forcing the other to squat (but no arm wheeling). Now children heard either 2a or 2b, and children's looking time at each screen was measured.

(2) a. The duck is gorping the bunny!
    b. The duck and the bunny are gorping!

The first thing to notice is that both test sentences involve a novel verb, *gorp*, and two nouns (bunny and duck). If children know the two nouns and nothing else, then no matter what the test sentence, they should look equally at both video screens. However, if children know that transitive verbs imply affectedness, then when they hear the transitive sentence 2a, they should prefer to look at the scene in which the duck is pushing the bunny down into a squatting position (since this involves affectedness). In the intransitive condition, because intransitive verbs do not involve affectedness, children should prefer to look at the scene in which the two characters are independently spinning their arms. In fact, children who heard 2a spent more time looking at the screen in which one character forced the other to squat, and children who heard 2b spent more time looking at the screen where both characters were wheeling their arms. (This experimental methodology, known as the **intermodal preferential looking paradigm** [Golinkoff, Hirsh-Pasek, Cauley & Gordon, 1987], measures how long children look at a scene that either matches or does not match a linguistic prompt. On average, children will look *longer* at the scene that matches the sentence. See appendix B.)

	Likely event	Unlikely event
The duck is gorping the bunny!	forcing-to-squat	arm-wheeling
The duck and the bunny are gorping!	arm-wheeling	forcing-to-squat

Importantly, both sentences contain two noun phrases, so children were not simply counting the number of nouns and responding on that basis. Rather, children were using argument structure frames (transitive vs. intransitive sentence) to interpret the meaning of the unknown verb.

In another study, Fisher, Gleitman, and Gleitman (1991) showed 3- to 5-year-old children videos of two characters interacting. Their interaction involved one character feeding the other one. This scene could be described with either a transitive sentence (*The rabbit is feeding the elephant*) or an intransitive sentence (*The elephant is eating*). The experimenters assigned the children to one of three conditions: a transitive condition, in which they heard *The rabbit is gorping the elephant!*, an intransitive condition, in which they heard *The elephant is gorping!*, or a control condition, in which they heard *Look! Gorping!* (i.e., there was no sentence frame information). Following the presentation of the scene and the sentence, children were asked, “What does *gorping* mean?”

The researchers found that children in the transitive condition tended to answer “feeding,” children in the intransitive condition tended to answer “eating,” and children in the control condition did not display any preference (they answered both “feeding” and “eating” at approximately equal rates).

More recently, tests of children’s ability to use sentence frame information to restrict the meanings of verbs has been extended to even younger children. For example, Fisher and her colleagues conducted a series of studies in which they presented children as young as 21 months with videos of people performing an action alone or an action affecting another person, along with audio input containing a made-up verb in a transitive (*He’s gorping*

Table 5.3
Example dialogue

Speaker	Intransitive	Transitive
Speaker A	Guess what? Jane blicked!	Guess what? Jane blicked the baby!
Speaker B	Hmm, she blicked?	Hmm, she blicked the baby?
Speaker A	And Bill was blicking.	And Bill was blicking the duck.
Speaker B	Yeah, he was blicking.	Yeah, he was blicking the duck.

Source: Yuan and Fisher (2009).

him!) or an intransitive sentence (*He's gorping!*). In one study, toddlers spent more time looking at the two-person event when they heard the transitive sentence, indicating they matched that sentence with the video (Yuan et al., 2007). In another study, researchers showed 2-year-olds a video of two people having a conversation that used a made-up verb in both transitive and intransitive sentences. Rather than watch an accompanying scene of a solo or dual action, toddlers merely saw two people talking and listened to their conversation (Yuan and Fisher, 2009). An example conversation is given in table 5.3.

After the conversation ended, those same toddlers were shown a video of a solo action (e.g., a person raising one arm) or an action that involved two participants (e.g., a person swinging another person's leg while they sat in a chair), and they were told to "Find blicking!" Even though the conversations and video events had not been presented simultaneously, 2-year-olds preferred to look at the two-person action if they had heard the transitive conversation and the one-person action if they had heard the intransitive conversation. This suggests that by age 2, children can draw inferences about the likely meaning of a verb simply by hearing it used in a particular argument structure.

Where does the knowledge of how argument structure frames relate to verb meanings come from? This knowledge is taken to be innate. In fact, when we look across languages, we find a great deal of uniformity in how verb meanings correspond to their argument structure frames. There can be variation in whether certain arguments are pronounced or silent (for example, many languages allow the subject of the sentence to be left unpronounced, as in Spanish: *(Yo) te buscaba* '(I) was looking for you'), but the rough correspondence between the number of arguments and general types of verb meanings is crosslinguistically robust. Verbs that are intransitive in English (e.g., *arrive*, *go*, *sleep*) are also intransitive in other languages (e.g., *fika*, *enda*, and *lala*, respectively, in Swahili); verbs that are transitive in English (*hit*, *follow*) are also transitive in other languages (*piga*, *fuata*, respectively, in Swahili); verbs that are ditransitive in English (*give*, *send*) are also ditransitive in other languages (*-pa*, *peleka*, respectively, in Swahili); and verbs that take a whole clause as their complement (e.g., *think*) are universally so (e.g., *sema* in Swahili).

So far in this chapter we have described noun learning in terms of meaning that can be derived from conceptual constraints, and we've described verb learning in terms of meaning that can be derived from syntactic constraints. But the constraints of syntactic bootstrapping can be applied to learning noun meanings as well. For example, while nouns for individuated inanimate objects can be used with the indefinite article *a* or the definite article *the* (a/the stone, a/the block, a/the car), nouns for substances cannot be (*a dirt, *a sand) but instead can be used with the determiner *some* (some dirt, some sand), and proper names for animate entities can't be used with either kind of determiner or article (*a Mary, *some Mary, *the Mary). In a classic study by Brown (1957), it was shown that children use these restrictions to draw inferences about what a novel word refers to. Three- to five-year-old children were shown a drawing of two hands kneading some spaghetti-like stuff in a bowl. When they were asked to "point to the blick," children pointed to the bowl, but when they were asked to "point to some blick," they pointed to the spaghetti-like stuff.

Katz, Baker, and Macnamara (1974) investigated this issue in a different way. These authors showed children either (i) a pair of dolls or (ii) a pair of blocks. The dolls differed from each other in their hair color but were otherwise identical, and the blocks differed from each other in their color but were otherwise identical. Some children were shown the dolls; other children were shown the blocks. One group of children was introduced to the dolls/blocks by the experimenter saying something like, *Look what I brought you: this is* ... At this point, one of the dolls/blocks was labeled either as a common noun or a proper noun using nonsense nouns, for example, *This is a zav* (common noun) or *This is Zav* (proper noun). The other doll/block was not labeled or even referred to. Later, children were asked to perform some actions involving the object, like *Put the zav / Zav under the blanket*, in which children had to choose between the target doll/block and the other object that had not been discussed or labeled. Researchers coded what proportion of the time children picked the doll or block that had been labeled earlier and performed the expected action.

The thinking here is that if the child knows that the indefinite article (*a*) is used with common nouns, then they should treat the doll or block that was introduced as *a zav* pretty much the same as the other doll or block that had not been introduced—they are both just *zavs*. But if introduced to the object with the proper noun (*Zav*), if they know that the absence of an article on a count noun means it is a proper noun, then they should assume that *Zav* is actually the name of that particular entity, and when asked to do something to *Zav*, they should only select the object that they were introduced to, not the unlabeled one. A further issue is that dolls, like animate things, can have proper names, while blocks typically cannot. If children know this restriction, then they should not limit *Zav* to referring only to the particular block that had been introduced.

And this is exactly what was found (see table 5.4). The results show that when a doll or block was introduced with an article (*a zav*), children treated the target object as interchangeable with its like-pair object. So a doll introduced as *a zav* is considered just that—a doll among many dolls. And when the child was asked to perform an action on *a zav*, because there was no difference between the target doll or the other doll, they picked (essentially) at chance. As expected, this was also true when the target object was a block, no matter how it was labeled.

Table 5.4
Percentage of selection of original object (results for girls)

	A *zav*	*Zav*
Doll	48%	75%
Block	44%	48%

Source: Katz et al. (1974).

However, when the object is a doll and is introduced as a proper noun without the article, children preferred to select the very same doll that had been introduced to them earlier 75% of the time, not the other (otherwise similar) doll. This means that children assumed the article-less noun was a proper noun. This effect did not play out when the object was a block, because children know that inanimate objects typically do not get proper names. So this experiment shows that children use the presence or absence of the determiner to give them some meaning of the new noun.

Sidebar 5.5: Gender Differences in Katz et al. (1974)

An interesting wrinkle in this result was that it held only for girls, not boys. The researchers suggest that the reason for this is that boys don't take dolls to be surrogate people, but girls do. So this is essentially an effect of early childhood socialization. In a subsequent experiment, researchers changed the protocol to force boys to pay attention to the humanlike nature of dolls, and they obtained similar results with boys as they did with girls in this original experiment.

Recent studies by Ferguson, Graf, and Waxman (2014, 2018) illustrate how very young toddlers can use a known verb to figure out some semantic properties of its subject. If they are shown a picture of an unfamiliar animal (e.g., a hedgehog) and an unfamiliar object and are told, "The dax is crying," 19-month-olds can later pick out the hedgehog as the "dax" instead of the inanimate object (because only animate things can cry). But if they are told, "The dax is right here," they pick either the hedgehog or the object at chance, because being "right here" is a property that can apply to animate or inanimate things.

Taken together, these studies provide examples of young children using semantic or syntactic information within a sentence to inform the acquisition of the meanings of individual words.

5.5 Summary

The acquisition of the lexicon is viewed by some as being relatively uninteresting in that it is often thought that Universal Grammar (UG) does not have anything to say about word meanings. Each language has its own vocabulary, so how could lexical knowledge be innate? Certainly, no one proposes that children are born knowing the meanings of particular words. However, the issues that motivate UG are very much present in the area of lexical acquisition. In this chapter, we saw that children face significant logical challenges in learning the meanings of words, which we referred to as the mapping problem (which included both the Gavagai problem and the hot stove problem) and the categorization problem. We saw that in the absence of some kind of structured learning system, a naïve child would simply not be able to acquire the lexicon through mere exposure to words in the input.

We saw, however, that there is strong evidence for several learning principles that children use in acquiring the meanings of words. These principles are simply biases that guide children in navigating an otherwise infinite hypothesis space. By invoking these learning principles, children are able to reduce the hypothesis space to something manageable. Errors are still predicted to occur, but those errors are restricted in scope and easy to overcome. The three principles/constraints we reviewed (the **principle of reference**, the **whole object constraint**, and the **principle of mutual exclusivity**) provide children with mechanisms to overcome the primary problems of lexical acquisition.

Finally, we saw that children may make use of syntax to help them in acquiring the meanings of some words, in particular verbs. This is referred to as syntactic bootstrapping and is a very productive area of research in the field. Putting all of this together, we see that the acquisition of word meaning is far from the simple process of learning by ostension, as we often think of it. Instead, learning the meanings of words is complex stuff, and we need some pretty sophisticated learning mechanisms to get the job done. Children, it seems, have those sophisticated learning mechanisms in place, so the errors they end up making are not uninterpretable or bizarre, but predictable and understandable.

5.6 Further Reading

Bloom, Paul. 2000. *Learning, Development, and Conceptual Change: How Children Learn the Meanings of Words*. Cambridge, MA: MIT Press.

Carey, Susan. 2009. *The Origin of Concepts*. New York: Oxford University Press.

Gleitman, Lila, and Barbara Landau (eds.). 1994. *The Acquisition of the Lexicon*. Cambridge, MA: MIT Press.

5.7 Exercises

1. The following utterances were produced spontaneously by two children, Adam at age 2;3 and Eve at age 1;7. Read each set of expressions and identify the lexical categories of the words: nouns, verbs, adjectives, adverbs, prepositions, and function words (include in this category any determiners and pronouns). What category has the most different words in it? What category has the fewest or is not used at all? Given what you have read in this chapter, is the distribution of lexical categories in these children's speech about what you would expect? Why or why not?

Adam (2;3)	Eve (1;7)
my suitcase.	more coffee.
spaghetti.	no.
Monroe suitcase.	Fraser's coffee
find dirt.	Cromer busy.
spaghetti.	at home.
yeah.	drink juice.
no.	Fraser cup.
look.	grape juice.
Adam glove.	get grape juice cup.
put ball.	a spoon.
put the ball.	drink coffee.
where ball go?	I see.

2. We described an experiment by Soja, Carey, and Spelke (1991) in which children appeared to assume that a novel label referred to an object as opposed to a substance (see sidebar 5.3). But, of course, we have nouns that label substances (*water, sand, dirt, gel*). In a subsequent experiment, these researchers modified their study by changing the *determiner* they used with the made-up word. For example, instead of saying, "This is my blicket," they now said either, "This is a blicket" or "This is some blicket." Then, children were asked in the test phase, "Which is the blicket?"

Make a prediction about how children would respond in this case. Do you think hearing "a blicket" versus "some blicket" would change how they interpret the meaning of *blicket*? Explain your prediction. If you predict a difference in this condition compared to the one we discussed earlier in the chapter, what would this tell you about additional cues children might use in word learning?

3. We saw that mutual exclusivity can sometimes conflict with the whole object constraint and that this conflict can help children learn the words for parts or characteristics of objects. How else can mutual exclusivity be helpful for learning? Let's look at one example: Imagine a child is presented with two objects, a familiar one (say, a ball) and an unfamiliar one (say, a turkey baster). The adult says, "Please hand me the flimmick!" Which object do you think the child will pick up? Explain your answer and how mutual exclusivity plays a role in learning in this situation.

4. Mutual exclusivity says that an object can have only one label. Now consider a bilingual child: to be bilingual means having (at least) two labels for each object. So what does this mean? Here are some possibilities: (i) bilingual children do not bother with mutual exclusivity because it cannot apply for them, (ii) bilingual children have great difficulty learning words for objects because of mutual exclusivity, or (iii) bilingual children allow mutual exclusivity to apply within each of their languages but not *across* them. How might you design an experiment to tease apart these possibilities?

5. The syntactic bootstrapping hypothesis makes the claim that children infer verb meanings on the basis of argument structure, and argument structure is determined by the number and syntactic positions of nouns in the sentence. Should this learning procedure work the same way in a language that allows null subjects? (Null subjects of a sentence are not pronounced; this happens in Italian, Spanish, and Mandarin, among other languages.) What about languages that allow null objects? (Objects are not pronounced, e.g., in Mandarin and Korean.) How could you test to see if syntactic bootstrapping works in these languages?

6. Verbs that label mental states (e.g., *think*, *believe*, *know*, *guess*, *suppose*) typically select a complement that contains a whole sentence (e.g., *Susan thinks that Fred is nice*, where *Fred is nice* is a whole sentence).

(i) What sentence frames might allow learners to use syntactic bootstrapping to figure out that a novel verb labels a mental state? Make up some novel verbs and use them in sentence frames that would be compatible with mental verbs.

(ii) Can you think of any situations children might experience in which the utterance of a verb in isolation (*Gorping!*) could refer to a mental state?

(iii) Given your answers to (i) and (ii), which type of cue (sentence frames or situations) is stronger for learning mental verbs? Explain your answer.

5.8 References

Baldwin, Dare A., and Ellen M. Markman. 1989. Establishing word-object relations: A first step. *Child Development* 60(2): 381–398.

Bates, Elizabeth, Virginia Marchman, Donna Thal, Larry Fenson, Philip Dale, J. Steven Reznick, Judy Reilly, and Jeff Hartung. 1994. Developmental and stylistic variation in the composition of early vocabulary. *Journal of Child Language* 21(1): 85–123.

Bloom, Paul. 2000. *How Children Learn the Meanings of Words*. Cambridge, MA: MIT Press.

Bloom, Paul. 2002. Mindreading, communication and the learning of names for things. *Mind and Language* 17(1–2): 37–54.

Booth, Amy, and Sandra Waxman. 2009. A horse of a different color: Specifying with precision infants' mappings of novel nouns and adjectives. *Child Development* 80(1): 15–22.

Brown, Roger. 1957. Linguistic determinism and the part of speech. *Journal of Abnormal and Social Psychology* 55: 1–5.

Carey, Susan. 1978. The child as word learner. In Joan Bresnan, George Miller, and Morris Halle (eds.), *Linguistic Theory and Psychological Reality*, pp. 264–293. Cambridge, MA: MIT Press.

Carey, Susan. 1985. *Conceptual Change in Childhood*. Cambridge, MA: MIT Press.

Carey, Susan. 2010. Beyond fast mapping. *Language Learning and Development* 6(3): 184–205.

Carey, Susan, and Elsa Bartlett. 1978. Acquiring a single new word. *Proceedings of the Stanford Child Language Conference* 15: 17–29.

Clark, Eve. 1978. Strategies for communicating. *Child Development* 49: 953–959.

Dapretto, Mirella. 1995. The development of word retrieval abilities in the second year: A new perspective on the naming explosion. Poster presented at the biennial meeting of the Society for Research in Child Development, Indianapolis.

Fenson, Larry, Philip Dale, J. Steven Reznick, Elizabeth Bates, Donna Thal, Stephen J. Pethick, Michael Tomasello, Carolyn B. Mervis, and Joan Stiles. 1994. Variability in early communication development. *Monographs of the Society for Research in Child Development* 59(5), serial no. 242.

Ferguson, Brock, Eileen Graf, and Sandra Waxman. 2014. Infants use known verbs to learn novel nouns: Evidence from 15- and 19-month-olds. *Cognition* 131: 139–146.

Ferguson, Brock, Eileen Graf, and Sandra Waxman. 2018. When veps cry: Two-year-olds efficiently learn novel words from linguistic contexts alone. *Language Learning and Development* 14(1): 1–12.

Fisher, Cynthia, Henry Gleitman, and Lila R. Gleitman. 1991. On the semantic content of subcategorization frames. *Cognitive Psychology* 23: 331–392.

Gelman, Susan, John D. Coley, and Gail M. Gottfried. 1994. Essentialist beliefs in children: The acquisition of concepts and theories. In Lawrence Hirschfeld and Susan Gelman (eds.), *Mapping the Mind: Domain Specificity in Cognition and Culture*, pp. 341–365. New York: Cambridge University Press.

Gentner, Dedre. 1978. On relational meaning: The acquisition of verb meaning. *Child Development* 49: 988–998.

Gleitman, Lila. 1990. The structural sources of verb meanings. *Language Acquisition* 1: 3–55.

Gleitman, Lila, Kimberly Cassidy, Rebecca Nappa, Anna Papafragou, and John C. Trueswell. 2005. Hard words. *Language Learning and Development* 1: 23–64.

Goldfield, Beverly A., and J. Steven Reznick. 1990. Early lexical acquisition: Rate, content, and the vocabulary spurt. *Journal of Child Language* 17: 171–183.

Golinkoff, Roberta, Kathryn Hirsh-Pasek, Kathleen Cauley, and Laura Gordon. 1987. The eyes have it: Lexical and syntactic comprehension in a new paradigm. *Journal of Child Language* 14(1): 23–45.

Golinkoff, Roberta M., Carolyn B. Mervis, and Kathryn Hirsh-Pasek. 1994. Early object labels: The case for a developmental lexical principles framework. *Journal of Child Language*, 21: 125–155. Also in Katharine Perera (ed.), *Growing Points in Child Language*, pp. 125–156. Cambridge, MA: Cambridge University Press.

Harley, Trevor A., and Siobhan B. G. MacAndrew. 2001. Constraints upon word substitution speech errors. *Journal of Psycholinguistic Research* 30(4): 395–418.

Hirsh-Pasek, Kathryn, and Roberta Golinkoff. 1996. *The Origins of Grammar*. Cambridge, MA: MIT Press.

Hoek, Dorothy, David Ingram, and Deborah Gibson. 1986. Some possible causes of children's early word overextensions. *Journal of Child Language* 13: 477–494.

Katz, Nancy, Erica Baker, and John Macnamara. 1974. What's in a name? A study of how children learn common and proper names. *Child Development* 45: 469–473.

Keil, Frank C. 1989. *Concepts, Kinds, and Cognitive Development*. Cambridge, MA: MIT Press.

Keil, Frank C., and Nancy Batterman. 1984. A characteristic-to-defining shift in the development of word meaning. *Journal of Verbal Learning and Verbal Behavior* 23: 221–236.

Landau, Barbara, Linda Smith, and Susan Jones. 1988. The importance of shape in early lexical learning. *Cognitive Development* 3: 299–321.

Macnamara, John. 1982. *Names for Things: A Study of Human Learning.* Cambridge, MA: MIT Press.

Markman, Ellen M. 1989. *Categorization and Naming in Children: Problems of Induction.* Cambridge, MA: MIT Press.

Markman, Ellen M., and Jean E. Hutchinson. 1984. Children's sensitivity to constraints on word meanings: Taxonomic vs. thematic relations. *Cognitive Psychology* 16: 1–27.

Markman, Ellen M., and Gwyn F. Wachtel. 1988. Children's use of mutual exclusivity to constrain the meanings of words. *Cognitive Psychology* 20: 121–157.

Naigles, Letitia. 1990. Children use syntax to learn verb meanings. *Journal of Child Language* 17: 357–374.

Nelson, Keith. 1973. Structure and strategy in learning to talk. *Monographs of the Society for Research in Child Development* 38(1/2), serial no. 149.

Petitto, Laura A. 1992. Modularity and constraints in early lexical acquisition: Evidence from children's first words/signs and gestures. In Megan Gunnar and Michael Maratsos (eds.), *Modularity and Constraints in Language and Cognition: The Minnesota Symposia on Child Psychology*, pp. 25–58. Hillsdale, NJ: Lawrence Erlbaum Associates.

Quine, W. V. O. 1960. *Word and Object.* Cambridge, MA: MIT Press.

Soja, Nancy, Susan Carey, and Elizabeth S. Spelke. 1991. Ontological categories guide young children's inductions of word meaning: Object terms and substance terms. *Cognition* 38: 179–211. Reprinted in Alvin I. Goldman (ed.), *Readings in Philosophy and Cognitive Science*, pp. 461–480. Cambridge, MA: MIT Press.

Taylor, Marjorie, and Susan A. Gelman. 1988. Adjectives and nouns: Children's strategies for learning new words. *Child Development* 59: 411–419.

Woodward, Amanda, Ann Philips, and Elizabeth Spelke. 1993. Infants' expectations about the motion of animate versus inanimate objects. In *Proceedings of the Meeting of the Cognitive Science Society, Boulder, CO*, pp. 1087–1091. Hillsdale, NJ: Lawrence Erlbaum Associates.

Yuan, Sylvia, and Cynthia Fisher. 2009. "Really? He blicked the cat?" Two-year-olds learn distributional facts about verbs in the absence of a referential context. *Psychological Science* 20(5): 619–626.

Yuan, Sylvia, Cynthia Fisher, Yael Gertner, and Jesse Snedeker. 2007. Participants are more than physical bodies: 21-month-olds assign relational meaning to transitive novel verbs. Paper presented at the meeting of the Society for Research in Child Development, Boston.

6 The Acquisition of Morphology

6.0 Introduction

Morphology is the study of the smallest meaningful units of language (morphemes) and how they are organized in languages. In this chapter, we begin with a brief discussion of various kinds of morphemes that language exhibits. We then discuss the very important contributions of Roger Brown, a major figure in the history of this field who contributed significantly to our understanding of the acquisition of morphology. This is followed by a discussion of some basic properties of how children acquire morphology, along with a description of the kinds of errors children make.

Morphemes can be categorized in several ways. One basic division in morphology is the division into **inflectional morphemes** and **derivational morphemes** (word-creating morphemes). Inflection involves different forms of the same basic word, in which the basic meaning of the 'host' word is retained but some additional layer of meaning is added. For example, the word *book* is a root, and it can have inflectional morphology added to it (e.g., *book-s*). Here, the plural *-s* does not change the basic meaning of the word but simply denotes a plurality of books. Similarly, the word *kiss* is a root verb; adding the past tense *-ed* inflectional morpheme, *kissed*, does not change the core meaning of the word but simply denotes that the kissing event occurred in the past. So inflection does not change the basic meaning of the word; rather, it adds some layer of meaning to the original meaning. Typical inflectional categories include grammatical number (singular vs. plural), tense (past, future), agreement (first, second, or third person), and gender (masculine, feminine).[1]

Inflectional morphology can be contrasted with derivational morphology. The addition of a derivational morpheme to a word results in a different word entirely, in which the meaning of the new word *can be* but *need not be* related to the meaning of the original word. For example, *blend* is a verb root that can be modified by the use of derivational morphology in a number of ways, such as *blender* (someone/thing that blends), *unblend* (to undo the blending), and *reblend* (to blend again). These are new words, derived from the original root, and each able to carry its own inflectional morphology, such as *blenders* (plural of *blender*), *unblended* (past tense), and *reblending* (progressive). So inflectional morphology adds meaning to lexical roots or stems, while derivational morphology changes words to create new words (which can, in turn, be inflected with normal inflectional morphology). In these examples, the new words (e.g., *blender*) are semantically related to the root word (*blend*); an example in which the two words are not semantically related (at least, not in a transparent way) would be *vacate* and *vacation*. The word *vacate* is a verb (as in, *You need to vacate the premises*) and *-ion* is a suffix that can be added to verbs to make a noun; in fact, *vacation* is a noun. But while you do "vacate" your home when you go on vacation, if you vacate a room or a building, you are not then on vacation (if only it were that simple!).

Furthermore, the word *blend* may be combined with other lexical roots, in a process called **compounding**, to create new words (e.g., *a fruit blend*). In sum, then, there are at least three kinds of processes at play here (inflection, derivation, and compounding). Because the bulk of research into the acquisition of morphology has focused on inflection, we will focus on inflection in this chapter.

Another important distinction in morphology relates to how closely the morphology is connected to its associates. For example, there is a sense in which the past tense *-ed* in English (e.g., *blended*) is more closely connected to the verb than, say, the definite article *the* is to a noun (e.g., *the cars*). The past tense morpheme is referred to as a **bound morpheme**, since it is tightly bound to the verb (they are usually adjacent to each other), cannot be easily separated from it (you typically can't stick things between the root and the past-tense morpheme: **blend-quickly-ed*), and cannot occur without the verb root (you can't just use the past-tense morpheme by itself—it needs a host). The definite article, on the other hand, is referred to as a **free morpheme**, since it has more independence than its bound counterpart. It can, for example, be separated from the host morpheme by adjectives (*the fast cars*), numerals (*the three cars*), or quantifiers (*the many cars*). In this chapter, we focus on bound morphology, though some discussion of free morphology occurs too.

There are various kinds of bound morphology, the most common of which are prefixes (occurring before the root or stem, e.g., untie, retie) and suffixes (occurring after the root or stem, e.g., kicked, kicking, kicker). Other kinds of bound morphology do occur, such as infixes (which are inserted into the root) and circumfixes (which go around the root). For an illustration of the different types of morphemes, see figure 6.1. The vast majority of research on inflection has been on prefixes and suffixes, so our attention will be focused on these types of affixes.

These various kinds of morphology (inflectional vs. derivational; bound vs. free; prefix vs. suffix vs. infix vs. circumfix), as well as differences in the manner in which languages integrate morphemes into the rest of the sentence,[2] all make the acquisition of morphology a tremendously interesting research area. Most of what we know about the acquisition of inflection comes from the acquisition of Germanic and Romance languages. The research

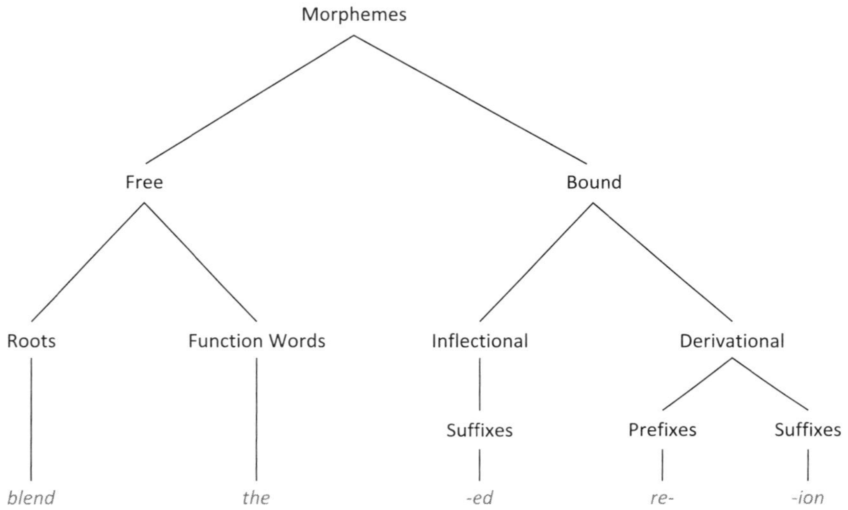

Figure 6.1
Types of morphemes found in English.

tradition on the acquisition of English dates back to the earliest days of the field (as reviewed below), but in recent years a large amount of research on other languages has been published, and we will include discussion of these languages as they become relevant. We begin with the foundational study on the acquisition of morphology—Brown's (1973) monograph on the acquisition of English morphology.

6.1 The Foundation: Roger Brown

Roger Brown is widely seen as a pioneer in the field of language acquisition. He developed the method commonly used to collect naturalistic data; he created a system to gauge children's linguistic maturity in a way that did not depend on chronological age; and he developed a system to assess whether morphemes had been acquired by children. In the 1960s, the context of all research was the new wave of Chomskian ideas. One of the ideas that was gaining traction at the time was that language acquisition is generally quite uniform from child to child. Brown was interested in assessing this claim, and he did so by investigating three English-acquiring children and their use of grammatical morphemes. He asked whether these three children acquired the morphemes in English in the same order.

He identified three children and recorded their interactions with caregivers on a regular basis for several years. These three children were given the pseudonyms Eve, Adam, and Sarah, and their data (freely available through the CHILDES system) are now part of the fabric of the field. Brown transcribed all their interactions and then analyzed them to determine whether these three children acquired fourteen English morphemes in the same order (thus confirming Chomsky's claim that language acquisition is uniform) or in a varied order. The fourteen morphemes he investigated are these:

(i) Present progressive (-*ing*)
(ii/iii) Prepositions *in*, *on*
(iv) Plural (-*s*)
(v) Past irregular (e.g., *went*)
(vi) Possessive (-*s*)
(vii) Uncontractible copula (*is*)[3]
(viii) Articles (*a*, *the*)
(ix) Past regular (-*ed*)
(x) Third person regular (-*s*)
(xi) Third person irregular (e.g., *has*, *does*)
(xii) Uncontractible auxiliary (*is*)
(xiii) Contractible copula (*'s*)
(xiv) Contractible auxiliary (*'s*)

6.1.1 Brown's Method for Establishing When a Morpheme Has Been Acquired

In the 1960s, there was no established method for determining when something had been acquired by a child. Brown and his colleagues set about to develop such a method, and the method they developed has now been accepted as the norm in the field. This method includes several important principles, reviewed here.

6.1.1.1 Obligatory Contexts

Brown's method begins with the recognition that the simple fact of a child using a particular morpheme is not sufficient to establish that the child has a deep and adultlike knowledge of that morpheme. For example, in a context where there are three cookies, if a child says, "I want the cookie," does the fact that the child used the definite article *the* mean the child has acquired the definite article? Well, maybe, but in this case, use of 'the' seems inappropriate: a better thing to say is "I want a cookie." And equally importantly, if the child said "ball is over there," could this be counted as evidence that the child had not acquired the definite article? Again, it depends on the context. If there was no previously mentioned ball in the context, then this is actually evidence that the child omitted the indefinite article (*a*/*an*), not the definite article So this is not an obligatory context for the definite article.

Sidebar 6.1: Obligatory Contexts

> Some other examples of obligatory contexts in English:
>
> (i) Third-person-singular subjects, in present tense, require *-s* on verbs:
> John think__ that he will win. (thinks)
>
> (ii) Progressive verbs require a form of auxiliary *be*:
> Mary __ swimming. (is, was)
>
> (iii) Past tense requires *-ed* on verbs (or relevant irregular past form):
> Yesterday, Sue climb__ a tree. (climbed)
>
> (iv) Nouns require plural marking when there are multiple referents:
> Alex bought three book__. (books)

Brown argued that when establishing whether a morpheme has been acquired, one must only consider obligatory contexts for that morpheme. If it is unclear what the context is, or if the context is compatible with more than one morpheme, then one cannot consider what the child says as evidence for or against knowledge of a particular morpheme.

6.1.1.2 90% Criterion

What Brown found was that in obligatory contexts, at early ages, children produce morphemes in very low proportions, sometimes producing the appropriate morphology in 0% of obligatory contexts (i.e., never producing the required form). But as they mature, that rate gradually climbs, until at some point, children reach an adultlike 100% (or close to it) supply of morphology in obligatory contexts. So at what point does one give credit to the child for having acquired the morpheme? Must we wait until the child has achieved 100% use in obligatory contexts? This seems overly stringent, since surely a child who produced a morpheme in 95% of obligatory contexts has significant knowledge of that morpheme. So Brown set the criterion at 90% of obligatory contexts. That is, a child has to produce a particular morpheme correctly in at least 90% of obligatory contexts before the child is considered to have acquired that morpheme. This number was an arbitrary number and has since been criticized as being overly stringent too. For example, if a child produces a morpheme in 80% of obligatory contexts (or even 60%), shouldn't we say that the child has some knowledge of that morpheme and how it is used? As such, some modern researchers set their own (lower) criterion (often between 70% and 90%), despite the fact that this 90% figure is now the standard in the field.

6.1.1.3 Consistency

Brown noticed that the rate of morpheme use in obligatory contexts varied greatly from transcript to transcript. In one transcript, the child might produce third-person-singular *-s* in 70% of obligatory contexts, while in the very next transcript, the child might produce *-s*

Table 6.1
Hypothetical data set of a child's production of third-person-singular -*s* in obligatory contexts

Transcript	Age of child	# utterances	Morphemes/ Obligatory contexts	% Morpheme use in obligatory context
1	1;10.04	1,357	68/123	55
2	1;10.28	45	4/4	100
3	1;11.08	1,022	70/97	72
4	1;11.27	973	85/105	81
5	2;0.03	1,102	120/130	92
6	2;0.25	985	102/110	93
7	2;1.05	1,001	132/139	95

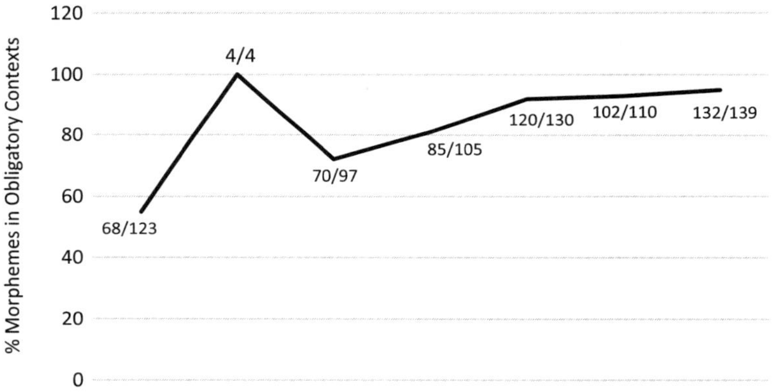

Figure 6.2
Hypothetical graph of child's production of third-person-singular -*s* in obligatory contexts.

in only 35% of obligatory contexts. And in the very next transcript, that figure may swing all the way up to 90%. This is in part due to changes in the child's mood/temperament, changes in who the interlocutors are, and changes in the size of the sample in each transcript.

If the sample size is small, there will be relatively few instances of obligatory contexts for third-person-singular -*s*, so the rate of use of -*s* may not be genuine but just a blip in the data. Consider table 6.1 (with a graphical representation in figure 6.2), which is a hypothetical data set showing the use of third-person -*s* in the speech of one child. In transcript 1, the rate of supply of -*s* in obligatory contexts is 55%, and in transcript 2, this rate rises to 100%. However, there were only four obligatory contexts in that transcript, and in the next two transcripts the rate of use is well below the 90% threshold. It is clearly incorrect to conclude that the child has acquired third-person-singular morphology by transcript 2.

Because of such cases, Brown included as part of the criterion for acquisition the requirement that the 90% threshold be met across three consecutive transcripts. Thus, if that threshold is crossed for one transcript, perhaps due to a sampling anomaly (transcript 2, above), and then drops below 90%, the child is not considered to have acquired the morpheme. But if the rate remains above 90% for three consecutive transcripts (as is the case in the hypothetical data set above in transcripts 5, 6, and 7), then the child is considered to have acquired the morpheme at the first transcript in which the threshold was met—that is, transcript 5.

6.1.1.4 Mean Length of Utterance

Remember that Brown was intent on comparing the language development of Eve, Adam, and Sarah (the three children in his study) to establish whether there was uniformity in the acquisition of the fourteen morphemes listed at the beginning of section 6.1. In order to do this, he needed a metric to compare the language of each child. One way to do this is to ask, "At what *age* does each child acquire each morpheme?" However, Brown and others (notably Brown, Cazden, and Bellugi-Klima, 1968; Brown and Bellugi, 1964) had discovered that age by itself is not always a good indicator of linguistic maturity, so he used the measure now known as **mean length of utterance** (MLU). This measure represents the average number of morphemes per utterance that a child produces. It is measured by counting up all the morphemes in a (randomly selected) sequence of 100 utterances and dividing that number by 100. This yields the MLU for that sample.

Sidebar 6.2: Rules for Calculating MLU

Brown laid out the following rules for how to calculate MLU:

(i) Take the first 100 utterances in a transcript (he advised starting on the second page in order to allow for the child to "warm up" to the interviewer).

(ii) Count each root and inflectional morpheme as 1 morpheme (e.g., *eat-s* and *eat-ing* would each count as 2 morphemes).

(iii) Conversational words like *yes*, *no*, and *hi* get counted as morphemes, but "nonword" expressions like *um* and *oh* do not.

(iv) Auxiliary verbs (*is*, *can*) and semi-auxiliaries (*gonna*, *wanna*) count as 1 morpheme each.

(v) Irregular forms (*feet*, *went*) are counted as single morphemes, unless there is clear evidence that the child uses the corresponding regular morpheme productively (in at least 90% of obligatory contexts).

(vi) Disregard any incomplete or (partially or fully) unintelligible utterances.

(vii) Add up the total number of morphemes and divide by the number of utterances.

Notice that it is recommended to use 100 utterances; since MLU is simply an average, it can be calculated over any number of utterances. If you are working with a number other than 100, simply divide the number of morphemes by the total number of utterances in your sample.

For example, in the following examples we give the number of morphemes that would be counted for each utterance for the purposes of MLU calculation.

(1) a. Mommy take that 3
    b. I doing 3
    c. What is that? 3
    d. That falled. 3

Notice that in 1d we count the regular *-ed* morpheme separately from the verb stem even though the form *falled* is not "correct" from the perspective of adult grammar. Nevertheless, the child used the past tense morpheme here, so it gets counted.

There are some problems with MLU. First, calculating MLU in morphemes makes it problematic to compare children across languages. For example, Italian children, just to get an utterance out, must use a lot more morphology than English children. So if MLU (calculated in morphemes) was used to compare Italian and English children (e.g., Valian, 1991), the MLU of the Italian children would be significantly higher than that of English children. In this case, instead, researchers calculate MLU in words (this is notated as MLUw, in contrast to MLUm, which is MLU calculated in morphemes).

Second, even within varieties of the same language, use of MLUm can be problematic. If MLU is used to assess children speaking different dialects of English, there is a danger that children acquiring one dialect will appear to have lower MLUs than others. For example, African American English (AAE) is different from standard American English in that it often does not have some inflectional elements where standard English does. The sentence *She is a teacher* in AAE is said (quite grammatically) as *She a teacher.* There are numerous other inflectional elements that AAE does not mark, such as agreement and possessives. So using our MLUm system, an AAE-speaking child that says, "She a teacher," would look like they are grammatically less developed than a standard American English–speaking child who says, "She is a teacher." This is wrong, since the AAE-speaking child has shown complete mastery of their target system, so using MLU in this way leads to the incorrect conclusion that AAE-speaking children (or those acquiring varieties of English that have reduced inflectional systems) are less developed. Nonetheless, when studying children acquiring a common language or dialect, MLUm remains widely acknowledged as a better measure of linguistic growth than chronological age.

6.1.2 Brown's Findings

With all these methodological innovations, what did Brown find? Quite amazingly, he found that across the three children he studied, the fourteen morphemes investigated were indeed acquired in close to the same order, given in table 6.2.

Several factors might be at play in determining this language-specific order of acquisition, such as frequency, semantic weight, and saliency, all of which have been shown to

Table 6.2
Order of acquisition of fourteen morphemes, and average ranks

Morpheme	Average Rank
1. Present progressive (*-ing*)	2.33
2./3. Prepositions *in, on*	2.50
4. Plural (*-s*)	3.00
5. Past irregular (e.g., *went*)	6.00
6. Possessive (*-s*)	6.33
7. Uncontractible copula (*is*)	6.50
8. Articles (*a, the*)	7.00
9. Past regular (*-ed*)	9.00
10. Third person regular (*-s*)	9.66
11. Third person irregular (e.g., *has, does*)	10.83
12. Uncontractible auxiliary (*is*)	11.66
13. Contractible copula (*'s*)	12.66
14. Contractible auxiliary (*'s*)	14.00

Source: Brown (1973).

play a part. But interestingly, no single factor is sufficient to explain this order. For example, the order of morphemes in table 6.2 is not in descending order of frequency in the input: the articles *the* and *a* are the most frequent items in the input to a child, but these are acquired later than possessive *-s*.

Furthermore, it is not the case that the equivalents of these morphemes are acquired in the same order across languages. For example, the progressive is not acquired before the past perfective in all languages, nor is plural marking universally acquired before definiteness. So semantic content or function is not sufficient to predict the order of acquisition of morphemes. It seems that all these factors combine, on a language-by-language basis, to produce the order of acquisition that any language exhibits. However, within any single language, morphemes do seem to be acquired in the same order.

6.2 Acquisition of a Rule, or Memorized Chunk: Jean Berko (Gleason)

One question that arises from Brown's method is: Are children actually acquiring the rules of morphology, or are they simply acquiring memorized pieces of language and very astutely recruiting them in the appropriate contexts? For example, does the child who correctly produces the word *cat* and its plural counterpart *cats* actually know that the rule of pluralization in English involves the suffixation of *-s* ([s], [z], or [əz], as the case may be), or have they simply memorized that when talking about one cat, *cat* is appropriate, and when talking about many cats, *cats* is appropriate? Maybe the child does not really know the abstract rule of plural formation in English but has a large inventory of memorized

Figure 6.3
Example of the Wug Test (Berko, 1958). (Image from Wikimedia Commons, https://commons.wikimedia.org/wiki/File:Wug.svg.)

singular-plural pairs. This kind of learning might result in correct production of plural morphology in obligatory contexts, but the nature of the knowledge being exhibited would be qualitatively different from that of adults. It turns out this question had been answered more than a decade earlier than Brown's work in the seminal work by Berko (1958). Berko (now more commonly known as Berko Gleason, which we refer to her as henceforth) is the innovator of the so-called **Wug Test**, a device whose basic principles continue to be used by researchers in modern experiments today.

The central idea behind the Wug Test (see figure 6.3) is to present children with a novel word (i.e., one that they have never heard before) and to elicit from them an inflected form of the word. Because children have never heard these words before, they could not possibly produce the correct form on the basis of memorization of previously heard forms. In the experiment, children saw a picture of a stylized creature—something that was vaguely familiar but certainly nothing identifiable—which was labeled as follows: "This is a wug." Children then saw a second picture with two of the novel creatures in it and were prompted as follows: "Now there is another one. There are two of them. There are two ______." Children were expected to complete the sentence with the correct noun form. If children were aware of the pluralization rule in English, this task should be easy: the correct answer is [wʌgz].

The findings showed that children generally did apply the rules of inflection from very early ages (as young as 4 years), though there was some variation across the three English **allomorphs**, or morpheme variants, of the plural. Specifically, children were quite good at adding the [-s] and the [-z] allomorphs (as in *pifs* [pɪfs] and *wugs* [wʌgz]) but not so good at adding the [-əz] allomorph (*tasses* [tæsəz]). Instead of adding the -əz morpheme, children in Berko Gleason's study simply repeated the stem (*tass* [tæs]). It is not known exactly why children had difficulty with that particular allomorph, but one possibility is that since many of the words it attaches to already end in a sound that is like the plural morpheme itself ([s] or [z]), children think the word is already plural. Another possibility is that children are aware that in English some nouns take a zero morpheme in the plural—*sheep* (singular) vs. *sheep* (plural)—and they are applying that rule in these cases as well.

More recently, research using similar methods has shown that knowledge of the rule of pluralization is in place (to some extent) at ages younger than even 3 years (Zapf and Smith, 2007; Lukyanenko and Fisher, 2014). Using this method, or variations of it, Berko Gleason and others have also tested children on their knowledge of a variety of other inflectional forms (progressive aspect, past tense, third-person-singular agreement, possessive, the agentive derivational morpheme *-er*, and compounding). While there is evidence that from very early on children do employ rules of grammar, there is also a wealth of evidence that the process of employing memorized, fixed forms (often referred to as **chunking**) is a significant part of children's productions (see, e.g., Peters, 1983; Pine and Lieven, 1993; Lieven, Salomo, and Tomasello, 2009).

6.3 General Properties of the Acquisition of Inflection

We turn now to some general properties of the acquisition of (inflectional) morphology that have been observed in a variety of different languages. While there are exceptions to each of these generalizations, such exceptions can usually be explained through some language-specific factors. When investigating the acquisition of any new language, divergence from one of these generalizations might be taken as evidence that there is something worthy of further investigation.

6.3.1 Rapidity and Accuracy

Children typically acquire their inflectional system before age 4 years, with some errors related to the rare and nonsystematic aspects of inflection remaining for another year or so (see section 6.3.4). There is some debate in the literature on the underlying nature of the learning that accounts for this rapid acquisition. Some researchers (e.g., Wexler, 2004) claim that the reason children's acquisition of inflection is relatively quick and maximally rule governed is that children are endowed with a disposition for rule learning, knowledge of the kinds of semantic categories typically encoded by inflection, and the kinds of morphological processes that languages typically exhibit.

In addition, as a whole, children are very accurate in their use of the systematic parts of the inflectional paradigm, even from quite early stages of development. Wexler (2004), in fact, refers to children, perhaps slightly hyperbolically, as "little inflection learning machines."

Saying that children are, by and large, accurate with their morphology does not mean that their speech is error-free. In fact, the inflectional errors that children make (e.g., *goed*, *foots*, *toothes*) are among the cutest errors children produce. Some researchers (e.g., Rubino and Pine, 1998) emphasize these errors and the variability across children in their morphological productions and point out that while inflection is acquired by age 4, the preceding few years are not quite the clean, predictable process that rule-based systems might predict. But interestingly, the kinds of errors children make can be seen as evidence of knowledge of the inflectional system of their language, not ignorance of that system. And all sides agree that the *systematic* parts of any inflectional system are acquired quickly, while the rare and nonsystematic parts of inflection tend to lag behind.

We will return to the types of morphological errors children make, but first we discuss two additional aspects of morphology.

6.3.2 Prefixation versus Suffixation

The two most common kinds of bound inflectional morphology found in the languages of the world are prefixes and suffixes. Children appear to find suffixation easier than prefixation, an observation first noted in the operating principle of Slobin (1973), in which he says children "pay attention to the ends of words." This has been shown in many different kinds of languages, even those that have rich sets of both prefixes and suffixes. For example, Deen (2005) showed that children acquiring Swahili omit the obligatory prefixes in as many as 80% of verbs at certain stages of acquisition, while the obligatory suffixes are omitted in less than 1% of cases during the same time period. No stage was ever detected in which obligatory suffixes are omitted at significant rates, suggesting that inflectional suffixation is acquired (in Swahili at least) before the onset of multiword utterances. This is true in the many languages that have rich inflectional affixation, for example, Seotho (Demuth, 1992), Siswati (Kunene, 1979), Zulu (Suzman, 1996), Quechua (Courtney, 1998), and Georgian (Imedadze and Tuite, 1992).

As for other kinds of morphology (infixes and circumfixes), there is relatively little research. Segalowitz and Galang (1978) tested three age groups of children on their acquisition of focus morphology (also sometimes referred to as voice) in Tagalog: 3-year-olds (3;1–3;11, mean 3;6), 5-year-olds (5;1–5;9, mean 5;6) and 7-year-olds (7;1–7;5, mean 7;4). Verb morphology in Tagalog indicates the thematic role of the argument bearing the focus marker *ang*. When the agent of a transitive (2) clause is focus-marked with *ang*, the verb takes an agent-focus infix <um>; in contrast, when the theme of a transitive clause is focus-marked with *ang* (3), the verb takes a theme-focus infix <in>.

(2)	K<um>akain	ng	mansanas	ang	lalake.
	<AF>IPFV~eat	NFOC	apple	FOC	boy

'The boy is eating an/the apple.'

(3)	K<in>akain	ng	lalake	ang	mansanas.
	<TF>IPFV~eat	NFOC	boy	FOC	apple

'A/the boy is eating the apple.'

Segalowitz and Galang (1978) found that in an elicited production task, even their youngest age group exhibited knowledge of this focus morphology, correctly producing both focus types in more than 70% of responses. This result has been replicated, most recently by Tanaka et al. (2014) for children as young as age 3;10. This suggests that infixes are not as problematic as prefixes, but this is data from just one language.

6.3.3 Rich versus Impoverished Morphology

Intuitively, one might think that a language with less morphology would be easier to acquire than one with a rich set of morphemes. English, for example, has a relatively meager inventory of bound inflectional morphology, while languages like Spanish and Italian have much richer inventories. As adults, we find such complex inflectional systems daunting and challenging—how many of us had a stack of flash cards or a flash-card app to learn verb conjugations? But study after study has shown that children acquiring morphologically rich languages acquire the inflectional system of their language at much earlier ages than children acquiring more impoverished languages like English. Some English-acquiring children continue to struggle with the present-tense third-person-singular *-s* morpheme after age 3, while Italian-speaking children have been shown to master Italian morphology significantly earlier (Hyams, 1986; Valian, 1991). A quick glance at the literature reveals that this is not limited to Italian—children acquiring languages from a variety of language families with rich morphology exhibit control over their morphology at very young ages: Turkish (Aksu-Koc and Slobin, 1985), Greek (Tsimpli, 1992; Stephany, 1995), Arabic (Aljenaie, 2010), Hungarian (Gábor and Lukács, 2012), Sesotho (Demuth, 1992), Zulu (Suzman, 1996), Hebrew (Schaeffer and Ben-Shalom, 2004), Malagasy (Hyams et al., 2006), Swahili (Deen, 2005).

The reason for this is likely the role such rich morphology plays in the respective languages. As discussed by Dressler (2007), morphology in so-called rich morphological languages often does much of the work done by syntax (e.g., word order) in morphologically impoverished languages like English. It is also the case that inflection in morphologically rich languages like Italian is more predictable than in languages like English. For example, agreement in English is not particularly systematic. In fact, it really is a pain since the suffix *-s* on verbs occurs *only* with third-person-singular subjects and nowhere else (see the verb paradigm in 4a), whereas in Spanish and Italian, agreement shows up uniquely for each part of the paradigm in 4b.

(4) a. English

	Singular	Plural
First (I/we)	-0	-0
Second (you)	-0	-0
Third (he/she/they)	-s	-0

b. Spanish *-ar* (e.g., *cantar* 'to sing')[4]

	Singular	Plural
First	-o	-amos
Second	-as	-áis/-an
Third	-a	-an

Agreement in Spanish is more systematic, more predictable, and occurs in all verbs. This means that in languages like Spanish, Swahili, and Russian, inflection must be marked, whereas in languages like English and French, the role of inflection is less robust, and this may be one factor that contributes to the difference in speed of acquisition among the two kinds of languages.

6.3.4 Kinds of Morphological Errors

In section 6.3.1, it was claimed that children are, by and large, accurate in their production of inflection. However, that does not mean child language is completely error-free. Rather, children do produce errors, but the errors are not all errors of ignorance. Different kinds of errors arise for different reasons, as discussed next.

6.3.4.1 Errors of Commission versus Errors of Omission

There are three kinds of errors that children make with inflection: **errors of omission**, **errors of commission**, and **overregularization errors**. We'll focus on errors of omission and commission here, returning to overregularization in the next section. Errors of omission are cases in which a child simply leaves out a morpheme. For example, if a morpheme is required in a particular context (e.g., you are talking about an event that happened in the past) and the child doesn't use the required morpheme (e.g., *Yesterday I walk-0 to school*), this is an error of omission. Such an error could indicate a lack of morphological knowledge, or it could indicate knowledge of inflection but uncertainty about the precise contexts for the use of inflection. Errors of commission, on the other hand, are cases in which a child uses the *wrong form*, and these can be properly thought of as actual errors that result from an absence of knowledge. For example, if a child says *I walks to school*, this is a commission error because the child has used a morpheme (*-s*) but has done so incorrectly.

As we flesh out below, errors of omission are far more frequent than errors of commission, which are virtually unattested. This is important because it tells us about the kind of learner a child is. If children are knowledgeable about how language works, we expect them to navigate the complexity of morphology with relative ease. But if children are truly ignorant of the basic properties of morphology, then we expect acquisition of morphology to be messy, unsystematic, and highly variable from child to child. What we will show here is that the former is definitely closer to the truth than the latter: children's bedrooms might be messy, but their knowledge of morphology is not.

We just mentioned that errors of omission are the more frequent type, and they have been documented in a variety of languages. Some examples are provided in 5, where Ø = missing morphemes and IND = indicative mood.

(5) Errors of omission

a. He bite-Ø me — Agreement Omission (Brown, 1973)

b. Ø–qet–il–e — Agreement Omission, Sesotho (Demuth, 1992)

target:	ke–	qe–	il–	e
	1sg–	finish–	PAST–	IND

'I finished'

c. alafu a–Ø–rud–i — Tense Omission, Swahili (Deen, 2005)

target:	alafu	a–	li–	rud–	i
	then	3sg–	PAST–	return–	IND

'Then he returned.'

d. I Ø in the kitchen — Copula Omission (Becker, 2000)

e. baby Ø talking — Auxiliary Omission (Radford, 1990)

f. Paula Ø play-Ø ball — Determiner Omission (Radford, 1990)

target: Paula is playing with the ball

The particular types of morphemes that children omit can vary from language to language. For example, while children acquiring Italian or Spanish almost never omit agreement morphemes on the verb (they don't produce **cant-0* instead of *canta* 'he is singing'), they often omit articles on nouns (e.g., *casa* 'house' instead of *la casa* 'the house'). This may have to do with the bound or free nature of these morphemes, or it may have to do with the fact that agreement is a suffix in these languages, while articles occur before nouns. So in languages like Italian and Spanish, the omission of inflection occurs but is not as obvious as in some other languages.

In many other languages, bound morphemes are omitted quite frequently. Sano and Hyams (1994) reported that in the speech of the children Eve (age 1;6–1;10), Adam (age 2;3–3;0), and Nina (2;4–2;5) (all data available on CHILDES; MacWhinney, 2000), the rate of omission of *-s* in third-person-singular contexts was 78%, 81%, and 75%, respectively. Deen (2005) reported similar percentages for errors of omission in Swahili, in which the verb is minimally inflected for subject agreement (SA), tense (T), and mood, as shown in 6. In the speech of two children (Haw, aged 2;2–2;6, and Mus, aged 2;0–2;3), SA was omitted 72% and 54% of the time, respectively, and T was omitted 70% and 40% of the time, respectively. Thus, the omission of inflection is both widely attested crosslinguistically and can be very frequent within each language (depending on the language).

(6) Swahili minimal verbal complex: SA–T–V–Mood

Example:	ni–	li–	anguk–	a
	1sg–	PAST–	fall–	ind
	'I fell.'			

In contrast to errors of omission, errors of commission (also known as errors of substitution) are remarkably rare in child language. Harris and Wexler (1995) investigated the speech of ten English-speaking children (age range 1;6–4;1) and identified 1,724 verbs that occurred in the first-person-singular context, of which only three occurred with the incorrect third-person-singular *-s* suffix—an error rate of only 0.17%. Contrast this to the more than 70% error rate for errors of omission described above.

Similarly, in the speech of two Swahili-acquiring children (aged 2;10–3;0 and 1;8–2;1), Deen (2005) found low rates of errors of commission. Of the 224 verbal utterances produced by the older child, only three agreement errors were found (an error rate of 1.3%), and of the 197 verbal utterances produced by the younger child, only one error was found (an error rate of 0.5%). Table 6.3 (adapted from Sano and Hyams, 1994) shows the rate of agreement errors in a number of children acquiring various languages.

Moreover, in languages with agglutinating affixes (this means they get "stacked" onto the root), it is often the case that children must produce multiple affixes in any single verbal utterance. For example, in Swahili, the minimal verbal complex consists of three inflec-

Table 6.3
Rate of agreement commission errors in a range of languages

Child	Language	Age	Utterances	Percentage error
Simone (Clahsen and Penke, 1992)	German	1;7–2;8	1,732	1.0
Martina (Guasti, 1993/94)	Italian	1;8–2;7	478	1.6
Diana (Guasti, 1993/94)	Italian	1;10–2;6	610	1.5
Guglielmo (Guasti, 1993/94)	Italian	2;2–2;7	201	3.3
Claudia (Pizzuto and Caselli, 1992)	Italian	1;4–2;4	1,410	3.0
Francesco (Pizzuto and Caselli, 1992)	Italian	1;5–2;10	1,264	2.0
Marco (Pizzuto and Caselli, 1992)	Italian	1;5–3;0	415	4.0
Gisela (Torrens, 1995)	Catalan	1;10–2;6	81	1.2
Guillem (Torrens, 1995)	Catalan	1;9–2;6	129	2.3

tional affixes (subject agreement, tense, and mood) plus the verb root. A further three inflectional affixes (e.g., object agreement in some transitive clauses) and a further half dozen or so derivational suffixes may occur. In the data from Deen (2005), not a single error of affix ordering was observed—children omitted morphemes, sometimes to large extents, but they never placed them in the wrong order. The same has been found in numerous other languages with similar systems (e.g., Clancy, 1986, for Japanese; Kim, 1997, for Korean; Aksu-Koc and Slobin, 1985, for Turkish; and Courtney and Saville-Troike, 2002, for both Navajo and Quechua).

Errors of commission are rare in child language, especially when compared to the rate of errors of omission. However, when one inspects the transcripts of children in this age range (roughly 1;10 to 3;6), one finds numerous morphological errors that look like errors of commission because a morpheme is being used incorrectly, such as the use of *mans* to mean *men*, or *mouses* to mean *mice*, or *foots* to mean *feet*. Such errors are referred to as errors of overregularization. There is a sense in which such errors cannot be characterized as errors of ignorance (as is the case with the errors of commission), which is why we distinguish them here in their own category, separate from errors of commission.

6.3.4.2 Overregularization and U-Shaped Development

Overregularization is not a true error in the sense that it does not indicate ignorance of the inflectional system. Rather, overregularization results from children correctly applying a rule but overextending that rule to words that just *happen to be exceptions*. Children apply grammatical rules generously, but unfortunately for them, languages are not always well behaved and there can be exceptions to basic rules of morphology. The English plural rule is to add *-s* to the end of the noun (book/books, computer/computers, lion/lions, glass/glasses), but exceptions to this rule abound (man/men, mouse/mice, foot/feet, deer/deer, child/children). When a child figures out what the rule is, they can be forgiven for simply applying that rule to all nouns. When they do that to one of the exceptions in the language, an overregularization error arises: the child is making the language more regular in its morphological system than it actually is, hence the term **overregularization**.

An interesting fact about such overregularization errors is that they don't occur when children are very young, and (perhaps counterintuitively) such errors become frequent at a later point in development (Cazden, 1968; Marcus et al., 1992). So a child at an early age might initially use the correct morphological forms, like *men*, *went*, *feet*, *ran*, only to later stop using these correct forms and start producing overregularized errors, like *mans*, *goed*, *foots*, and *runned*. At an even later stage, children eliminate such errors from their speech and go back to the correct forms. This developmental path is referred to as **U-shaped development**, since high accuracy is followed by a period of low accuracy, which is followed by a return to high accuracy (see figure 6.4).

Let's stop for a moment here and think about why this might be. We understand why children produce overregularization errors (the dip in the U-shaped curve)—because they

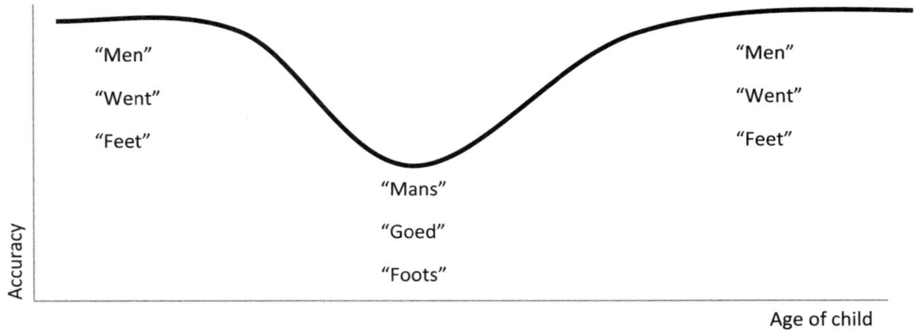

Figure 6.4
Hypothetical U-shaped development.

are applying a rule to the language without considering exceptions. We also understand why that curve swings upward later in development—children are learning each exception on a case-by-case basis. But why do you think the accuracy rate is so high at early ages? Shouldn't accuracy be worst at young ages?

Well, let's remember that errors of overregularization are evidence that children have acquired the underlying system in the language but have not yet acquired the exceptions. So at the point that the curve initially starts to bend downward into the U-shape, children are just beginning to learn the regular rule in the language. But before that, children may not have acquired the regular rule. Instead, they are using words that they hear in their input without having analyzed them as root+inflection. So a word like *books* is not *book*+plural for the child but simply 'plural book'. It's a single lexical item with no internal morphology to speak of. We referred to this as an unanalyzed chunk earlier (they are also sometimes called **formulaic items**). But this is true for regular nouns like *books* and *cups* as well as for irregular nouns like *men* and *feet*. Later, when the child begins to notice the regularity, they begin to apply the rule of inflection, and in the case of regular nouns, this yields accurate forms: the child still produces things like *books* and *cups*. But in the case of irregular nouns, the child incorrectly applies the plural rule to the irregular noun, yielding something like *mans*, *mens*, *feets*, *foots*. From the perspective of adult grammar, these forms are not accurate, so the rate of accuracy drops. However, when accuracy begins to fall because of overregularization, this should be taken as evidence of a child having acquired the rules governing a particular inflectional paradigm, not an absence of knowledge. Thus, errors of overregularization, while perhaps similar to errors of commission in that both involve the use of non-target morphology, are fundamentally errors that involve the overapplication of a rule rather than errors of ignorance.

Sidebar 6.3: Overregularization

> How do you think the following words would be overregularized (if at all)?
>
> fish
> break
> bring
> mouse

6.4 The Role of Input

There has been a significant amount of research on the relationship between input and acquisition in terms of word learning and syntactic development but relatively little specifically on the acquisition of morphology. The one area in which input frequency has been investigated is the phenomenon of overregularization. There have been numerous analyses of the input to children to ascertain the number of correct irregular forms (e.g., *went*, *feet*, *mice*) a child must hear before the overregularized form (e.g., *goed*, *foots*, *mouses*) is eliminated from the child's language. The two most prominent approaches in the literature may be dubbed the **blocking approach** (Marcus et al., 1992) and the **competition approach**. The former contends that once the child hears the irregular form in the input, the overregularized form is blocked and therefore immediately removed from the child's grammar. Input, therefore, will quickly purge the system of the overregularization. Competition models (e.g., Rumelhart and McClelland, 1986) predict that children initially postulate the overregularized form, and when they hear the irregular form for the first time, they entertain both options for a while. As children hear the irregular form time and time again and fail to hear the overregularized form, the irregular form gains strength and the overregularization is slowly purged from the system. As Maratsos (2000) points out, this is precisely why the more common irregulars (e.g., *went* and *men*) are acquired earlier than the less frequent irregulars (*geese*). This approach assumes that a far larger amount of input is required in order to cleanse the system of overregularization errors.

More generally, though, how much input is required for an inflectional morpheme to be acquired? Are hundreds of tokens required, or just a handful? Kim et al. (2014) investigated how much input (and what kind of input) is required in order for a child to acquire a very rare morpheme. Korean has a plural marker, *-tul*, that occurs to the right of the root to which it attaches but to the left of the case marker: [root-*tul*-case]. The first noun in example 7 exemplifies this plural marker, referred to as the intrinsic plural marker (IPM). This plural marker is very frequent in Korean. However, Korean also has a far rarer plural form, referred to as the extrinsic plural marker (EPM), also pronounced *-tul*, exemplified in the second nominal in 7. Note that the EPM is homophonous with the IPM, but it occurs to the right of the case marker: [root-case-*tul*]. The EPM is not in fact a marker of plurality but a marker of distributivity—with the EPM, the sentence must have a distributive meaning, such as the one provided in the gloss. In the absence of the EPM, the sentence could have a collective meaning—that is, 'The students all (as a group) gave the children money.' The question Kim et al. asked is, when do Korean children acquire the EPM, and what kind of input is required?

(7)	Haksayng-tul-i	ai-eykey-tul	ton-ul	cwu-ess-ta.
	student-IPM-NOM	child-DAT-EPM	money-ACC	give-PST-DECL
	'Students each gave the children money.'			

Kim et al. found that in a corpus of speech, the EPM is virtually nonexistent. They then tested twenty Korean children aged 5;3–6;9 (mean 6;1) using a **Truth Value Judgment Task** (see appendix B) and found that all the children failed to show knowledge of the distributive requirement of EPM. This is not surprising, given the near-absence of tokens in the input. But this raises an important question: If the EPM is so rare that children aged 6 years have not yet acquired it, how could they ever acquire this morpheme? How many tokens do they need to acquire EPM?

Kim et al. then exposed children to scenarios involving interaction between a mother and child in which the distributive meaning was exemplified, and the children were then retested for knowledge of EPM. They found that fourteen of the twenty children acquired the distributive meaning of EPM after *just one exposure* to EPM. A subsequent testing two weeks later showed that all fourteen of the children retained knowledge that EPM carries a distributive function.

This shows that the amount of exposure required to acquire some properties of morphosyntax need not be large—as little as a single, meaningful exposure is enough for children to acquire some aspects of morphosyntax. Kim et al. refer to this as **syntactic fast mapping**, akin to lexical fast mapping (see chapter 5), since just one exposure is sufficient to produce long-lasting knowledge of the properties of EPM in Korean.

While frequency has been the most widely discussed factor in the acquisition of morphology, several other factors have been shown to have an effect on the accuracy and speed of acquisition, for example, the phonological complexity of individual morphemes (phonologically complex morphemes may be harder to acquire) and semantic transparency (whether a morpheme marks something that has a communicative function, like tense, versus something that is less obviously useful in communication, such as agreement). The relative role of these additional factors is still under investigation, and their impact on the acquisition of morphology remains unclear.

Sidebar 6.4: Counting Morphemes in English

> Using the transcript sample.cha, how often does the possessive *'s* occur in child-directed speech (all speaker lines that start with *MOT)? There is another homophonous form that occurs in the child-directed speech: the contracted *'s* form (as in *it + is = it's*). How often does this occur? You can use a simple search command and count each token by hand.

6.5 Summary

In this chapter we reviewed several clear generalizations in the acquisition of morphology. While on a general level, children acquire morphology quickly and accurately, there are numerous caveats we must consider. Children find prefixes more challenging than suffixes; they omit some types of morphology (in some languages) often and overregularize rules; but they rarely commit true errors of commission. Moreover, children acquiring morphologically rich and systematic languages (like Italian) acquire the morphology of their language earlier than those acquiring unsystematic or impoverished languages (like English). These generalizations have emerged from decades of research on well-known languages like English, German, Italian, and Spanish. Despite this, crosslinguistic research has grown in recent years, and we continue to learn more about the acquisition of morphology.

Many unanswered questions remain, however, which perhaps will be addressed by further crosslinguistic research. What are the relative roles of frequency, phonology, saliency, and grammatical function in the acquisition of morphology? One thing that would help answer this question would be more data from a variety of languages that show the relative effect of each of these factors in each language. Additionally, remarkably little is known about the acquisition of infixes and circumfixes. While these kinds of affixes are rare in the languages of the world, one wonders whether they pose a difficulty for children to acquire, more so than even prefixes. The little research that there is on Tagalog suggests that children do not have difficulty with infixes, but this is just one language. Moreover, the vast majority of research on child morphology has focused on languages that are isolating, weakly inflected, or agglutinating. There has been relatively little research on the acquisition of synthetic or polysynthetic languages (although see Kelly et al., 2014, for an excellent overview). This represents, in our view, an important avenue of future research.

6.6 Further Reading

Guijarro-Fuentes, Pedro, María Pilar Larrañaga, and John Clibbens. 2008. *First Language Acquisition of Morphology and Syntax: Perspectives across Languages and Learners*. Amsterdam: John Benjamins.

Slobin, Dan I. (ed.). 1985. *The Crosslinguistic Study of Language Acquisition*, vol. 1: *The Data*. Hillsdale, NJ: Lawrence Erlbaum Associates.

6.7 Exercises

1. Write an example of a sentence in English that contains an inflectional morpheme (underline the inflectional morpheme):

2. a. Write an example of a word in English that contains a prefix (underline the prefix):
b. Write an example of a word in English that contains a suffix (underline the suffix):

3. Categorize the following into errors of commission, omission, overregularization, or non-errors:

(i) *Yesterday, I bring some books home.*
(ii) *Yesterday, I bringed some books home.*
(iii) *Yesterday, I brought one books home.*
(iv) *Yesterday, I brought one book home.*

4. Consider the following utterances from Sammy, age 2;7. First, calculate the number of morphemes in each utterance and write the number on the blank line on the right. Words in parentheses were not uttered by Sammy but were inserted by the person transcribing to help you understand what Sammy was trying to say; these words should not be counted as part of the MLU calculations.

Utterance	Number of morphemes
1. Daddy coming	________________
2. (I) want more milk.	________________
3. Me want more milk.	________________
4. Him like that.	________________
5. Daddy home.	________________
6. The doggie's outside.	________________
7. Him eats mud.	________________
8. He loves hamburgers!	________________
9. Her riding bike.	________________
10. Me wearing new clothes.	________________
11. He see snow outside.	________________
12. (She) sleeping.	________________
13. Need go in pool.	________________
14. The car is blue.	________________
15. The baby's sleeping.	________________
16. He's a big dog.	________________
17. That's mines.	________________
18. This is my cracker.	________________
19. (She) have a new baby.	________________
20. He doesn't likes that.	________________
Total:	________________

(i) What is this child's MLU? _______________

(ii) What kind of pronoun errors does this child make?

____

(iii) Should the pronoun errors affect the MLU count? Yes / No

5. A child aged 3;1 is saying words like *feets* (plural of *foot*), *runned* (past tense of *run*), and *bringed* (past tense of *bring*). Can you predict what forms the child would be producing at the following ages?

	2;0	3;1	4;6
Plural of *foot*		Feets	
Past tense of *run*		Runned	
Past tense of *bring*		Bringed	

6.8 References

Aksu-Koc, Ayhan, and Dan Slobin. 1985. The acquisition of Turkish. In Dan I. Slobin (ed.), *The Crosslinguistic Study of Language Acquisition*, vol. 1: *The Data*, pp. 839–878. Hillsdale, NJ: Lawrence Erlbaum Associates.

Aljenaie, Khawla. 2010. Verbal inflection in the acquisition of Kuwaiti Arabic. *Journal of Child Language* 37(4): 841–863.

Ambridge, Ben, and Elena Lieven. 2011. *Child Language Acquisition: Contrasting Theoretical Approaches*. Cambridge: Cambridge University Press.

Ambridge, Ben, Julian M. Pine, Caroline F. Rowland, Franklin Chang, and Amy Bidgood. 2013. The retreat from overgeneralization in child language acquisition: Word learning, morphology and verb argument structure. *Wiley Interdisciplinary Reviews: Cognitive Science* 4: 47–62.

Argus, Reili. 2009. The early development of case and number in Estonian. In Ursula Stephany and Maria Voeikova (eds.), *Development of Nominal Inflection in First Language Acquisition: A Cross-Linguistic Perspective*, pp. 111–152. New York: Mouton de Gruyter.

Austin, Jennifer. 2012. The case-agreement hierarchy in acquisition: Evidence from children learning Basque. *Lingua* 122: 289–302.

Babyonyshev, Maria, and Stefania Marin. 2006. Acquisition of pronominal clitics in Romanian. *Catalan Journal of Linguistics* 5: 17–44.

Bavin, Edith, and Timothy Shopen. 1985. Children's acquisition of Warlpiri: Comprehension of transitive sentences. *Journal of Child Language* 12(3): 597–610.

Becker, Misha. 2000. The development of the copula in child English: The lightness of *be*. PhD diss., University of California, Los Angeles.

Berko, Jean. 1958. The child's learning of English morphology. *Word* 14: 150–177.

Berman, Ruth A. 1981. Language development and language knowledge: Evidence from the acquisition of Hebrew morphophonology. *Journal of Child Language* 8(3): 609–626.

Blom, Elma. 2007. Modality, infinitives and finite bare verbs in Dutch and English child language. *Language Acquisition* 14(1): 75–113.

Blount, Ben. 1988. Cognition and phonology in acquisition of plurals and possessives by Luo children. *Language Sciences* 10(1): 225–240.

Brown, Roger. 1973. *A First Language*. Cambridge, MA: Harvard University Press.

Brown, Roger, and Ursula Bellugi. 1964. Three processes in the child's acquisition of syntax. *Harvard Educational Review* 34: 133–151.

Brown, Roger, Courtney Cazden, and Ursula Bellugi-Klima. 1968. The child's grammar from I to III. In John P. Hill (ed.), *Minnesota Symposia on Child Psychology*, 28–73. Minneapolis: University of Minnesota Press.

Cazden, Courtney. 1968. The acquisition of noun and verb inflections. *Child Development* 39: 433–448.

Choi, Soonja, and Alison Gopnik. 1995. Early acquisition of verbs in Korean: A cross-linguistic study. *Journal of Child Language* 22(3): 497–529.

Clahsen, Harald. 1986. Verb inflections in German child language: Acquisition of agreement markings and the functions they encode. *Linguistics* 24(1): 79–121.

Clahsen, Harald, and Martina Penke. 1992. The acquisition of agreement morphology and its syntactic consequences: New evidence on German child language from the Simone-corpus. In Jürgen Meisel (ed.), *The Acquisition of Verb Placement: Functional Categories and V2 Phenomena in Language Acquisition*, pp. 181–223. Dordrecht: Kluwer.

Clancy, Patricia. 1986. The acquisition of Japanese. In Dan I. Slobin (ed.), *The Crosslinguistic Study of Language Acquisition*, vol. 1: *The Data*, pp. 373–524. Hillsdale, NJ: Lawrence Erlbaum Associates.

Courtney, E. H., and M. Saville-Troike. 2002. Learning to construct verbs in Navajo and Quechua. *Journal of Child Language* 29: 623–654.

Courtney, Ellen. 1998. Child acquisition of Quechua morphosyntax. Unpublished doctoral dissertation, University of Arizona, Tucson.

Dasinger, Lisa. 1997. Issues in the acquisition of Estonian, Finnish and Hungarian: A crosslinguistic comparison. In Dan I. Slobin (ed.), *The Crosslinguistic Study of Language Acquisition*, vol. 4, pp. 1–86. Hillsdale, NJ: Lawrence Erlbaum.

Deen, Kamil Ud. 2005. *The Acquisition of Swahili*. Amsterdam: John Benjamins.

Demuth, Katherine. 1992. The acquisition of Sesotho. In Dan I. Slobin (ed.), *The Crosslinguistic Study of Language Acquisition*, vol. 3, pp. 557–638. Mahwah, NJ: Lawrence Erlbaum Associates.

Dressler, Wolfgang. 2007. Introduction. In Sabine Laaha and Steven Gillis (eds.), Typological perspectives on the acquisition of noun and verb morphology, *Antwerp Papers in Linguistics*, vol. 112, pp. 3–9. Antwerp: University of Antwerp.

Erbaugh, Mary. 1992. The acquisition of Mandarin. In Dan I. Slobin (ed.), *The Crosslinguistic Study of Language Acquisition*, vol. 3, pp. 373–455. Mahwah, NJ: Lawrence Erlbaum Associates.

Fortescue, Michael, and Lennert Olsen. 1992. The acquisition of West Greenlandic. In Dan I. Slobin (ed.), *The Crosslinguistic Study of Language Acquisition*, vol. 3, pp. 111–220. Hillsdale, NJ: Lawrence Erlbaum Associates.

Gábor, Bálint, and Ágnes Lukács. 2012. Early morphological productivity in Hungarian: Evidence from sentence repetition and elicited production. *Journal of Child Language* 39(2): 411–442.

Gagarina, Natalia, and Maria D. Voeikova. 2009. Acquisition of case and number in Russian. In Ursula Stephany and Maria Voeikova (eds.), *Development of Nominal Inflection in First Language Acquisition: A Cross-Linguistic Perspective*, pp. 179–216. New York: Mouton de Gruyter.

Galang, Rosita. 1982. Acquisition of Tagalog verb morphology: Linguistic and cognitive factors. *Philippine Journal of Linguistics* 13(2): 1–15.

Gordishevsky, Galina, and Jeannette C. Schaeffer. 2008. The development and interaction of case and number in early Russian. In Pedro Guijarro-Fuentes, María Pilar Larrañaga, and John Clibbens (eds.), *First Language Acquisition of Morphology and Syntax: Perspectives across Languages and Learners*, pp. 31–59. Amsterdam: John Benjamins.

Grinstead, John. 2000. Case, inflection and subject licensing in child Catalan and Spanish. *Journal of Child Language* 27(1): 119–155.

Grinstead, John (ed.). 2009. *Hispanic Child Languages: Typical and Impaired Development*. Amsterdam: John Benjamins.

Guasti, Maria Teresa. 1993/94. Verb syntax in Italian child grammar: Finite and nonfinite verbs. *Language Acquisition* 3(1): 1–40.

Gxilishe, Sandile, Peter de Villiers, and Jill de Villiers. 2007. The acquisition of subject agreement in Xhosa. In Alyona Belikova et al. (eds.), *Proceedings of the 2nd Conference on Generative Approaches to Language Acquisition North America (GALANA)*, pp. 114–123. Somerville, MA: Cascadilla Proceedings Project.

Hamann, Cornelia, and Kim Plunkett. 1998. Subjectless sentences in child Danish. *Cognition* 69: 35–72.

Harris, Tony, and Kenneth Wexler. 1996. The Optional-Infinitive stage in child English: Evidence from negation. In H. Clahsen (ed.), *Generative Perspectives on Language Acquisition: Empirical Findings*, pp. 1–42. Amsterdam: John Benjamins.

Henden, Åge Kristian. 2013. Early lexical and grammar development in Norwegian language acquisition. Unpublished thesis, Norwegian Institute of Science and Technology.

Hyams, Nina. 1986. *Language Acquisition and the Theory of Parameters*. Dordrecht: Reidel.

Hyams, Nina, Dimitris Ntelitheos, and Cecile Manorohanta. 2006. Acquisition of the Malagasy voicing system: Implications for the adult grammar. *Natural Language and Linguistic Theory* 24(4): 1049–1092.

Ilic, Tatjana, and Kamil Ud Deen. 2003. Object raising and cliticization in Serbo-Croatian child language. In Jaqueline van Kampen and Segio Baauw (eds.), *Proceedings to the GALA Conference*, pp. 235–243. Utrecht: LOT.

Imedadze, Natela, and Kevin Tuite. 1992. The acquisition of Georgian. In Dan I. Slobin (ed.), *The Crosslinguistic Study of Language Acquisition*, vol. 4, pp. 39–109. Mahwah, NJ: Lawrence Erlbaum Associates.

Josefsson, Gunlög, Christer Platzack, and Gisela Håkansson. 2004. *The Acquisition of Swedish Grammar*. Amsterdam: John Benjamins.

Kelly, Barbara, Gillian Wigglesworth, Rachel Nordlinger, and Joseph Blythe. 2014. The acquisition of polysynthetic languages. *Language and Linguistics Compass* 8: 51–64.

Ketrez, F. Nihan, and Ayhan Aksu-Koc. 2009. Early nominal morphology in Turkish: Emergence of case and number. In Ursula Stephany and Maria Voeikova (eds.), *The Development of Number and Case in the First Language Acquisition: A Cross-Linguistic Perspective*, pp. 15–48. Berlin: Mouton de Gruyter.

Kim, Young-Joo. 1997. The acquisition of Korean. In D. I. Slobin (ed.), *The Crosslinguistic Study of Language Acquisition*, vol. 4, pp. 335–435. Hillsdale, NJ: Lawrence Erlbaum Associates.

Kim, Chae Eun, Willian O'Grady, and Kamil Ud Deen. 2014. The Korean extrinsic plural marker. *Korean Linguistics* 16(1): 1–17.

Kovačević, Melita, Marijan Palmović, and Gordana Hržica. 2009. The acquisition of case, number and gender in Croatian. In Ursula Stephany and Maria Voeikova (eds.), *Development of Nominal Inflection in First Language Acquisition: A Cross-Linguistic Perspective*, pp. 153–178. New York: Mouton de Gruyter.

Kunene, Lwandle. 1979. The acquisition of SiSwati as a first language: A morphological study with special reference to noun prefixes, noun classes, and some agreement markers. Unpublished doctoral dissertation, University of California, Los Angeles.

Laalo, Klaus. 2009. Acquisition of case and plural in Finnish. In Ursula Stephany and Maria Voeikova (eds.), *Development of Nominal Inflection in First Language Acquisition: A Cross-Linguistic Perspective*, pp. 49–90. New York: Mouton de Gruyter.

Lakshmanan, Usha. 2006. Assessing linguistic competence: Verbal inflection in child Tamil. *Language Assessment Quarterly* 3(2): 171–205.

Lieven, Elena, Dorothé Salomo, and Michael Tomasello. 2009. Two-year-old children's production of multiword utterances: A usage-based analysis. *Cognitive Linguistics* 20: 481–507.

Lukavský, Jiří, and Filip Smolík. 2009. Word order and case inflection in Czech: On-line sentence comprehension in children and adults. In Niels. A. Taatgen and Hedderik van Rijn (eds.), *Proceedings of the 31st Conference of the Cognitive Science Society*, pp. 1358–1363. Austin, TX: Cognitive Science Society.

Lukyanenko, Cynthia, and Cynthia Fisher. 2014. 30-month-olds use verb agreement features in online sentence processing. In Will Orman and Matthew James Valleau (eds.), *Proceedings of the 38th Annual Boston University Conference on Language Development*, pp. 292–305. Boston: Cascadilla Press.

MacWhinney, Brian. 2000. *The CHILDES Project: Tools for Analyzing Talk*. 3rd ed. Mahwah, NJ: Lawrence Erlbaum Associates.

Maratsos, Michael. 2000. More overregularizations after all: New data and discussion on Marcus, Pinker, Ullman, Hollander, Rosen and Xu. *Journal of Child Language* 27(1): 183–212.

Marcus, Gary, Steven Pinker, Michael Ullman, Michelle Hollander, John Rosen, and Fei Xu. 1992. *Overregularization in Language Acquisition*. Chicago: University of Chicago Press.

Murasugi, Keiko, and Koji Sugisaki. 2008. The acquisition of Japanese syntax. In Shigeru Miyagawa (ed.), *The Oxford Handbook of Japanese Linguistics*, pp. 250–286. Oxford: Oxford University Press.

Narafshan, Mehry Haddad, Firooz Sadighi, Mohammad Sadegh Bagheri, and Nasrin Shokrpour. 2014. The role of input in first language acquisition. *International Journal of Applied Linguistics and English Literature* 3(1): 86–91.

Nirmala, Chervela. 1983. Development of plural in Telugu children. In *Osmania Papers in Linguistics*, vols. 9–10, pp. 1–20. Hyderabad, India: Hyderabad Department of Linguistics: Osmania University Press.

Ochs, Elinor. 1982. Ergativity and word order in Samoan child language. *Language* 58: 646–671.

Olbishevska, Olesya. 2004. Acquisition of aspect (lexical vs. grammatical) by Ukrainian children. *Cahiers linguistiques d'Ottawa* 32: 66–86.

Otsu, Yukio. 1994. Case marking particles and phrase structure in early Japanese. In Barbara Lust, Margarita Suner, and John Whitman (eds.), *Syntactic Theory and First Language Acquisition: Cross-linguistic Perspectives*, pp. 159–169. Hillsdale, NJ: Lawrence Erlbaum Associates.

Otsu, Yukio. 1999. First language acquisition. In Matsuko Tsujimura (ed.), *The Handbook of Japanese Linguistics*, pp. 378–397. Oxford: Blackwell.

Peters, Ann M. 1983. *The Units of Language Acquisition*. Cambridge: Cambridge University Press.

Peters, Ann M. 1997. Language typology, prosody and the acquisition of grammatical morphemes. In Dan I. Slobin (ed.), *The Crosslinguistic Study of Language Acquisition*, vol. 5: *Expanding the Contexts*, pp. 136–197. Hillsdale, NJ: Lawrence Erlbaum Associates.

Pine, Julian M., and Elena Lieven. 1993. Reanalysing rote-learned phrases: Individual differences in the transition to multi-word speech. *Journal of Child Language* 20: 551–571.

Pires, Acrisio, and Jason Rothman. 2009. *Minimalist Inquiries into Child and Adult Language Acquisition*. Berlin: Mouton de Gruyter.

Pizutto, Elena, and Cristina Caselli. 1992. The acquisition of Italian morphology: Implications for models of language development. *Journal of Child Language* 19(3): 491–557.

Prévost, Philippe. 2009. *The Acquisition of French: The Development of Inflectional Morphology and Syntax in L1 Acquisition, Bilingualism, and L2 Acquisition*. Amsterdam: John Benjamins.

Pye, Clifton. 1992. The acquisition of K'iche' Maya. In Dan I. Slobin (ed.), *The Crosslinguistic Study of Language Acquisition*, vol. 3, pp. 221–308. Hillsdale, NJ: Lawrence Erlbaum Associates.

Radford, Andrew. 1990. *Syntactic Theory and the Acquisition of English Syntax: The Nature of Early Child Grammars of English*. Oxford: Basil Blackwell.

Raghavendra, Parimala, and Laurence B. Leonard. 1989. The acquisition of agglutinating languages: Converging evidence from Tamil. *Journal of Child Language* 16(2): 313–322.

Ragnarsdóttir, Hrafnhildur, Hanne Gram Simonsen, and Kim Plunkett. 1999. The acquisition of past tense morphology in Icelandic and Norwegian children: An experimental study. *Journal of Child Language* 26(3): 577–618.

Rose, Yvan, and Julie Brittain. 2011. Grammar matters: Evidence from phonological and morphological development in Northern East Cree. In Mihaela Pirvulescu (ed.), *Proceedings of the 4th Conference on Generative Approaches to Language Acquisition in North America (GALANA)*, pp. 193–208. Somerville, MA: Cascadilla Press.

Rubino, Rejane, and Julian Pine. 1998. Subject agreement in Brazilian Portuguese: What low error rates hide. *Journal of Child Language* 25: 35–59.

Rumelhart, David, and James L. McClelland. 1986. On learning the past tenses of English. In *Parallel Distributed Processing*, vol. 2, pp. 216–271. Cambridge, MA: MIT Press.

Rus, Dominic. 2008. The acquisition of verbal inflection in child grammars in a variability model of early morphosyntactic development: A biolinguistic perspective. PhD diss., Georgetown University.

Sano, Tetsuya. 1995. Roots in language acquisition: A comparative study of Japanese and European languages. PhD dissertation, University of California, Los Angeles.

Sano, Tetsuya, and Nina Hyams. 1994. Agreement, finiteness, and development of null arguments. In Mercè Gonzalez (ed.), *The Proceedings to NELS 24*, pp. 543–558. Amherst, MA: Graduate Linguistic Student Association.

Sanusi, Issa, and Bolanle Arokoyo. 2011. Acquisition of early verbs by Yoruba children: An analysis of a longitudinal study. *Ilorin Journal of Linguistics, Literature and Culture* 2: 1–26.

Sarma, Vaijayanthi. 2000. Case, agreement and word order: Issues in the syntax and acquisition of Tamil. PhD diss., Massachusetts Institute of Technology.

Schaeffer, Jeannette, and Dorit Ben-Shalom. 2004. On root infinitives in child Hebrew. *Language Acquisition* 12(1): 83–96.

Schieffelin, Bambi. 1985. The acquisition of Kaluli. In Dan I. Slobin (ed.), *The Crosslinguistic Study of Language Acquisition*, vol. 1: *The Data*, pp. 525–592. Hillsdale, NJ: Lawrence Erlbaum Associates.

Segalowitz, Norman S., and Rosita Galang. 1978. Agent-patient word-order preference in the acquisition of Tagalog. *Journal of Child Language* 5(1): 47–64.

Slobin, Dan I. 1966. The acquisition of Russian as a native language. In Frank Smith and George A. Miller (eds.), *The Genesis of Language: A Psycholinguistic Approach*, pp. 129–148. Cambridge: Cambridge University Press.

Slobin, Dan I. 1973. Cognitive prerequisites for the development of grammar. In Charles A. Ferguson and Dan I. Slobin (eds.), *Studies of Child Language Development*, pp. 175–208. New York: Holt, Rinehart, and Winston.

Smoczynska, Magdalena. 1985. The acquisition of Polish. In Dan I. Slobin (ed.), *The Crosslinguistic Study of Language Acquisition*, vol. 1: *The Data*, pp. 595–686. Hillsdale, NJ: Lawrence Erlbaum Associates.

Somashekar, Shamitha. 1999. Developmental trends in the acquisition of relative clauses: Cross-linguistic experimental study of Tulu. PhD diss., Cornell University.

Spencer, Andrew, and Arnold Zwicky (eds.). 2001. *The Handbook of Morphology*. Oxford: Blackwell.

Stephany, Ursula. 1995. The acquisition of Greek. In Dan I. Slobin (ed.), *The Crosslinguistic Study of Language Acquisition*, vol. 4, pp. 183–333. Hillsdale, NJ: Lawrence Erlbaum Associates.

Stephany, Ursula, and Anastasia Christofidou. 2009. The emergence of nominal inflection in Greek. In Ursula Stephany and Maria Voeikova (eds.), *Development of Nominal Inflection in First Language Acquisition: A Cross-Linguistic Perspective*, 217–264. New York: Mouton de Gruyter.

Stoll, Sabine, Balthasar Bickel, Elena Lieven, Netra Paudyal, Goma Bandaje, Toya Bhatta, Martin Gaenszle, Judith Pettigrew, Ichchha Purna Rai, Manoj Rai, and Novel Kishore Rai. 2012. Nouns and verbs in Chintang: Children's usage and surrounding adult speech. *Journal of Child Language* 39(2): 284–321.

Suzman, Susan. 1991 Language acquisition in Zulu. PhD diss., University of the Witwatersrand.

Suzman, Susan. 1996. Acquisition of noun class systems in related Bantu languages. In Carolyn E. Johnson and John H. V. Gilbert (eds.), *Children's Language*, vol. 9, pp. 87–104. Mahwah, NJ: Lawrence Erlbaum Associates.

Tanaka, Nozomi, William O'Grady, Kamil Ud Deen, Chae Eun Kim, Ryoko Hattori, Ivan Bondoc, and Jennifer Soriano. 2014. Acquisition of Tagalog relative clauses. In Will Orman and Matthew

James Valleau (eds.), *Proceedings of the 38th Annual Boston University Conference on Language Development*, pp. 463–470. Somerville, MA: Cascadilla Press.

Torrens, Vicenç. 1995. L'adquisició de la sintaxi en català i castellà: La categoria functional de flexió. PhD diss., Universitat de Barcelona.

Tran, Jennie. 2011. The acquisition of Vietnamese classifiers. PhD diss., University of Hawaii at Manoa.

Tsimpli, Ianthi-Maria. 1992. Functional categories and maturation: The prefunctional stage of language acquisition. PhD thesis, University of London.

Valian, Virginia. 1991. Syntactic subjects in the early speech of American and Italian children. *Cognition* 40: 21–81.

Varlokosta, Spiridoula, Anne Vainikka, and Bernhard Rohrbacher. 1998. Functional projections, markedness, and "root infinitives" in early child Greek. *Linguistic Review* 15: 187–208.

Westergaard, Marit. 2008. Verb movement and subject placement in the acquisition of word order: Pragmatics or structural economy? In Pedro Guijarro-Fuentes, Pilar Larranaga, and John Clibbens (eds.), *First Language Acquisition of Morphology and Syntax: Perspectives across Languages and Learners*, 61–86. Language Acquisition and Language Disorders 45. Amsterdam: John Benjamins.

Wexler, Kenneth. 2004. Beauty and awe: Language acquisition as high science. Plenary address at BUCLD 29, November 6.

Wijnen, Frank, and Maaike Verrips. 1998. The acquisition of Dutch syntax. In Steven Gillis and Annick De Houwer (eds.), *The Acquisition of Dutch*, pp. 223–300. Amsterdam: John Benjamins.

Zapf, Jennifer, and Linda Smith. 2007. When do children generalize the plural to novel nouns? *First Language* 27(1): 53–73.

IV Module 4: The Sentence Level

7 Syntactic Development

7.0 Introduction

Syntax relates to how words in a sentence are organized into phrases and sentences that convey meaning. Part of the study of syntax concerns mapping out the rules that govern how verbs combine with noun phrases and other constituents to yield coherent meanings; this is the domain of **argument structure**. Syntax is also concerned with the expression of **functional** aspects of language such as tense (e.g., present vs. past), negation, and question formation. A third main focus of syntax is how pronouns (*him, her*) and other variables are interpreted. This chapter focuses on each of these domains in turn, with respect to the question of what children produce and comprehend and how children gain adultlike knowledge.

Two things should guide the reader's attention in the following material, both discussed in chapter 1: the Logical Problem of Language Acquisition (LPLA) and the Developmental Problem of Language Acquisition (DPLA). Recall that the LPLA concerns the remarkable fact that children come to know things that they have no external evidence for. What we mean is that neither the incoming linguistic data nor paralinguistic information (like eye gaze, hand gestures), nor correction, nor other inductive patterns in the input are enough for the child to acquire syntactic knowledge in the manner that they do. And the DPLA concerns the precise developmental path that children take in the acquisition of a particular phenomenon: Why do children develop along the particular path that they do, as opposed to countless other possible paths? Both of these problems are brought into sharp focus in the study of the acquisition of syntax, which is why this area has dominated the field for decades.

7.1 Bootstrapping into Syntax: Semantic Bootstrapping

Imagine you hear a sentence of an unknown language:

(1) Sasa hivi nitakula kitu kikubwa kwa samamu ninajaa sana.

This is a sentence of an actual language, Swahili, and it means 'Right now, I am about to eat something big because I am very hungry'.[1] You hear (or read) the words in a linear order, but if you don't know the language yet, you don't know which words are the nouns and which are the verbs nor how the words group together in a structure. This is the situation all children find themselves in before they have bootstrapped into the grammatical system of their language. In the first part of this chapter, we will look at one proposal for how children do this.

The term **bootstrapping** refers to the process of using the resources you have at hand to solve a problem by yourself, without explicit help from any external source (after all, parents don't explain the rules of syntax to their children). We saw in chapter 3 an example of how children can use information from prosody, rhythm, and phonotactics to solve the segmentation problem by bootstrapping, which was called **prosodic bootstrapping**. In chapter 5 we saw another kind of bootstrapping, **syntactic bootstrapping**, in which children exploit the semantic properties of argument structure to narrow down the meanings of verbs.

One explanation for how children begin to figure out some of the basic properties of sentence structure is called **semantic bootstrapping**. Semantics is the study of meaning, and the idea of semantic bootstrapping is that learners first learn some basic semantic properties of words, and then they can use this semantic (meaning) information to bootstrap into syntax. How exactly does this work?

Pinker (1982, 1984, 1987) proposed that although linguists define lexical categories in terms of grammatical function (nouns are words that can be preceded by a determiner, verbs can be marked for tense) rather than meaning (nouns don't necessarily label people, places, and things [e.g., *decision*] and verbs don't always denote actions [e.g., *evoke*]), "it is plausible that, when speaking to infants, parents refer to people and physical objects using nouns, that they refer to physical actions and changes of state using verbs, that they communicate definiteness using determiners, and so on" (Pinker, 1984, p. 39). The idea is that there are consistent and reliable correspondences between lexical categories (e.g., nouns, verbs) and semantic features (e.g., objects, events). Therefore, assuming that young children can perceptually and conceptually differentiate objects from events, for example, children should be able to determine that there is a category of words that generally label objects (we call them nouns) and another category of words that generally label events (verbs), and so on for other lexical categories.

Furthermore, Pinker suggested that children come to the language-learning task with some assumptions about sentence structure. For example, subjects are prototypically animate (they denote living and sentient things, such as people and animals) and they are agents of an action. An agent of an action is one that volitionally causes an event or a change of state to occur. If you are able to determine through perception which entity is acting upon which other entity in a given situation, you will be able to associate the acting entity (the **do-er** or **agent**) with one word in a sentence describing the situation and the acted-upon entity (usually called the **patient** or **theme**) with a different word. Imagine that you see a scene in which one animal is pursuing another animal. Then, when you hear a sentence such as

(2) The dog is chasing the rabbit

if you already know the meanings of the nouns *dog* and *rabbit*, you can infer that the word *dog* maps onto the one doing the chasing (i.e., the agent [subject]), that *rabbit* maps onto the one being chased (i.e., the patient [object]), and that the verb *chase* denotes some kind of pursuing action. Furthermore, because agents are typically sentence subjects, the learner can also infer that the sentence has subject-verb-object (SVO) word order. This is semantic bootstrapping: you have figured out something about your language's syntax (in this case, it has SVO word order) by using semantic notions of agency, and you've identified some grammatical categories by using semantic features (things vs. actions).

Importantly, Pinker extended this approach to cases in which nouns do not refer to people and concrete objects, verbs do not label observable events, and the subject is not an agent, as in 3.

(3) The decision evoked a harsh response.

Here the semantic cues to lexical categories and sentence structure are unavailable. However, with enough evidence from sentences like 2, the child will have been able to build up a mini-grammar with rules that can now be extended to less transparent cases. For example, one rule the child might have discovered is that nouns can be preceded by determiners (*the*, as in, *the dog*, *the rabbit*), so that when they hear *the decision*, the know the word *decision* must be a noun. Similarly, children could notice that words that label events, like *chase*, indicate through different morphology whether the event is ongoing (*is chasing*) or is completed (*chased*). Now when they hear *evoked* with a past-tense marker, they know that *evoke* must be a verb. Another rule the child might have figured out is that the subject

Sidebar 7.1: Semantic vs. Syntactic Bootstrapping

The astute reader might recall from the end of chapter 5 (section 5.5) that the argument for syntactic bootstrapping is that children can use syntax to learn verb semantics, but semantic bootstrapping appears to make the opposite argument: children can use semantics to learn syntax. Isn't this a contradiction?

In some sense, semantic and syntactic bootstrapping are at odds with each other. Semantic bootstrapping *assumes* that children can infer something about verb meanings, enough to know whether the first noun is the do-er or the done-to argument, while syntactic bootstrapping says that's not a fair assumption. Pinker and Gleitman, the main proponents of semantic and syntactic bootstrapping, respectively, have engaged in a lively debate about these ideas (Pinker, 1987; Gleitman, 1990). But in another sense the two approaches are trying to answer slightly different questions (how you learn basic word order vs. how you learn verb meanings), and they agree on a number of important points: children can map some concrete nouns to object meanings, and children expect sentences to have a hierarchical structure with predicates and arguments.

always comes before the word in the sentence that is marked for tense—*is* in 2, *evoked* in 3—so *the decision* must be the subject in 3 even though it's not an agent. The procedure of using the rules of the mini grammar to determine the sentence structure of a nontransparent sentence is known as **structure-dependent distributional learning.** In this way, children can begin to bootstrap into the word order and syntactic rules of their language.

There is evidence that children learn the word order of their language quite early. Hirsh-Pasek et al. (1985) showed that children who produce only one word at a time (i.e., they are at the one-word stage) at age 16–19 months correctly interpret sentences like 4 as meaning that Big Bird is acting upon Cookie Monster, not that Cookie Monster is acting upon Big Bird (Hirsh-Pasek and Golinkoff, 1996).

(4) Big Bird is tickling Cookie Monster.

If children can use semantic bootstrapping and structure-dependent distributional learning to correctly interpret word order in their language, do they also use correct word order in their own productions? To a very large extent, the answer is yes. Occasionally one finds utterances like 5, displaying an incorrect order of words, but most utterances at the two- and three-word stage exhibit adultlike word orders, as in 6.

(5) a. Come Cromer? (Adam 2;4)
 b. Bit me dog. (Adam 2;5)

(6) a. Block broke. (Eve 1;6)
 b. Drink juice. (Eve 1;7)
 c. Find dirt. (Adam 2;3)
 d. Cromer paper. (Adam 2;5)
 e. Give that Cromer (Adam 2;6)
 f. What doing? (Adam 2;6)

7.2 Functional Structure and Optional Infinitives

Figuring out basic word order is an important achievement for a child, but it takes them only so far. After all, we don't merely speak in sentences like *Mom go*, *Teddy fall down*, or *Mary kick ball.* In addition to knowing the order of words in sentences, the child must also acquire what is commonly referred to as **functional structure**. Functional structure provides information about things like (i) the time of the event (present or past), (ii) the polarity of the sentence (whether it is a negative or positive statement), and (iii) modality (whether an action may or could occur). This information is variably marked in sentences (by inflectional morphology, typically), depending on the intended meaning. The sentences in 7a–7e show that a simple transitive sentence may occur with any number of different inflectional patterns, all of which are perfectly grammatical. But 7f–7g show that, unlike functional structure, argument structure is not variable: the verb must have a subject and an object.

(7) a. Beth pulls Tony (present tense, third-person agreement)
b. Beth pulled Tony (past)
c. Beth did not pull Tony (negative, past)
d. Beth may pull Tony (modal)
e. Beth was pulling Tony (past progressive)
f. *Beth pulled (missing object)
g. *Pulled Tony (missing subject)

Notions such as tense, agreement, negation, and so on are classic functional categories that are often (though not always) marked with explicit inflectional morphology. Other categories include number (singular, dual, plural), aspect (perfective, imperfective), mood (indicative, subjunctive, imperative), and, within the nominal domain, definiteness, number (singular, plural), and case (this relates to the syntactic role of a nominal in a sentence). Such information is very important for a child to acquire, since in many ways these categories are the essence of natural human language. The functional categories let us take words for actions and objects and transform them into the complex web of meanings we readily convey to one another in normal speech; the functional categories of language form the glue with which the words of a sentence are bound together. Omission or incorrect use of such categories by speakers older than about 4 years is very noticeable and can in fact be an indicator of a language disorder (see chapter 8, section 8.4); because of this, the acquisition of functional structure has received a tremendous amount of attention within the fields of theoretical language acquisition and language pathology.

We'll begin with an inventory of functional categories and discuss how such information is encoded in traditional generative syntax. One of the hallmarks of children's early sentences is that many words and morphemes that mark functional categories are missing—this property is known as **telegraphic speech**. Focusing on the category of tense, we'll then address the questions of why children omit morphemes that mark tense and whether children represent this functional category in their underlying grammar even if they do not produce these morphemes. After that, we will turn to other aspects of syntax such as negation, question formation, and passivization.

7.2.1 Functional Categories and Structure

Functional categories can be part of either the verbal domain or the nominal domain. The bulk of research in the field has focused on verbal functional categories, which are the categories we focus on here. Table 7.1 lists some functional categories in English expressed via bound morphemes (e.g., *-ed*, *-ing*) or free morphemes (words such as *that*, *may*, *is*).

Table 7.1
Some functional categories in English, with examples

Category label	English examples
Complementizer	that, because, whether, if, since
Conjunction	and, or, but, nor
Wh-words	who, what, where, when, why, how
Auxiliary verbs	is, has, does
Modal auxiliaries	can, may, will, might
Tense	-ed
Agreement	-s (in present tense only)
Negation	not, no
Aspect	-ing (progressive)

All languages express at least some functional categories, and some categories are found in all languages, such as negation and wh-words. Other categories exhibit more variation across languages: for example, Mandarin does not have agreement. Some languages like Swahili, Russian, Cree, Kaqchikel, Quechua, among many others, mark a rich array of functional categories using morphology. Here is an example from Swahili:

(8) Jana mimi nilimwona raisi

Jana	mimi	ni-	li-	mw-	on-	a	raisi
yesterday	I	1SA	past	3OA	see	IND	president

'Yesterday I saw the president'

In 8, the subject is *mimi*, meaning 'I', and the object is *raisi*, meaning 'president'. We see the following categories expressed on the verb: subject agreement (SA: first person singular, agreeing with *mimi*), tense (past tense), object agreement (OA: indicating third person singular, agreeing with *raisi*), and mood (the *a* vowel on the verb, indicating that the event was a real one, as opposed to a hypothetical or imagined event). Swahili is a language in which functional categories run rampant, but there are other languages in which fewer functional categories are expressed using morphology. Mandarin is one such language, but of course Mandarin too marks various functional categories.

(9)

我	吃了	番石榴
wo	chi-le	fanshiliu
1SG	eat-Perf	guava

'I ate the guava'

Here we see that right after the verb is a particle *le*. This particle indicates aspect, in this case perfective aspect, which expresses the notion that the event was completed already. If this particle were omitted, the sentence would mean something like 'I am eating the guava.'[2]

In order to understand functional structure, let us begin with English to see where and how such information is encoded in the syntactic tree. We will then move on to other languages and see how different categories are represented in syntax in those languages. This process allows us to see the complexity and variation in human language that children are faced with.

In a simple transitive sentence in English like *Beth pulled Tony*, the argument structure of the verb is clear to see: the verb *pull* requires a subject (*Beth*, the agent) and an object (*Tony*, the theme). A simple tree might be constructed as follows:

(10)

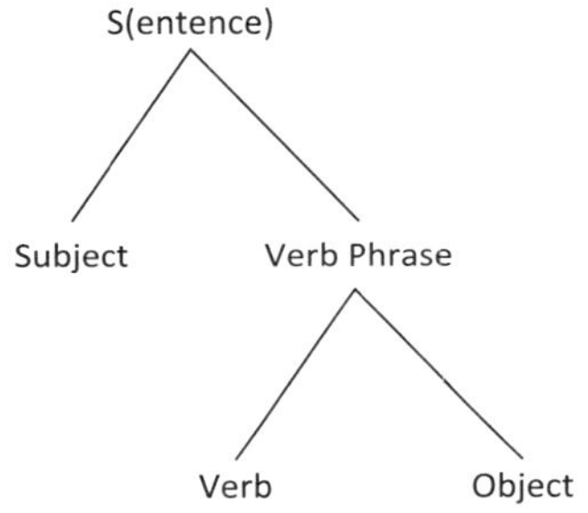

However, this tree does not encode any information about the tense of the verb (which is marked with *-ed*). Therefore, we need to introduce a new phrase within the tree that can incorporate such inflectional information:

(11)

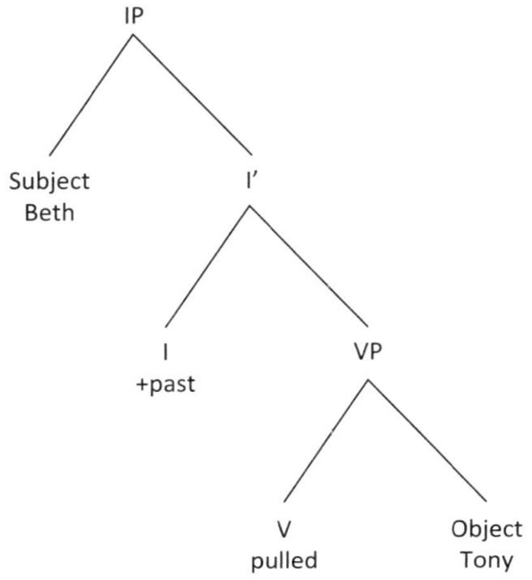

Sidebar 7.2: X-bar Structures

Before we go further, let's review some things about syntax and how we represent sentence structure. One difference you can see between the trees in 10 and 11 is that the tree in 11 has what we call a **bar level I-bar (I′)**. All phrases in syntax have this level, which allows us to create nested structures with asymmetrical relationships. In the abstract, these are referred to as **X-bar structures**, and we can use the symbol X to stand in for any category. Each phrase has the following type of structure:

(i)

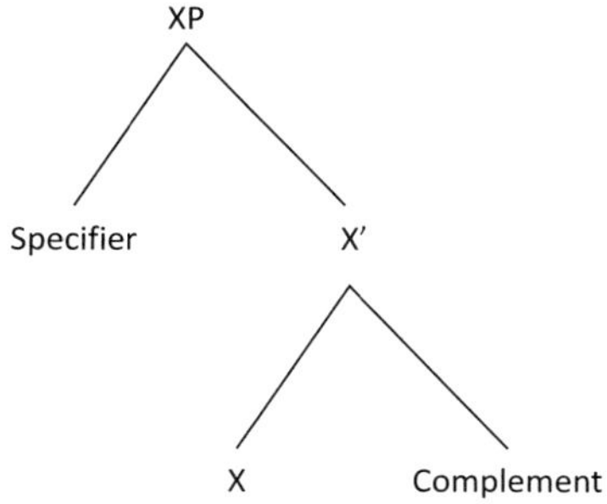

The node in the tree labeled 'X' is the **head** of its phrase. The **specifier** is a node that can host another phrase, and the **complement** of the head likewise hosts another phrase. What phrases go in these positions are determined by the head itself. For example, if a VP has as its head a transitive verb like *hug*, it will have a noun phrase (NP) complement. If there is an adverb in the VP, like *often*, it will go in the specifier of that VP:

(ii)

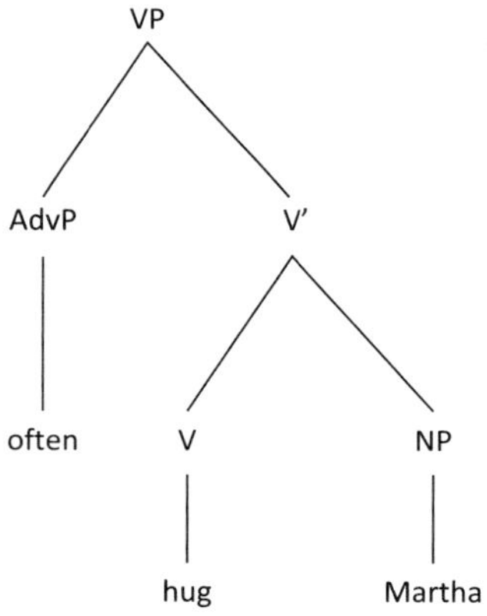

Sometimes, the specifier and complement positions in a phrase are not spelled out, as in the NP *Martha* or the adverb phrase (AdvP) *often* in tree (ii). What's important is that these positions are *potentially* available in every phrase, and they are filled (or not) depending on the requirements of the particular sentence. Furthermore, as we'll see, some functional phrases are found in some languages but not in others. Another word for a phrase is **projection**.

The first thing to notice is that the top node has changed from S(entence) to IP (inflectional phrase). Within this phrase is the inflectional head (I, for inflection), which houses the inflectional morpheme that ultimately ends up on the verb. Below IP is the verb phrase (VP), which contains the verb and object. The question of how the inflection affix and the verb get attached to each other is complex and does not affect the point of all of this, which is that inflectional material is assumed to originate somewhere above the verb phrase and below the subject.

This is an important point: the reason that all functional positions occur above the VP is that inflectional information such as tense, agreement, mood, and so on are not in fact part of the argument structure of the verb.

As we saw in 8, other languages may express different functional categories, so the functional structure may be more articulated. Consider the Swahili sentence in 12a, with a morpheme-by-morpheme gloss provided in 12b:

(12) a. Beth alimvuta Tony 'Beth pulled Tony'

b. Beth	a–	li–	m–	vut–	a	Tony
Beth	SA3s–	past–	OA3s–	pull–	Mood_IND	Tony

In this sentence, there are four inflectional morphemes present. From left to right, *a-* is a third-person-singular subject-agreement morpheme (the verb agrees with the subject, *Beth*), *-li-* is a past-tense morpheme, *-m-* is a third-person-singular object-agreement morpheme (the verb agrees with the object, *Tony*), and the suffix *-a* marks the mood of the verb as indicative (as opposed to subjunctive or negative). Thus a simple tree that includes IP as the only phrase for functional structure would not suffice, so some researchers postulate a more articulated structure, along the lines of 13, in which separate phrases are postulated for subject agreement (S-Agr), tense, object agreement (O-Agr), and mood (mood relates to whether a sentence is indicative or subjunctive).

(13)

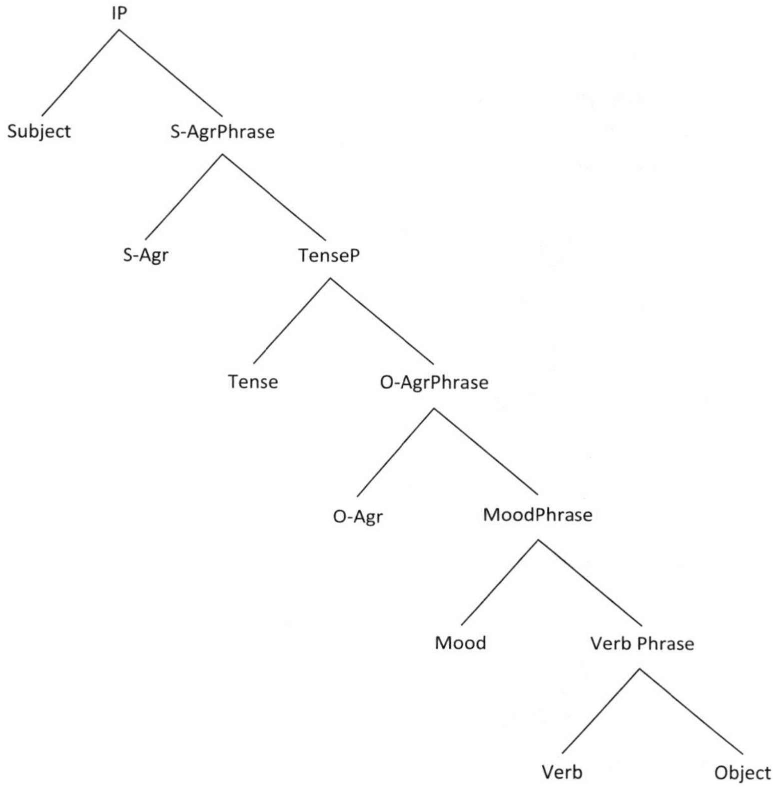

Once again, notice that all the inflectional material is housed in functional categories that are higher in the tree than the VP. This articulated structure is sometimes referred to as an **expanded IP structure**.

If we must assume four categories for Swahili (and presume the Swahili child must acquire this richer functional structure) but only one category for English, does this mean that functional structure varies from language to language? According to Cinque (1999), on the basis of a survey of over 100 languages, there may be as many as fifty distinct functional categories that could, in principle, be part of Universal Grammar (UG). So across human language, we have dozens of different functional categories, but no single language exemplifies all these categories. What this means for children acquiring language is that because the full inventory of functional categories is part of their linguistic endowment, the child's job is to figure out the system that their ambient language actually employs. A child acquiring English will need to work out that their language does not encode object agreement, but a child acquiring Swahili will need to learn that their language does. How do children do this? And how do we know that the options for representing functional structure are really provided by UG, as opposed to being inferred on the basis of the language children hear?

The first argument against an input-driven theory of functional categories is the well-known problem of induction. As we saw in chapter 2, being exposed to a series of examples and having to induce abstract structure from those examples is fraught with difficulty (and is widely thought to be impossible). A second argument may be derived from a careful analysis of children's errors, which we turn to in the next section. The gist of this argument is as follows: Children make a certain kind of error during the telegraphic speech stage, and the particular patterns associated with this error reveal that even while producing morphological errors, children still show evidence of abstract syntactic structure within the functional domain (i.e., above the VP). This shows that those abstract positions exist, even at very early ages.

7.2.2 Telegraphic Speech

Having discussed the range of different functional categories possible in human language and where they are projected in syntactic structure (above the VP), let's return to child English and look at some syntactic properties of the earliest multiword utterances. The following sentences were produced by some very young children (Brown, 1973). What is missing from each sentence? That is, how would an adult express each of these ideas, and what is the difference between what the child said and what an adult would say?

(14) a. Where book? (Adam 2;4)
b. Sitting chair. (Adam 2;4)
c. Adam need one. (Adam 2;5)
d. Big doggie eating. (Adam 2;7)
e. Apple drop. (Adam 2;9)
f. That my box. (Eve 1;10)

The type of speech given in the examples in 14 is called **telegraphic speech** (discussed in the chapter on the acquisition of morphology) because it resembles the kind of language people used in writing telegrams. These were electronic messages that could

Sidebar 7.3: Looking for Functional Categories

> Look back at the functional categories listed in table 7.1. Which categories are expressed in the sentences in 14? Which categories are missing? (You might notice functional categories missing that were not listed in table 7.1, like determiners [*the*] or prepositions.)

be sent long distances from about the mid-nineteenth century through the twentieth century; crucially, telegraph services charged by the word. In order to keep the cost of your telegram low, you would naturally omit any words that were not essential to your message: typically the function words.

Another property of adult language when people sent telegrams (or even when they send postcards or texts) is the omission of the grammatical subject of a sentence. The use of subjects is not absolutely necessary for comprehension, though subjectless sentences are allowed in English only under certain circumstances, such as in these kinds of abbreviated communications. For example, a typical telegram might look like this: "Arrived safely. Missing you. Will be home next week." Note that all three of these sentences are missing the subject, but we can easily reconstruct what the intended subject is. And it turns out that this very same phenomenon occurs in child telegraphic speech: children learning languages like English often produce sentences in which the subject of the sentence is missing.

The fact that this speech is referred to as telegraphic suggests that the underlying motivation for children who are producing this reduced form of speech is the same as for people who sent telegrams: it is costly to produce words. On a certain level, this makes a lot of sense: the idea that children have limited production resources feels right. And because function words do not refer to objects or events in the world, they are more expendable than nouns or verbs. However, as we will see shortly, this is too simplistic an analysis of child telegraphic speech. In fact, telegraphic speech is far more complex than one might initially think, and it reveals a tremendous amount about the system that children bring to bear on the acquisition of language. Moreover, as we will also see shortly, telegraphic speech reveals that children do in fact have functional categories in their abstract syntax.

7.2.3 Optional Infinitives

The telegraphic utterances in 14 are missing various types of functional categories: determiners, auxiliary verbs, verb tense marking, and so on. Here we'll focus on auxiliary verbs and verb tense marking, which together express the abstract notion of **finiteness**. We can think of finiteness as a more abstract version of tense. Grammatical tense serves to locate an event or state in the present, past, or future with respect to the moment the sentence is uttered (see table 7.2).

A clause (sentence) has finiteness if it *can* be located in time, regardless of whether that location is before, during, or after the moment of speech. All of the sentences in table 7.2 are finite. They are distinguished from nonfinite clauses, which cannot be located in time in this way. Examples of nonfinite clauses are the ones in brackets in 15; notice that nonfinite clauses can't stand by themselves, as indicated by the starred sentences in 16. Instead, in adult grammar they have to be embedded under another (finite)

Table 7.2
Tenses and their meanings

Sentence	Tense	Meaning
John needed help	past	John's state of needing help held before the sentence was spoken
John needs help	present	John's state of needing help holds while the sentence is spoken
John will need help	future	John's state of needing help will hold after the sentence is spoken

clause (and to the extent that they have a temporal interpretation, they get that interpretation from the finite clause).

(15) a. I want [John to get help] (infinitive)
b. I saw [John get help] (bare verb)
c. I saw [John helping his friend] (bare participle)

(16) a. *John to get help
b. *John get help
c. *John helping his friend

The reason this distinction is important for child grammar is that when children begin to produce multiword sentences, they go through a phase of producing nonfinite sentences, like those in 17.

(17) a. Apple drop (= (the) apple drop(ped)) (bare verb)
b. Big doggie eating (= (the) big doggie (is) eating) (bare participle)

Children acquiring English produce bare verbs and bare participles. In languages like German, Dutch, and French, children at this stage produce sentences with verbs that are infinitives (INF), as seen in 18.

(18) a. Thorsten das hab-en German (Andreas, 2;1 [Wagner, 1985])
Thorsten that have-INF
'Thorsten has that'
b. Mama radio aan do-en Dutch (Peter, 2;0 [Wijnen and Verrips, 1998])
mummy radio on do-INF
'Mummy switch on radio'
c. Ferm-er yeux French (Daniel, 1;11 [Lightbown, 1977])
close-INF eyes
'(I have) closed (my) eyes'

Interestingly, at the same point in development that children produce nonfinite sentences, they also produce adultlike, *finite* ones:

(19) a. Hanna braucht das nicht German (Johanna, 2;5 [Becker, 1995])
Hanna need-3sg that not
'Hanna doesn't need that'
b. Elle roule pas French (Natalie, 1;9 [Pierce, 1989])
It roll-3sg not
'It doesn't roll'

Because children at this stage appear to alternate between producing nonfinite and finite sentences, the nonfinite forms seem to be optional, so they have come to be called **optional infinitives** (OI) (Wexler, 1994; Rizzi, 1993/1994). The questions researchers have tried to answer about these forms is why children optionally produce nonfinite sentences and what these forms tell us about children's underlying syntactic structure. In particular, when a child produces a nonfinite sentence like those in 18, does this mean that the syntax tree that children use is missing the functional structure needed to express tense, namely IP? Could it be that children's basic syntax is different from that of adults? Let's look at one highly influential theory for how we might reconcile the non-adultlike speech of children with a theory that postulates innate knowledge of functional structure.

7.2.3.1 The Truncation Hypothesis

While telegraphic speech is characterized by the omission of functional categories, this omission is not absolute. Save for the very first few words in a child's production, there isn't really a stage in which each and every functional element in children's language is missing. Rather, children sporadically (though not randomly) produce functional elements. In one sentence, tense might be missing, but in the very next one, tense is produced. How does a theory capture such apparent variability?

Rizzi (1994) put forward a theory that accounts for the optionality of inflection in child language and has a mechanism to deal with observed crosslinguistic variation. In section 7.2.1, we said that the top node of a sentence was called IP (inflectional phrase). In fact, we need to expand the tree a little to include at least one phrase above IP in all languages (not just those that have lots of rich inflection, like Swahili). Rizzi argued that for every adult sentence, the top node of the syntactic tree is a **complementizer phrase** (CP). The CP is typically the position that introduces words such as *that* and *for* in complex sentences such as 'I think **that** John is happy.' Such words are referred to as complementizers (hence the label). The CP position is also the part of the tree that wh-words (*who*, *what*, etc.) move to in wh-questions. Notice that question words such as *who*, *what*, *when* (the so-called wh-question words) typically occur at the beginning of a sentence in many languages. However, they are often interpreted in some other position. Consider the sentence 'What did John eat?' The question word *what* is interpreted as the object of the verb *eat*. This suggests that, at some level, this word originates in that object position. Because it is pronounced at the front of the sentence, it must move from the object position to a position that is structurally higher. This movement is shown in 20 (the *t*s in the tree are **traces** indicating where the wh-word *what* and the auxiliary verb *did* moved from; subscripts match the trace and the word that moved—e.g., *what* and the *t* after *eat* share the subscript *i*).

(20)

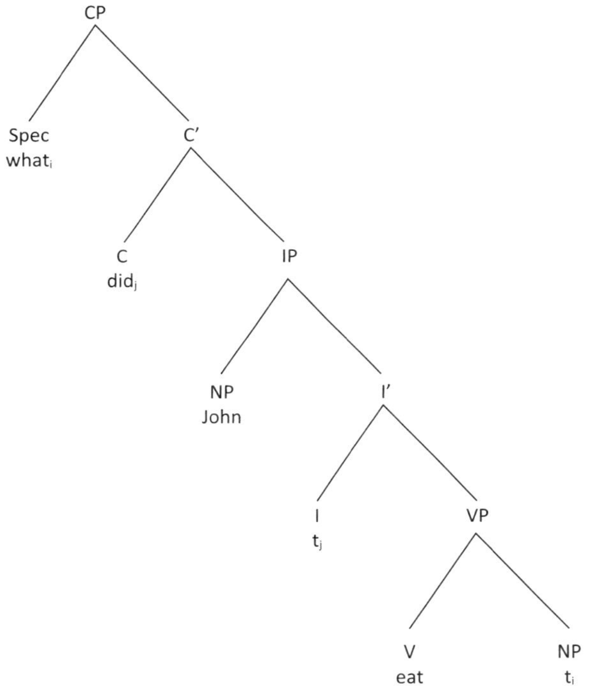

Rizzi argued that in every sentence in adult language, the top node (sometimes referred to as the root node) of the structure is always a CP, even if there is no question word or complementizer; that is, this structure is always there in the abstract. He referred to this as an **axiom** of language: *root* = *CP*. Children, on the other hand, have not set this axiom yet, so they may specify any node as the top node of the tree. For them, rather than having the axiom *root* = *CP*, they have something like *root* = *XP*, where X can be any node in the syntax tree. This means that children have the option (subconsciously, of course) of having a structure that only projects up to, say, the VP, and no higher. Thus, a sentence in which the top node is a VP is in fact a grammatical sentence for children; such a sentence would not contain any tense marking (and would therefore be nonfinite) because it would lack an IP node. Rizzi referred to this as the **truncation hypothesis** because everything above the node that is specified as the top node is (or can be) truncated (i.e., not projected).

Two crucial features of the truncation hypothesis must be noted. The first is that from utterance to utterance, any node may be specified as the top node of the tree. In some utterances, the tree may project all the way up to CP (just like an adult), but in other utterances it may project only as high as VP or somewhere in between (e.g., IP or other intermediate positions for languages that have such functional projections). The second crucial feature is that once a particular node is specified as the top node of the tree, everything below that node must be present in the structure. For example, it is not possible for the child to specify CP as the top node of the tree and then omit the IP from projection. So the trees in 21a and 21b are permissible, but the tree in 21c is not, because it has an intervening projection missing.

(21) a.

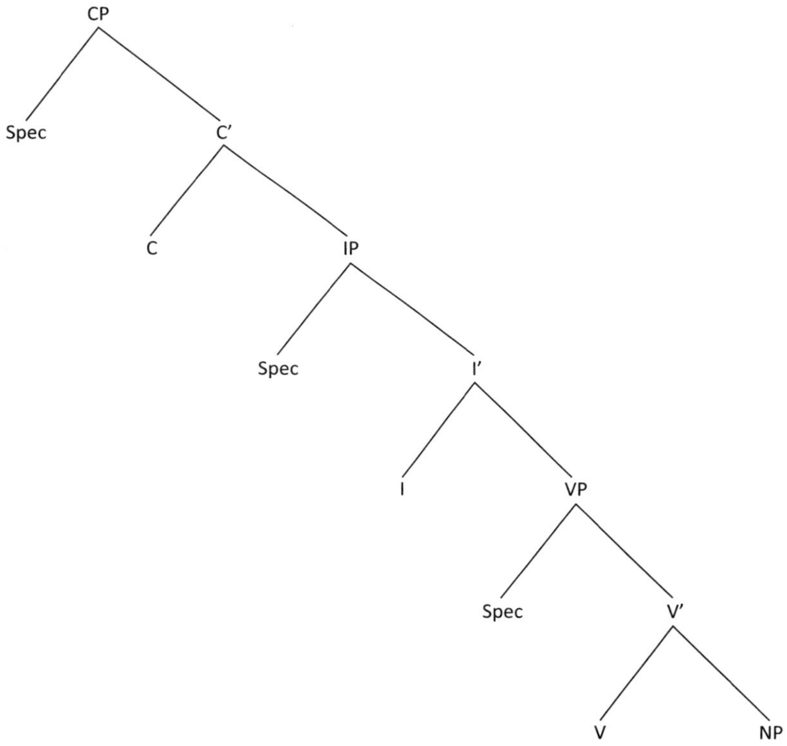

b.

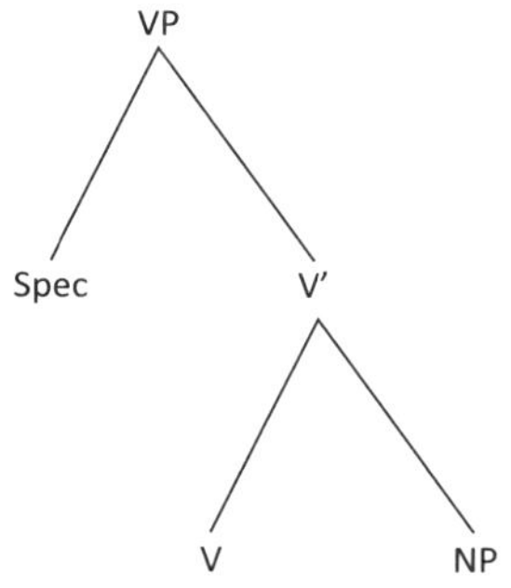

c.

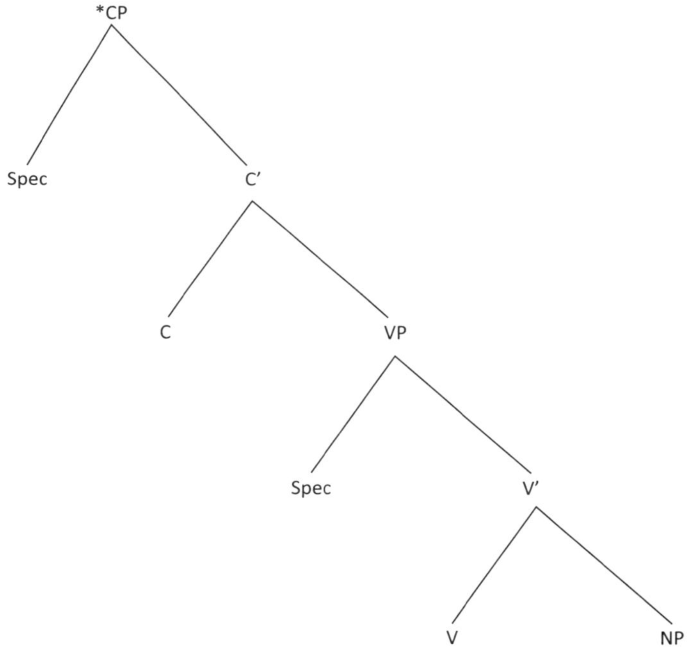

There are many benefits of this system (see Guasti, 2002, for a thorough and more technical overview of truncation and its merits). First, because the specification of the top node of the tree is variable, the child may sometimes specify the top node as either CP, IP, VP, or NP (that is, the object of the verb). This elegantly explains why the omission of functional elements is so sporadic in child speech: children have the option to project their tree to any node, on an utterance-by-utterance basis, which naturally leads to variability in production. Thus, the truncation hypothesis has a mechanism to account for the optionality of inflection. Moreover, if the top node is specified as anything below IP, then the grammatical

Sidebar 7.4: Summary of Truncation Hypothesis

(i) For children: *Root=XP*
(ii) Selection of 'X' varies from utterance to utterance
(iii) Whatever node is selected as root, everything below that is projected fully without exception.

subject may be omitted too, thus accounting for the fact that children often omit the subject of the sentence (noted in section 7.2.2).

But the virtues of this theory are far more profound than what we have discussed so far. Some of the most striking evidence for this approach comes from what are referred to as **contingencies** in the data. It has been observed that correlations exist in the production of child speech that seem rather random at first, but when understood within a generative framework of syntax, along with the truncation hypothesis, they suddenly make a whole lot of sense. These contingencies are *not* random but rather are a consequence of the generative system that children bring to bear on the acquisition of their language. Let's consider three such contingencies here, all of which involve the optional infinitive phenomenon we introduced in section 7.2.3.

7.2.3.2 Form-Position Contingencies in Optional Infinitives

Let's begin with French, a language in which there is a very clear distinction between the structural position that finite and infinitive verbs occupy. This can most easily be seen when the sentence is negative: in adult French, finite verbs always go before negation, while infinitive verbs always go after negation. (Note that the word indicating negation in French is *pas* 'not', not *ne*.)

(22) a. Jean n'**aime** **pas** ce livre — Finite verb > pas
John like-finite not that book
'John doesn't like that book'

b. Jean **veut pas** **perdre** le bus — Pas > nonfinite verb
John want-finite not miss-infinitive the bus
'John doesn't want to miss the bus'

The syntax of this is quite elegant, actually: linguists have shown that in finite sentences in French, the verb moves from its base position within the VP up to the higher functional projection in which tense occurs, namely IP. This verb raising moves the verb up to a position higher (and leftward) of the position for negation, as in 23. However, the nonfinite verb has no reason to raise (since tense is not active, the sentence being nonfinite); therefore, it remains in its base position, low in the VP. This means that the nonfinite verb occurs to the right of negation.

Sidebar 7.5: The Bare VP Hypothesis

A different idea for why children produce telegraphic speech is that their underlying syntax is different from adults' in that children have syntax trees that are drastically smaller and less complex than adults'. Specifically, Radford (1988, 1990) proposed that at early ages, the child's tree is missing all functional structure. His theory, which is referred to as either the **bare VP hypothesis** or the **small clause hypothesis**, claims that the child's tree consists of the VP structure and nothing above that, as in (i).

(i)

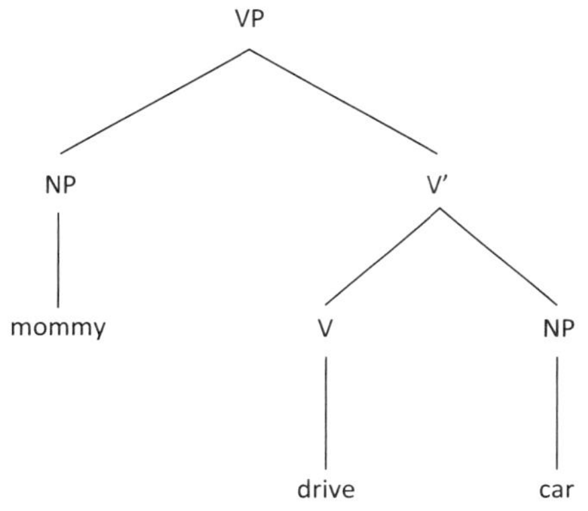

In a structure like (i) there are slots to host the agent of an action, the action itself, and the patient of the action. Furthermore, Radford proposed that other lexical categories could be projected, like NP and PP (prepositional phrase). Such structures are certainly not equivalent to adults' full clauses, but they are similar to adult syntax in that they possess an X-bar structure. In this way, while childlike structures are different from adultlike structures, they are not *fundamentally* different (as they would be if, for example, they had a flat structure or lacked specifiers).

One problem with the small clause hypothesis is that children acquiring morphologically rich languages do not appear to go through anything like the small clause stage that Radford hypothesized. Italian-speaking children, for example, almost never produce bare verbs—in fact, they produce large amounts of inflectional morphology at early ages, as described previously. Similar facts have been reported for numerous other languages, including Spanish, Catalan, Korean, Japanese, German, Dutch, Inuktitut, Quechua, Swahili, Sesotho, and Zulu. Thus, the theory that functional categories (and therefore inflectional morphology) are completely absent in child language at early stages is not supported by crosslinguistic data. Rather, the crosslinguistic data are more consistent with a model that allows for the optional omission of functional elements, such as the truncation hypothesis.

(23)

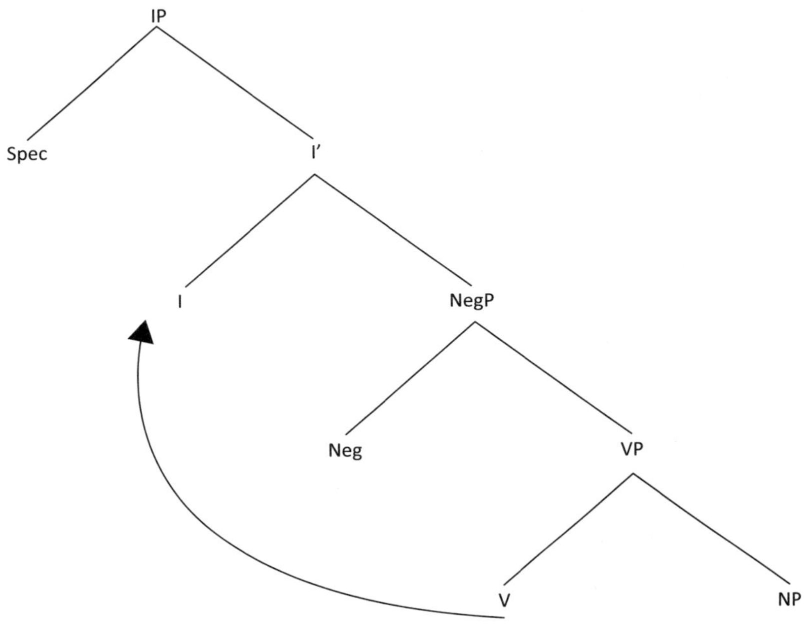

Now let's turn to child French. In an analysis of the speech of two French-acquiring children, Amy Pierce (1989, 1992) showed that children obey this form-position contingency in that when they produce finite verbs with negation, the verb occurs before negation 95% of the time, but when children produce negative optional infinitives (OIs), the infinitive verb occurs after negation 99% of the time (see table 7.3). Thus, French children seem to obey the distributional properties of the verb as dictated by its syntactic features, even while seemingly violating the requirement in adult language that all main clauses be finite. The conclusion is that whatever causes the error in child French that leads to OIs, it is certainly not a lack of knowledge of the syntax or morphology of the verb.

Similar findings have been reported for child German (Poeppel and Wexler, 1993) and child Dutch (Wijnen, 1997), in which the form-position contingency is slightly different: the finite verb occurs in second position in the sentence, while the nonfinite verb occurs at the end of the sentence. In both these languages, children's differential placement of verbs in their sentences according to whether the verb is finite is quite robust.

These results show at least two things. First, OIs are not an error of commission in the sense that the child lacks knowledge of the inflectional system. If that were the case, we would expect a more random distribution of the placement of these types of verbs. Second, the fact that children obey the distributional properties of finite and nonfinite verbs shows that they have knowledge of the functional structure in their language and they

Table 7.3
Placement of finite and nonfinite verbs with respect to negation (Pierce, 1989, 1992)

	Finite	Nonfinite
verb ... *pas*	121 (95.3%)	1
pas ... verb	6	118 (99.2%)
Total	127	119

Sources: Pierce (1989, 1992).

are using this functional structure to determine the position of the verbs. This is evidence that even though there appears to be an error in marking tense through morphology, this does not mean that the abstract representation of tense is missing from children's syntax.

How does the truncation hypothesis account for (i) the generation of OIs and (ii) the form-position contingency we see in French, German, and Dutch? If the child specifies the root of the clause as anything below IP (or TP [Tense Phrase], in a more articulated structure like we need for Swahili), then an OI will occur. And if the child specifies the top node as IP or higher, then a fully finite clause will occur. This has the rather desirable result of explaining why OIs are optional: that root node is specified variably, which means that OIs sometimes occur and sometimes do not. That is a big positive for the truncation hypothesis: it quite naturally explains the optionality of OIs.

As for the form-position contingencies that we noted above, if the top node is specified as something below IP, then there is no position to which the French verb may raise, and so it must remain in its base position within the VP. But if the top node is specified as IP (or higher), then there is a position to which the verb must raise (in order to get its finite features), and this results in movement of the verb to the left of negation. So truncation very neatly explains the position of finite and OI verbs in child language.

7.2.3.3 Null Subject Contingencies in Optional Infinitives

A second intriguing fact about OIs is that they have been found to have higher rates of null subjects than finite verbs. Sano and Hyams (1995), for example, find that in a range of languages, children produce significantly more overt subjects when the verb is fully inflected. Why would this be? The canonical position for subjects is the Specifier of IP. Although it is thought that subjects are probably generated within VP and then raise into the Specifier of IP position, this is where they occur in the adult language. If IP is missing because the child has projected VP as the highest node in a given sentence, then there is no real place for the subject. This means there is a greater likelihood of a null subject in OIs than in finite clauses.

Sidebar 7.6: Null Subject Explanations

There are several ways we could understand why children exhibit a null subject stage. The first is that perhaps this is telegraphic speech at its best: children are pressed for mental resources, so they drop the things that they don't absolutely have to say in order to be understood. Grammatical subjects are often old information—that is, information known to both the speaker and the listener, so it's easy to figure out who or what the intended subject is. That means dropping the subject won't lead to confusion in the conversation, which makes subject dropping a relatively risk-free thing for the child. Imagine if the child dropped the verb—what sense would that make? The adult would struggle to understand what was being said, and all kinds of confusion would occur. So null subjects make sense from that perspective. In fact, this has been proposed in the literature (see, e.g., Bloom, 1990).

Another way to understand null subjects begins with the observation that when English-speaking children drop subjects, they are speaking in a way that adult speakers of other languages speak, e.g., Italian or Spanish speakers. In Italian and Spanish, null subjects are grammatical—*(Io) ti cercavo* or *(Yo) te buscaba* 'I was looking for you'. Hyams (1986) proposed that English-speaking children (and perhaps all children) begin by assuming that their language is more like Italian in allowing null subjects. The child would need to detect evidence in the input that their language is actually not a null subject language, and then they would switch to a non–null subject setting (like English) for their language. As intriguing as this idea is, it has some serious challenges. Besides some technical problems too lengthy to discuss here, it was found that children acquiring English produce null subjects at significantly different rates than children acquiring Italian (Valian, 1991), suggesting that the underlying mechanisms that allowed for the null subjects were different in the two languages.

7.2.3.4 Wh-question Contingencies with Optional Infinitives

Finally, and perhaps most strikingly, there is the remarkable finding known as **Crisma's Effect**, named after the researcher Paola Crisma (1992), who first discovered that in the speech of one French-acquiring child (Philippe, 2;1–2;3, and since replicated in Dutch, German, and Swedish) who produced OIs at a rate of 17% among regular non-interrogative clauses, there was not even a single OI among this child's wh-questions. That is, OIs seem to never occur with wh-questions (see table 7.4). What is it about wh-questions that prevents OIs from occurring?

Recall that, according to the truncation hypothesis, OIs occur only when the top node of the syntax tree is specified as something below IP. However, in a wh-question, the top node *must* be specified as CP because that is where the wh-question must move to. Having specified the top node as CP, everything below CP must also be fully represented, so the absence of OIs in wh-questions is a very natural consequence of the truncation hypothesis.

7.2.3.5 Summary of Optional Infinitives and Truncation

We have seen that the truncation hypothesis is a remarkably successful and elegant theory. It is not perfect, however: there are a number of things that truncation cannot explain. For example, while OIs are quite common in child French, German, Dutch, Swedish, Russian, and Icelandic, for example, they never occur in numerous other languages. For example, Italian children do not produce infinitives as main clauses, nor do Japanese or Swahili children. Truncation does not explain why this split in the languages of the world occurs—although to be fair, there is no theory in the field that adequately explains this fact.

Table 7.4
Summary of Crisma's (1992) finding that OIs do not occur with wh-questions

	+finite	-finite	Total
Declarative	921	195 (17%)	1116
Wh-question	114	0	114

Source: Crisma (1992).

Second, and importantly, English is a very unusual language with respect to OIs. Several researchers (e.g., Wexler, 1994) have argued that the "bare" verbs, as in 24, produced by English-speaking children (which are typical child English utterances during the telegraphic speech stage) are the equivalent of OIs.

(24) a. Eve sit floor. (Eve, 1;7)
b. Papa have it. (Nina, 2;5)

On this view, the bare verb produced by the child in the utterance *Eve sit floor* to mean 'Eve is sitting on the floor' is essentially a syntactically nonfinite verb. English bare verb productions are clearly similar to OIs in that they lack tense marking, but the English language makes it difficult to tell for sure if these verbs are *truly* equivalent to French and German OIs: English bare verbs have no visible infinitive morphology, and because the position of the verb in English does not change according to whether it is finite or infinitive, we can't use the verb's position in the child's sentence to discern a difference between finite and nonfinite verbs. Moreover, the correlation of OIs and null subjects does not hold in English, nor does Crisma's Effect. Thus, the syntactic status of bare verbs in child English remains an open question.

So while truncation does not explain everything, a fair assessment of the model has to come up favorably. The model is simple yet covers a great deal of empirical ground. As such, it's a theory well worth considering in the explanation of many aspects of the acquisition of syntax during the telegraphic speech stage.

Most importantly for our purposes, this model postulates that the omission of inflectional and functional elements does not in fact mean that those syntactic categories are absent from the child's knowledge. In fact, the child knows everything about the syntax of their language, but what is missing is the axiom that the top node must always be CP. That small piece of missing knowledge has some rather consequential effects on the child's speech, but these effects do not need to be remedied by a process of inductive learning of abstract categories. As such, truncation is a far more realistic model than one based on learning purely from the input.

7.3 Other Aspects of Functional Structure

While the acquisition of tense and verbal inflection occupies a central area of focus in the study of syntactic development, it is not the only thing that syntax is concerned with. Here we track children's language development in three other domains of syntax: negation, questions, and passives. For these domains we will focus mainly on data from children acquiring English.

7.3.1 Negation

Negation changes the polarity of a sentence. In English, the most basic way to negate an utterance is to add the word *not* (*Jill is tall* → *Jill is not tall*). Other forms of negation include the adverb *never* and the determiner *no* (as in, *No burglars came to the party*). There are also negative NPs, such as *no one* and *nothing*, and the negative adverb *nowhere*.

When children begin expressing negation, they often simply use the form *no*. But used in isolation, its exact meaning can be difficult to know. Some of the ways children use *no* and *not* include talking about the lack of something, as in 25a, denial of what something is, as in 25b, and rejection of an object, as in 25c (Choi, 1988).

(25) a. no/all gone milk (= there's no more milk)
  b. no truck (= that's not a truck, that's a car)
  c. no dirty soap (= I don't want the dirty soap)

In terms of the use of negation as a syntactic device, an early study of children's syntax by Klima and Bellugi (1966) identified three stages in the development of negation. The first stage involves the word *no* or *not* at the beginning of an utterance or *no* at the end of an utterance. Some examples from their study are given in 26.

(26) a. No ... wipe finger
  b. No a boy bed
  c. No mitten
  d. No the sun shining
  e. Wear mitten no

This stage of negation development is referred to as **external negation** because the negation element is found externally to the rest of the sentence.

The primary shift from stage 1 to stage 2 is that the negation word moves *inside* the sentence and is expressed not only with the word *not* but also with the negative auxiliary verbs *can't* and *don't*. Some examples are given in 27 (from Klima and Bellugi), and these demonstrate children's advancement to **internal negation**.

(27) a. He no bite you
  b. I can't catch you
  c. I don't like him

d. Don't leave me.
e. He not little, he big.

As with many aspects of child language development, children don't wake up one day having left behind stage 1 completely and suddenly speaking only in stage 2–like sentences. Instead, there can be overlap, so that while children are producing utterances like those in 27, they are still producing sentences with external negation, such as *Touch the snow no.*

Although the stage 2 expressions are more like adult negation, they are not completely adultlike, since children's negative modals and auxiliary verbs like *can't* and *don't* do not have an affirmative counterpart: that is, children at this stage do not produce the forms *can* and *do*. Thus, we cannot be sure that they understand that *can't* is the contraction of *can* and *not*.

Finally, stage 3 shows the development of affirmative auxiliaries and modals alongside the negative ones, as well as a wider range of contracted auxiliaries (*isn't, didn't, won't*) and uncontracted negation. Some examples are given in 28.

(28) a. I didn't did it.
b. 'Cause he won't talk.
c. This not ice cream.
d. That was not me.
e. I am not a doctor.

What accounts for this development? Klima and Bellugi proposed that children's syntactic representation undergoes a radical change during this period, from an extremely limited, pared-down structure that admits only lexical categories and a single rule for appending a negation word to the beginning or end of the sentence, to a complex set of rules for combining negation with auxiliaries within the sentence in an adultlike way. A more recent explanation, more in line with current thinking about how much syntactic knowledge children have from early on, is that children already know that negation is projected between VP and IP, but they allow subjects, as well as verbs and objects, to remain inside VP (Déprez and Pierce, 1993). Normally the subject must raise into IP, but if IP is not projected, or if the subject is allowed to remain low in the structure, it would come after the negation word.

7.3.2 Questions

Questions, or interrogatives, are a marvelous function of human language that permit us to ask for more information, request verification, and find out about the world and people around us. All languages have a way of asking questions, and universally there are two types of questions we can ask: linguists call one type **yes-no questions**, since these questions take a *yes* or *no* (or *maybe* or *I don't know*) answer, and the other type **wh-questions**. Wh-questions ask *who* or *what* did something and *when*, *where*, *why*, or *how* an event took place.

Sidebar 7.7: Importance of Questions

To get a feel for how important questions are in language, try the following exercise: In pairs, try to have a conversation with another student without using questions of any kind. How far can you get in your "conversation"? What does your conversation end up consisting of? Does it feel awkward? If so, can you identify whether you were hampered more by the inability to ask yes-no questions or wh-questions?

Here we will focus on the development of questions in English. It is important to remember that, unlike negation, which involves an extra functional projection between VP and IP, questions involve the projection of additional structure above IP (namely, CP) as well as the **movement** of words and/or morphemes within a sentence. In adult English, we form yes-no questions by moving the auxiliary verb (or inserted auxiliary *do*) from the I (inflection) position into C (complementizer), as in 29.

(29)

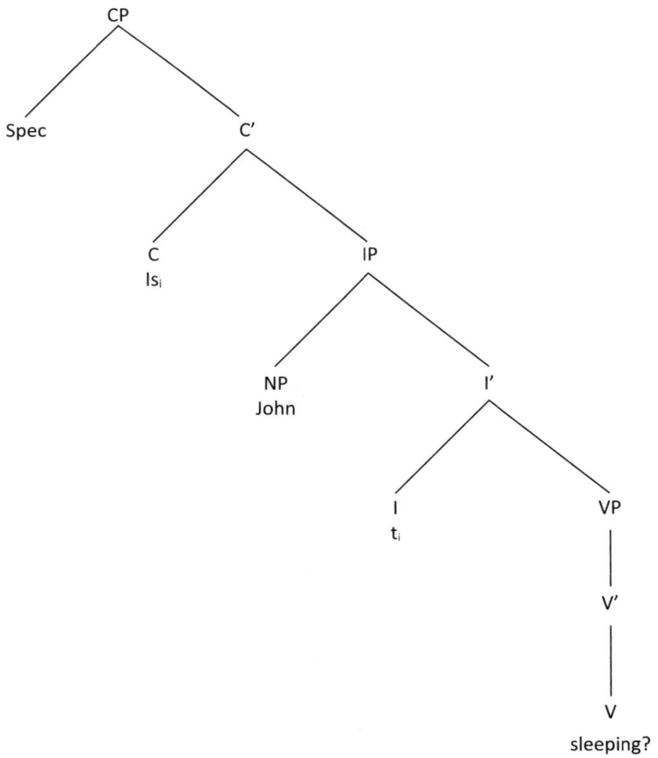

Wh-questions in English involve a similar type of movement, but in addition, the wh-word is moved from its base position into the specifier of CP, as we see in 30.

(30)

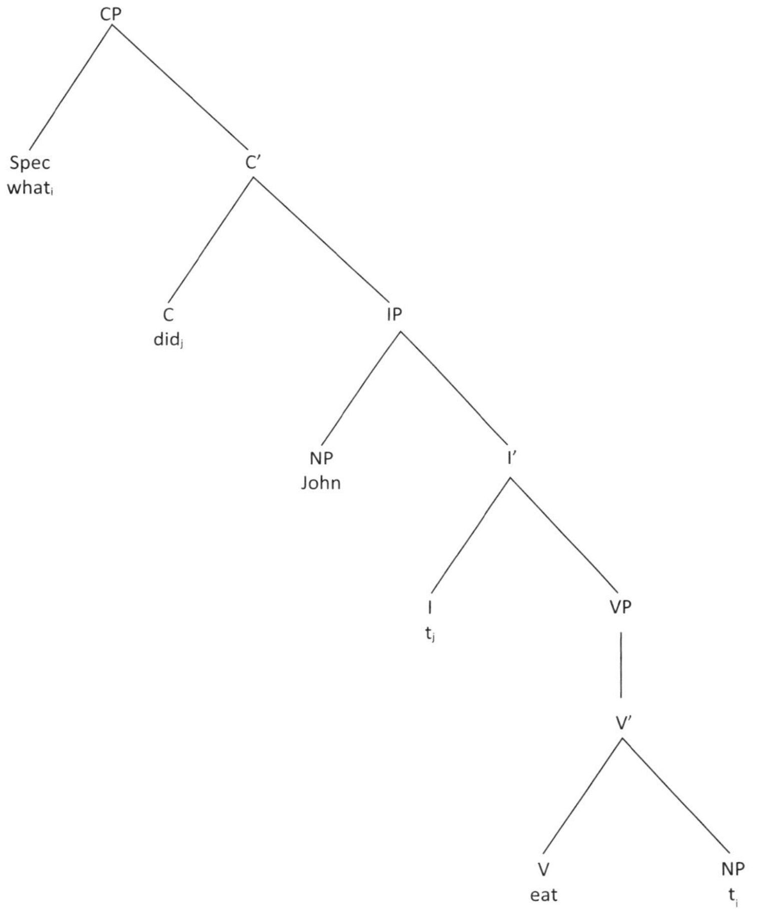

Thus, in order for children to learn how question formation works, they have to learn both auxiliary movement (called **subject-auxiliary inversion**, or SAI, because the auxiliary verb inverts its position with the subject) and wh-movement. Klima and Bellugi, who studied children's development of negation, also looked at how children start asking questions in English, and they identified three to four stages of question formation.

In stage 1, questions are telegraphic statements with rising intonation. That is, rather than using any syntactic operation of movement, children simply keep the words in their basic order and raise their voice pitch at the end of the sentence.

(31) a. Fraser water?
b. Mommy eggnog?
c. See hole?
d. I ride train?
e. What doing?
f. Where kitty?

While there are some wh-questions (31e–31f) and the wh-word always appears at the beginning of those sentences, they are quite limited and never employ auxiliary verbs as required in adult English (*What* ***are*** *you doing?* or *Where* ***is*** *the kitty?*).

In stage 2 children begin to produce auxiliary verbs, but these do not always invert with the subject. We see the beginnings of SAI, but it is inconsistent. Examples are given in 32.

(32) a. Why not . . . me can't dance?
b. You can't fix it?
c. This can't write a flower?

Such questions occur alongside questions that appear to simply involve rising intonation, as in stage 1:

(33) a. Mom pinch finger?
b. Where my mitten?
c. What book name?

For some children there are two additional distinct stages, while for other children there is only a single further stage. For children who distinguish two additional stages, the difference is that SAI becomes consistent first with yes-no questions, as in 34, and only later with wh-questions, as in 35. For other children, SAI becomes consistent with both types of questions at about the same time.

(34) a. Does the kitty stand up?
b. Will you help me?
c. Oh, did I caught it?
d. What I did yesterday?

(35) Where's his other eye?

As with their account of the development of negation, Klima and Bellugi's explanation for how questions develop involves a radical change and growth in syntactic structure across these stages. According to them, children start out with a simple rule for creating a question by intonation (and by appending a wh-word to the beginning for wh-questions). Gradually children's syntactic rules become more complex as they allow for auxiliary verbs and movement in their syntax trees.

One important question we can ask about wh-questions is whether children simply think they are formed by sticking a wh-word at the beginning of the sentence (plus SAI) or whether children understand that there are syntactic restrictions on how wh-movement takes place. In fact, wh-movement can cross several clauses, as in 36, but it cannot move in just any old way, as we see in 37, with explanation below.

(36) What$_i$ did [$_{IP}$ John think [$_{CP}$ that [$_{IP}$ Mary said [$_{CP}$ that [$_{IP}$ Jim wanted to buy t$_i$?]]]]]

(37) *Who$_i$ do [$_{IP}$ you wonder [$_{CP}$ why$_j$ t$_j$ [$_{IP}$ t$_i$ left?]]]

When a wh-word moves in a simple one-clause question, it ends up in Specifier of CP (SpecCP). One of the constraints on wh-movement is that when it moves across multiple clauses, it has to be able to stop off in the SpecCP position in each of the clauses it passes through. If something else is sitting in that position, such as 'why' in 37, then movement is impossible. How do we know that 'why' is in that SpecCP position? Well, we know that 'why' is a wh-word, and SpecCP is the position that holds wh-words. Roeper and De Villiers (1994) conducted an experiment to see if children understood this constraint on wh-movement.

They told children a story about a boy who climbed up a tree one afternoon and, while climbing, fell from the tree and hurt himself. That night when the boy was taking a bath, his father noticed a bruise on the boy's arm and asked about it. The boy responded by saying, "Oh, I must have hurt myself when I fell out of the tree this afternoon." This story has two important events: the falling-hurting event, which happened in the afternoon, and the telling event, which happened in the evening during his bath. Following the story, they asked each child one of the following questions:

(38) a. When did the boy say he hurt himself?
b. When did the boy say how he hurt himself?

Children who were asked 38a answered one of two ways: half of them said "when he fell out of the tree," and half of them said "when he was in the bath." In fact, both answers are available, because *when* could have moved from either the upper clause (asking about when he *said* it) or from the lower clause (asking about when he *hurt himself*). Children who were asked 38b, however, answered uniformly: in the bath. This question can only be asking about when the boy *said* it, not when he got hurt. This is because *how*, a wh-word, is sitting in the SpecCP position of the lower clause, so *when* could not have moved up from that lower position. This shows that the children, who were 3–5 years old, understood implicitly the constraints on wh-movement.

A separate but related strain of research has looked at what are called **long-distance wh-questions**. To understand what these are, remember that sentences come in many types, two of which are *simple* and *complex*. Simple sentences are those that have just one verb. Complex sentences, however, are those that have two verbs, one *embedded* within the other.

For example, the sentence *Ali tricked the thieves* is a simple sentence, but *Jade thinks Ali tricked the thieves* is a complex sentence, with the main sentence (*Jade thinks*) taking an embedded sentence (*Ali tricked the thieves*). Importantly, we can ask a wh-question about many parts of complex sentences, including subjects and objects of the embedded sentence:

(39) a. Who [t] thinks Ali tricked the thieves? — Main subject wh-question
b. Who does Jade think [t] tricked the thieves? — Embedded subject wh-question
c. Who does Jade think Ali tricked [t]? — Embedded object wh-question

In sentence 39a, the wh-word simply moves into the specifier position of the CP in its clause, as we have seen several times in this chapter already. However, in 39b–39c, the wh-word has moved from its original position in an embedded sentence to the SpecCP of the main clause, and importantly, as we saw above, when the wh-word moves out of its own sentence into the SpecCP in the higher sentence, it can't just move all the way in one big move. Instead, the wh-word needs to move in steps, first moving into the SpecCP in its own clause and then jumping up into the Spec of the next clause. Example 40 shows this movement for sentence 39b.

(40)

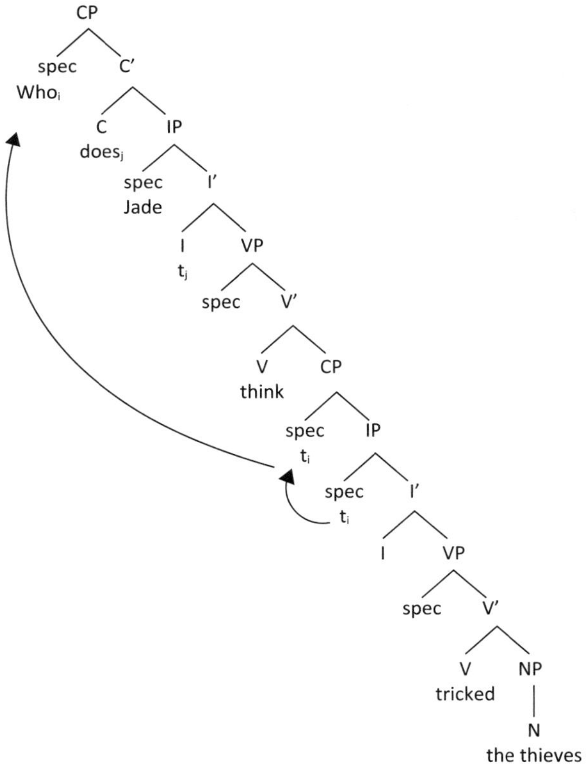

Thornton (1990) investigated children's production of such long-distance wh-questions. She used an elicited production task in which the experimenter played with a child and a bunch of toys. The child's job was to ask a puppet about various scenarios, as directed by the experimenter. For example, in one scenario, the child and the experimenter hid a marble in a box, and the child then had to ask the puppet a question. To accomplish this, the experimenter whispered in the child's ear, "We know there is a marble in the box. Ask Ratty [the puppet] what he thinks." They tested children aged 3 to 5 years and found that they generally did quite well in producing adultlike long-distance wh-questions (in this case, something like *What do you think is in the box?*).

However, interestingly, some of the children made a curious kind of error. Instead of saying, *What do you think is in the box?*, some children said something like *What do you think what's in the box?* This is referred to as a **medial wh-error** since children produced a wh-word in that medial SpecCP position. There are two interesting things about this finding. First, this error did not come from the input, since English-speaking children never hear medial wh-questions (that is, the question **What do you think what's in the box?* is ungrammatical in English).[3] Second, the production of medial wh-words is reflective of the fact mentioned above: in order to produce a wh-word, you need to move it through the SpecCP position in the embedded clause before moving it up to the main sentence CP position.

So it seems that the error children are making is that they are pronouncing the trace in the medial position, whereas adults don't pronounce it. But importantly, the fact that children are producing that medial wh-word indicates that they have knowledge of the computation required to produce long-distance wh-questions. This is intricate, abstract knowledge that does not have much (if any) evidence in the input, so this suggests that the underlying computational system for wh-question formation is something children bring to the table in acquiring a language.

7.3.3 Passive Construction

The third type of syntactic transformation we will consider is the passive construction. Passives are like questions in that they involve movement, but the type of movement is different. In passive sentences, the subject is the thing affected by the verb's action rather than the thing performing the action. Notice the difference between the **active** sentence in 41 and the corresponding **passive** in 42.

(41) The boy read the book.

(42) The book was read (by the boy).

Passives essentially take the typical, canonical order of arguments of a verb (which correspond to participants in an event) and turn them around so that the patient of the verb becomes the subject. The agent of the verb—*the boy* in 41 and 42—is optionally expressed in a prepositional phrase (PP) (also called the **by-phrase**). Thus, the movement involved in this construction is movement of the NP that is underlyingly the verb's object—*the book* in 41–42—from inside the VP up to the subject position in IP.

(43)

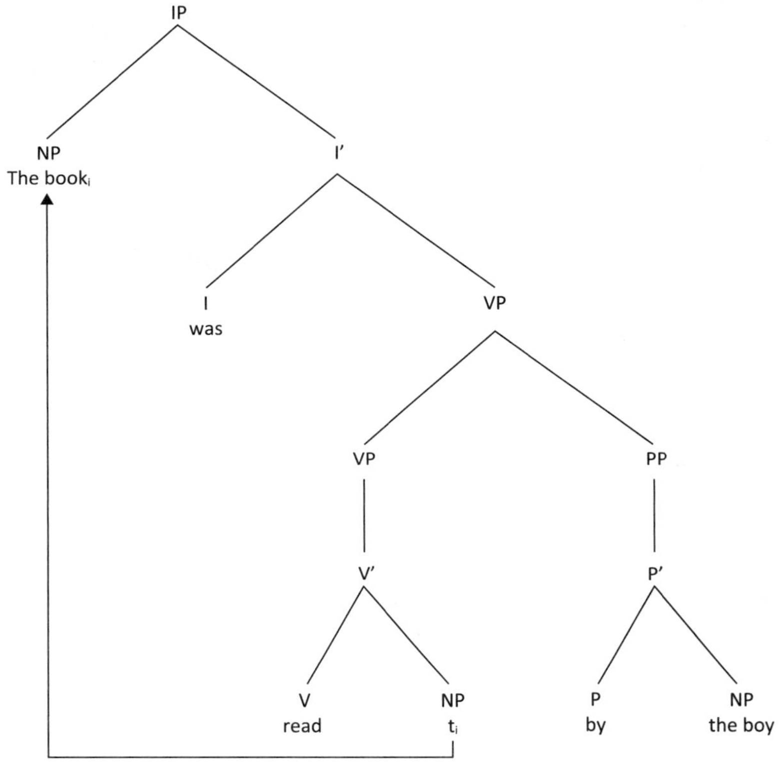

There are many reasons to expect the passive pattern to be acquired late by children. First, the passive is a very rare pattern in child-directed speech in many languages, including English. Gordon and Chafetz (1990) find that of all the verbal utterances children hear in English, only 0.4% of them are in the passive voice. Furthermore, we know that children acquire canonical word order (SVO) very early in development (e.g., as early as 16 months of age [Hirsh-Pasek and Golinkoff, 1996]). This canonical word order encodes the agent of the action (the do-er, or the one that instigates the action) as the subject and the patient or theme of the action (the one that is acted upon by the do-er) as the object. So in 41, the agent is the boy and the patient is the book. The passive reverses that. In sentence 42, the subject of the sentence is the patient of the action (the book) while the agent of the action (the boy) is (optionally) at the end of the sentence. Because children have mastered the canonical pattern so early, we can imagine that a reversed pattern like the passive would be tricky for them. Further complications involving the passive might make it even trickier, but for now, let's assume that there is good reason to suspect that the passive poses challenges for children.

In fact, a long history of research has examined children's acquisition of the passive (see Becker and Kirby, 2016, for an overview), with the overall finding being that passives do present some challenges for learners. While passives may be produced quite early on (see, e.g., Crain, Thornton, and Murasugi, 2009), accurate comprehension of them appears to remain elusive until as late as age 5 or 6 (in fact, some studies, such as Maratsos et al., 1985, have found that comprehension of some kinds of passives is not robust until as late as age 9). However, comprehension of passives depends very much on what *kind* of passive one is talking about, as a number of factors weigh in on the relative ease or difficulty of this construction.

We can distinguish four types of passives according to whether the arguments (recall, these are the participants in the event) can be reversed, meaning that either argument can be the subject or the object, and whether the verb denotes an action or a state. These types are exemplified in table 7.5.

The easiest passive sentences for children to understand are non-reversible actional passives. These are sentences in which the subject and the object are not semantically reversible. For example, a sentence like *The book was read by the boy* is not reversible because if you reverse the position of the subject and the object (*The boy was read by the book*), the result makes no sense. It is a grammatical sentence, but it is semantically ill formed, given all that we know about the verb *read* and the animacy of *book*. However, a sentence like *The boy was kissed by the girl* is reversible, since if the girl was the subject (*The girl was kissed by the boy*), the sentence would be perfectly normal. So non-reversible sentences give children a clue as to what the grammatical role of each noun is without the child having to use syntax at all: children can figure out the meaning of such sentences purely on the basis of the meanings of the nouns and the verb. This might be why children understand them so well, and because of that, researchers typically do not use non-reversible sentences in their experiments. We're interested in whether children use their syntax to figure out a sentence, and only reversible sentences require the use of syntax.

Reversible passives are harder than non-reversible sentences, but even within this type, some are harder and some are easier for children. The other major division is the actional and

Table 7.5
Types of passive constructions

	Actional	Non-actional
Reversible	The boy was kissed by the girl	The boy was seen by the girl
Non-reversible	The book was read by the boy	The book was seen by the boy

non-actional passives. Actional verbs are those that denote actions and events, such as *push, kiss, hit, grab, lift*. Such actions may be observable (you can see someone pushing someone or something else), and they often imply that the patient is somehow affected by the action.

Non-actional verbs are those that denote a state or something less visible, and they include verbs of perception or emotion, such as *see, love, hate, forget*. For example, in a sentence like *The boy saw the girl*, there is no action you can point to nor any visible outcome of the action. Passives involving such verbs are significantly harder for young children. For example, *The girl was seen by the boy* is often interpreted by children as *The girl saw the boy.*

In addition, we can distinguish between *short* and *long* passives. Short passives are those in which the *by*-phrase is absent. For example, *The girl was kissed* is a short passive. A long passive is one in which the *by*-phrase is present: *The girl was kissed by the boy.* It has generally been found that children have some difficulty with long passives. For example, Horgan (1978) found that young children rarely produce the *by*-phrase in naturalistic data, and when they produce long passives, they often substitute *by* with some other preposition (e.g., *from*). In experimental studies too, children have been found to have more difficulty with long passives than short passives (e.g., Fox and Grodzinsky, 1996), though there is some contention around this finding.

Many theories attempt to explain the late acquisition of the passive (see Deen, 2011, for a more detailed discussion). Some suggest that the movement of the patient into subject position is the difficult part of the passive (e.g., Borer and Wexler, 1987; Snyder and Hyams, 2015). Others argue that the *by*-phrase is particularly difficult for children and the movement of the object is not problematic at all (e.g., Fox and Grodzinsky, 1998). Yet others argue that it is simply the infrequency of the passives in the input that causes the difficulty—if the passives were more frequent, then the passive would be acquired much earlier (e.g., Demuth, 1989). In fact, in crosslinguistic work that looks at languages in which the passive occurs more frequently, Demuth et al. (2010) found that children acquiring Sesotho understand passives of all types as early as 3;1. In Sesotho, the passive occurs in child-directed speech at a rate of 2.7%—significantly higher than the rates found in English.

More recently, however, developments in the field have led to a rethinking of the finding that the passive is a late acquisition. There are some hints that the passive might indeed be acquired as early as age 3, even in languages like English. Developments in testing techniques have yielded results that show early acquisition of even the most difficult passive patterns. For example, O'Brien et al. (2006) found that children as young as 3;6 understand long, non-actional passives quite well. Moreover, experiments involving a technique called *priming* (see appendix B) show that children around age 4 years can produce passives of various kinds as well. It may be that artifacts of previous experiments (such as difficulty of the task) masked children's ability to interpret these sentences. Thus, significant discovery is still to be done in the area of the acquisition of passives.

7.3.4 Relative Clauses

The final type of syntactic construction we will consider is relative clauses. A relative clause is a clause that modifies a noun. It is a sentence that adds information about that noun, so in a sense it is a little like an adjective, except that unlike an adjective (which is usually just one word), the relative clause is a full sentence.

Typically, the relative clause itself is missing something within it, and that missing element corresponds to the noun that it modifies. Take 44a as an example. The noun that is modified is *The boy*, and we refer to this as the **head** of the relative clause. The head is modified by the sentence *the boy saw the girl*, except in this sentence, *the boy* is missing (shown in 44b, where the underscore ___ indicates the missing element within the relative clause). The relative clause itself gives us information about the head that we would not have otherwise.

(44) a. The boy that saw the girl
b. The boy that [___ saw the girl]

Importantly, the missing part in the relative clause refers to the same entity as the head noun. And note that the relative clause is preceded by *that*, which introduces the relative clause. This is referred to as a **relativizer**, or a relative pronoun. Without the relativizer, the sentence would sound like a regular subject-verb-object sentence, so *that* is required in order to mark this as a relative clause.

In the relative clause in 44, the head noun corresponds to the subject of the relative clause—that's why the underscore is in the subject position of the relative clause. This is referred to as a **subject relative clause** (RC). However, RCs come in lots of different types. Below are just a few examples:

(45) a. The boy that [the girl saw ___]	Object RC
b. The boy that [the girl gave the book to ___]	Indirect Object RC
c. The bat that [the girl hit the ball with ___]	Instrumental RC
d. The bridge that [the girl sat on ___]	Locative RC

In a wide-ranging and hugely influential study, Keenan and Comrie (1977) did a survey of languages all around the world and found that not every language allows all RC types.

Sidebar 7.8: Kinds of Relative Clauses

In languages like English, the relative clause is missing the element that corresponds to the head noun. This common pattern in the languages of the world is referred to as the **gap strategy**. That is, some languages leave a gap where the head noun would go, and a listener needs to fill in that gap as they hear the relative clause. However, many other languages (e.g., Hebrew, Cantonese) use what is called the **pronoun strategy**: instead of leaving a gap where the head noun would go, a pronoun is used (something like *the boy that the girl saw him*). Such pronouns are referred to as **resumptive pronouns**. And finally, many languages use a mix of gap and resumptive strategies, in which some kinds of relative clauses exhibit gaps and others exhibit pronouns.

In fact, they proposed something like a hierarchy of relative clauses.[4] They found that if a language allows RCs of any kind, then subject RCs are always possible. The second most common kind is object RCs, followed by indirect object RCs. Example 46 shows a partial list of RC types that Keenan and Comrie considered.[5]

(46) Subject > Object > Indirect Object > Instrumental/Locative > …

They refer to this as the **Noun Phrase Accessibility Hierarchy**. Importantly, this is not a frequency hierarchy—that is, subject RCs are not simply more common across the world's languages than object RCs. Rather, Keenan and Comrie claimed, this is an *implicational hierarchy*, which means that if a language exhibits anything on this scale, it will exhibit everything to the left of it. So if a language has object RCs, it *must* have subject RCs. If a language has indirect object RCs, it *must* have direct object RCs and subject RCs. Putting it another way, no language exists such that it contains indirect object RCs but not direct object RCs.

One way to understand this is that subject RCs are somehow privileged in human language. Because every language on the planet supposedly exhibits subject RCs, there must be something about them that gives them this special status.

In fact, when we consider how adults understand and produce relative clauses, we find that adults produce subject RCs far more often than any other kind. When we hear object RCs, we take longer to understand them than subject RCs. And finally, we respond to judgment and truth tasks more accurately with subject RCs than object RCs. All this leads to the idea that subject RCs are somehow advantaged within human language.

Turning to child language, this subject-RC advantage shows up quite clearly. Friedmann and Novogrodsky (2004) show that children (aged 4–5 years) acquiring Hebrew are able to correctly select a matching picture when tested with subject RCs more often than with object RCs. For example, when given two pictures, one showing a girl kissing her grandmother and the other showing the grandmother kissing the girl, children pick the correct picture when asked to point to the picture in which "this is the grandmother that ___ is kissing the girl" (SRC) more than when asked to point to the picture in which "this is the grandmother that the girl is kissing ___" (ORC). The accuracy rate in the SRC condition was 85%, while the accuracy in the ORC condition was 58%. This finding has been replicated many times in many different languages.

Sidebar 7.9: Animacy and Relative Clauses

A complicating factor with relative clauses is the animacy of the head noun and the nouns within the relative clause. Most studies have used reversible base sentences (see section 7.3.3 for reasons for this), but it has been claimed that relative clauses in which both nouns are animate (e.g., *the girl that the boy kissed*) are more natural in subject RCs than object RCs. Instead, object RCs are more natural when the direct object is inanimate, as in *the cart that the girl pushed*. Kidd et al. (2007) showed that when these animacy configurations are considered, children are able to produce object RCs on par with subject RCs.

This is interesting because of the way child language findings intersect so beautifully with psycholinguistic findings from adults as well as with typological facts from around the world's languages. In this way, child language acquisition helps us understand the degree to which such universal forces in human language manifest themselves.

7.4 The Problem of Variable Reference

In lexical semantics we talk about noun meaning involving something called **reference**. This simply means what in the world the noun *refers* to. For example, a noun such as *table* refers to an object that is a particular type of furniture. Proper names refer to specific individuals. But certain forms in language do not have a fixed reference, and these are called **variables**. Variables include traces of movement (see section 7.2), since the reference of the trace depends on the thing that has moved, and **pronominal forms** like *him*, *her*, *himself*, and *herself*, since the reference of the pronominal varies depending on the context. Pronominals like *himself* and *herself* are called **reflexive pronouns**, or just **reflexives**; pronominals like *him* and *her* are called **non-reflexive pronouns**, or sometimes just **pronouns**. In this section, we will focus on the linguistic (grammatical and pragmatic) constraints that influence how the various pronominal forms are interpreted.

To illustrate what it means for pronominals to have variable reference, an easy example comes from first- (*I*, *me*, *myself*) and second-person pronouns (*you*, *yourself*). First-person pronouns always refer to whoever is speaking, so as the speaker changes in a conversation, the referent of *I/me/myself* changes as well. The same goes for *you/yourself*. This means that the referent of the pronoun **varies** with the speaker.

The same is true with third-person pronominals, but here we have to deal with some further interpretive possibilities. Who does *himself* refer to in 47a? Who does *him* refer to in 47b?

(47) a. Bert likes himself.
     b. Bert likes him.

We can see that *himself* has to refer to the subject in 47a (Bert), but *him* cannot refer to the subject in 47b. Instead, *him* must refer to some other (male) individual who would have been mentioned earlier in a normal discourse. For example, *Ernie is a good friend of Bert's. Bert likes him* (him = Ernie).

In linguistics, we use **subscripts** (called **indices**) to indicate the reference of pronouns and nouns. When a pronoun and a full noun phrase share the same subscript (index), we say that they are **coindexed**, and normally this also means they **corefer** (refer to the same entity). Normally, indices start with the letter [i] and proceed alphabetically (so the next index is [j], then [k], etc.). (Examples 47′a–47′b show examples 47a–47b with their indices.) In 47′a, Bert is indexed with the subscript [i]. *Himself* is indexed with [i/*j]. This looks complicated, but it's actually quite simple. *Himself* refers to the entity that is indexed [i], and the [*j] means that anything that is indexed other than [i] is ungrammatical. So this means that *himself* ***must*** refer to Bert.

(47′) a. Bert_i likes $\text{himself}_{i/*j}$.
b. Bert_i likes $\text{him}_{*i/j}$.

Now consider 47′b, and notice that the indexing on *him* is the reverse of what we find on *himself.* The way to read the index on 47′b is that *him* cannot refer to anything indexed [i] (i.e., *Bert*) but can refer to anything else. Now consider who *him* and *himself* refer to in 48.

(48) a. Bert_i said that Ernie_j likes $\text{himself}_{*i/j/*k}$.
b. Bert_i said that Ernie_j likes $\text{him}_{i/*j/k}$.

The [k] subscript shows whether the pronominal can refer to an individual not mentioned in the sentence. We see that in 48a, *himself* must refer to Ernie—the subject of the clause in which *himself* occurs, but not the subject of the whole sentence—and it cannot refer to anyone else. In contrast, in 48b, *him* can refer to either Bert, the subject of the whole sentence, or to someone not mentioned in the sentence (e.g., Elmo, if Elmo is part of the discourse), but it cannot refer to Ernie, the subject of the clause containing *him.*

Notice that it does not matter (for English) whether the clause containing the pronominal is finite or nonfinite. We get the same pattern of coreference in 49, which contains an infinitive clause, as we did in 48.

(49) a. Bert_i wants Ernie_j to like $\text{himself}_{*i/j/*k}$.
b. Bert_i wants Ernie_j to like $\text{him}_{i/*j/k}$.

Finally, note that third-person-singular pronominals must agree in gender with their referent (called the **antecedent**). Thus, we get ungrammaticality when there is a gender mismatch.

(50) a. *Bert_i likes herself_i.
b. Bert_i likes her_j.
c. *Susan_i said that Bert_j likes herself_j.
d. Susan_i said that Bert_j likes $\text{her}_{i/*j}$.

In the following sections we will lay out how different types of pronominals can be interpreted and what children know (and at what ages) about the constraints on their interpretation. As we will see, there are two general types of constraints on how pronominals are interpreted: syntactic constraints and pragmatic constraints. Now we are in a position to formalize some of the syntactic constraints.

7.4.1 The Binding Theory

The part of linguistic theory that deals with these kinds of facts is referred to as **binding theory**. We will put the technical side of binding theory into a sidebar and focus here on outlining the phenomena and describing when children show knowledge of the underlying principles that govern variable reference. In order to do that, let us summarize what the phenomena are.

First, we saw above that reflexives must refer to the subject of that sentence. Second, reflexives cannot refer to any argument outside that sentence. For example, in 48a, *Bert said that Ernie likes himself*, the reflexive must refer to Ernie and not Bert (nor anyone else, such as Elmo). So the domain of the reflexive seems to be limited to its own sentence. And third, a reflexive must have a noun that it can refer *backward* to within that domain—that is, reflexives seem to look backward (or more accurately, upward in the structure) for a referent. We see this in the following example:

(51) *Bert said that [himself likes Ernie]

This sentence is ungrammatical because the reflexive must have a referent within its smallest clause, and remember that the complementizer *that* introduces a clause. This means that *Bert* is in a different clause than *himself*, thus ruling Bert out of contention as a referent. Furthermore, the reflexive needs its referent to be structurally higher than itself, so that rules Ernie out of contention. Because there is no potential referent for the reflexive, the sentence is ungrammatical. These three basic restrictions are summarized below:

Principle of Reflexives: A reflexive must have a referent that is structurally higher than it within its own (smallest) clause.

This is a very rough approximation to what is referred to in linguistics as **Principle A** of the binding theory. It is not technically correct since we do not fully define what 'structurally higher' means, nor what 'have a referent' means, but it suffices for our purposes (see sidebar 7.9 for more technical and precise definitions).

As for pronouns, we see a mirror image of the Principle of Reflexives. First, pronouns cannot refer to the subject of its smallest sentence or to any other noun phrase within its clause. Second, pronouns *can* refer to things outside its sentence, either something in a higher clause (e.g., *Ernie said that ...*) or something mentioned in the discourse but not in the actual sentence containing the pronoun. This can be summarized as follows:

Principle of Pronouns: A pronoun must not have a referent that is structurally higher than it within its own (smallest) clause.

This is a very rough approximation to what is referred to in linguistics as **Principle B** of the binding theory. It too is not technically correct, but it suffices for our purposes.

Finally, let us consider other kinds of nouns, referred to in the literature as **referring expressions** (or R-expressions). Such expressions include proper nouns and other "full" noun phrases (*the girl*, *the table*) that pick out an individual. These noun phrases inherently have the ability to refer. Sometimes it's clear who or what the R-expression picks out (e.g., Barack Obama—there is only one Barack Obama), while other times we might need more context to know who the intended referent is (e.g., John Smith—there are many John Smiths in the world). Similarly, for common nouns (*the/a girl*), there is more than one possible referent in the world. (In principle, there may be another Barack Obama in the world that we don't know about—there is no law that forbids people from naming their children after famous American presidents!) So although R-expressions pick out a referent, that referent might depend on context.

Do R-expressions work like pronouns or reflexives? It turns out that R-expressions behave differently from both reflexives and pronouns. While a reflexive must take a referent within its own clause, and while a pronoun cannot take a referent within its own clause (but may take a referent in a higher clause), R-expressions cannot take a referent anywhere within the sentence, even in a higher clause.

(52) a. He_i loves $Ernie_{*i}$
b. He_i said that $Bert_j$ loves $Ernie_{*i/*j/k}$
c. He_i said that Big $Bird_j$ believes that $Elmo_k$ insisted that $Bert_l$ loves $Ernie_{*i/*j/*k/*l/m}$

Example 52a shows that in a simple sentence, the R-expression *Ernie* cannot refer to the same person that *he* refers to. This means that the R-expression cannot take as its referent the preceding subject of that same sentence, making it different from a reflexive. Example 52b shows that even if there are referents in higher clauses, the R-expression *Ernie* still cannot refer to any of them, making it different from a pronoun. Not only that, R-expressions must not take referents anywhere in the sentence, even when they are separated by many clauses, as in 52c, in which *Ernie* can't take as a referent any of the previous pronouns or proper names mentioned in the sentence. This can be summarized as follows:

Principle of R-Expressions: An R-expression must not have a referent preceding it anywhere in the sentence, whether in the same clause or not.

This is a very rough approximation to what is referred to in linguistics as **Principle C** of the binding theory. Like the other principles already given, our definition is not technically correct, but it suffices for our purposes.

In the following sections we will look at experimental work designed to find out what children know about each of these binding principles, and when they know it.

7.4.2 Principle of Reflexives (Principle A)

Experiments investigating the Principle of Reflexives have generally shown that children respect restrictions on reflexive reference from quite early on. While some of the earliest work on reflexives put the age of acquisition around 5 years (Chien and Wexler, 1990), later work found success from as young as age 2;9 (McKee, 1992). In one study that used a Truth Value Judgment Task (see appendix B), McKee presented children with a story about two characters, Smurfette and a princess. In one condition Smurfette washed herself, and in the other condition Smurfette washed the princess. After the story, a puppet (operated by the experimenter) said, "I know what happened! Smurfette washed herself." Children were instructed to reward the puppet if it said the right thing or to "punish" the puppet if it said the wrong thing. The children in this study (ages 2;6 to 5;3) answered 'yes' 100% of the time in the "match" condition (in which Smurfette in fact washed herself) and 'no' 88% of the time in the "mismatch" condition (in which Smurfette washed the princess). This pat-

Sidebar 7.10: The Binding Principles

A more technically correct formulation of the binding principles is provided here.

(i) The Binding Principles

a. Principle A: A reflexive pronoun must be **bound** in its clause.[6]

b. Principle B: A non-reflexive pronoun must be **free** in its clause.

c. Principle C: A referring expression must be **free** everywhere.

The meaning of the word *free* in these principles is "not bound." To understand what this means, we need to introduce two formal notions. The first is called **c-command**. C-command is a relation between two nodes in a syntactic tree, which obtains under particular conditions (the *c* in c-command stands for *constituent*). The formal definition of c-command is given below, followed by explanation.

(ii) A node A c-commands a node B if and only if

a. the first branching node dominating A also dominates B, and

b. A does not itself dominate B.

Dominate means to be higher in the tree, and a *branching node* is a node with more than one line coming out of it. In the toy example in (iii), the nodes A and B are branching nodes, but none of the others are. Also in (iii), B dominates C and D, E dominates F, and A dominates everything.

(iii)

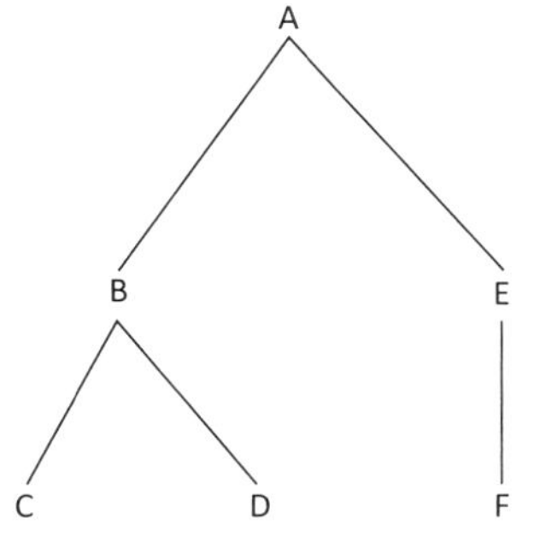

Question: Which node(s) in the tree does B c-command? Which node(s) in the tree does E c-command? Which node(s) in the tree does C c-command? Does A c-command anything?

(continued)

Sidebar 7.10 (continued)

We often talk about the relations between nodes in a syntactic tree in terms of female family relations (for example, in the tree in (iii), node B is the mother of node C, and C is the daughter of B, and C and D are sisters). One thing to remember about c-command is that sisters always c-command each other. An aunt also c-commands her nieces. Now let's look at a real example.

(iv)

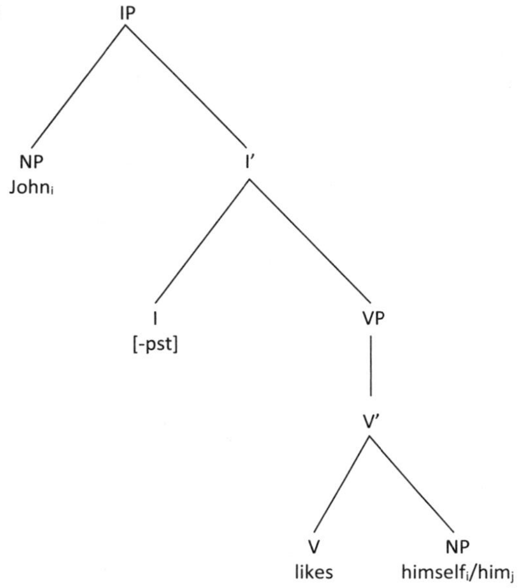

In (iv) we can see that the subject *John* c-commands the direct object of the verb. This is true because the first branching node dominating the noun phrase (NP) *John* is the inflection phrase (IP), and IP dominates the object NP; moreover, the NP *John* does not itself dominate the object NP.

C-command is a relation that holds no matter how close or far away two nodes are in the tree, as long as the conditions in (ii) are satisfied. Therefore, even if we have several embedded clauses, it is still possible for the subject of the main clause to c-command something in an embedded clause, such as a direct object. Thus, in (v) the NPs *John*, *Kenneth*, and *Bill* all c-command *him*.

(v)

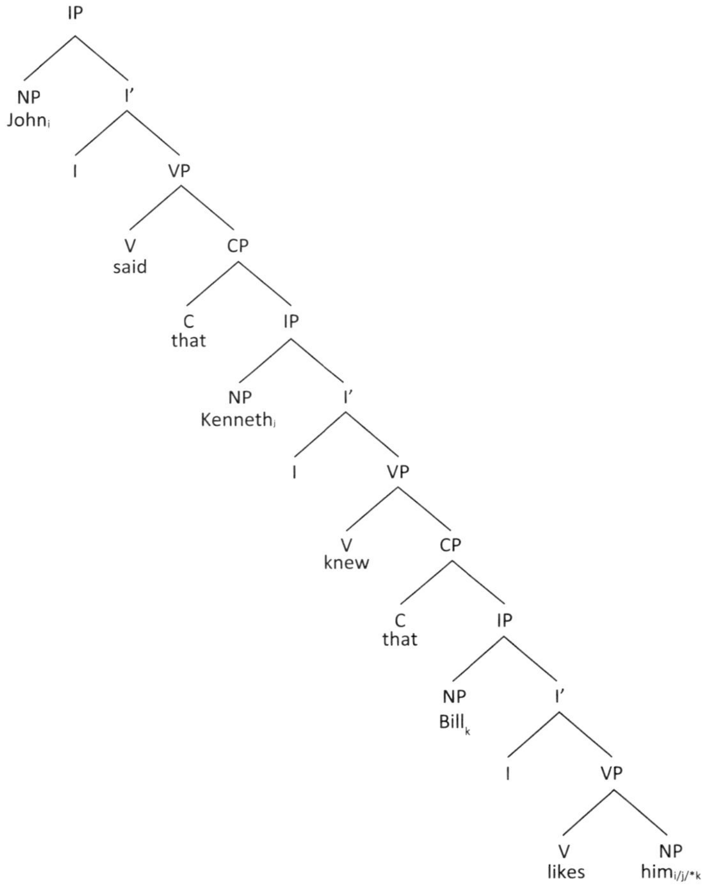

We can now introduce our second formal relation, known as **binding**.

(vi) A node A binds a node B if and only if

a. A c-commands B, and

b. A and B are coindexed.

Recall that coindexation simply refers to two nodes having the same index (subscript). Normally coindexation goes hand in hand with coreference, although we will see some cases in which coreference is possible without coindexation (see section 7.4.3). Going back to our example in (iv), we can now describe the relation between the subject, *John*, and the object pronominal in terms of binding: *John* binds the reflexive pronoun (*John* c-commands *himself* and they are coindexed) but not the nonreflexive pronoun (*John* c-commands *him* but they are not coindexed). Another way to say this is that

(continued)

Sidebar 7.10 (continued)

the reflexive is **bound** by *John*, its antecedent; whereas the non-reflexive pronoun is **not bound/free**. This matches our intuition that in (iv), *John* and *himself* refer to the same person, but *John* and *him* do not refer to the same person.

Finally, we must consider the question of why c-command is necessary in our definition of binding. Why can't we just say that the NP that *precedes* the pronominal is the antecedent? Consider the sentence in (vii).

(vii) John's sister invited him/*himself.

Why can't *John* bind *himself*, like it did in (v)? To understand why, we need to look at the syntactic structure of (vii), shown in (viii).

(viii)

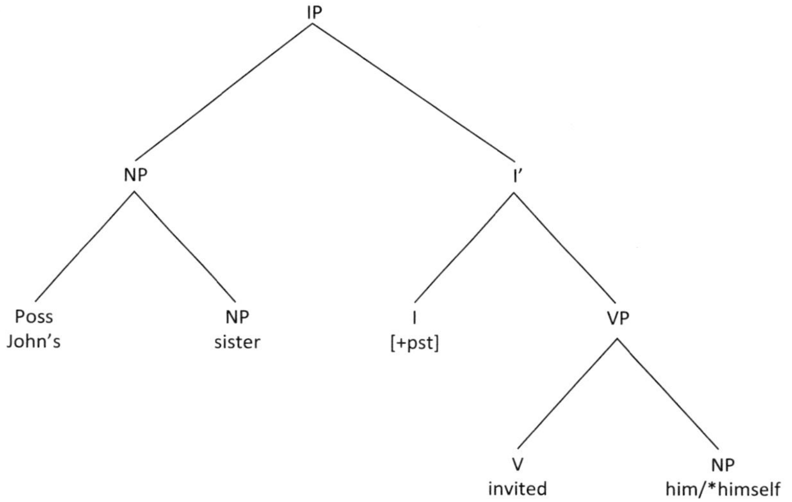

What is the first branching node dominating *John*? Here it is the higher NP, since *John* is a possessor within the subject NP. But that NP does not dominate the direct object, *himself*. Therefore, *John* does not c-command (and therefore does not bind) the direct object, and the reflexive, which must be bound within its clause, has no binding antecedent and the sentence is ungrammatical. The non-reflexive pronoun, *him*, on the other hand, must be free in its clause, and it is free since *John* does not c-command it (and therefore does not bind it).

tern of responses is correct and suggests that children limit *herself* to referring to something within its own clause.

In a language like English, in which the reflexive pronoun is marked for both gender and number (*himself* vs. *herself* vs. *themselves*), we might suppose that it is easier to learn what the reflexive is supposed to refer to, in comparison to a language in which the reflexive form does not show this morphological marking. Such a language is Italian, in which the third-person reflexive form *si* is the form for both masculine and feminine referents and for singular and plural third-person referents.

(53) a.

Il	ragazzo /	La	ragazza	si	lava.
the-masc	boy /	the-fem	girl	self	wash-3sg

'The boy/girl washes himself/herself.'

b.

I	ragazzi /	Le	ragazze	si	lavano.
the-masc-pl	boys /	the-fem-pl	girls	self	wash-3pl

'The boys/girls wash themselves.'

However, children acquiring Italian do not exhibit particular difficulties interpreting the reference of *si*. McKee's (1992) study included Italian-acquiring children (ages 3;7–5;5, slightly older than the English-speaking children). The Italian-speaking children performed almost perfectly in their responses to sentences with *si* in both the match and the mismatch conditions, as above.

7.4.3 Principle of Pronouns (Principle B)

The literature and conclusions about children's acquisition of the Principle of Pronouns is quite a bit more complex compared to the Principle of Reflexives. A large number of studies appear to show that children have great difficulty with the interpretation of non-reflexive pronouns. In general, the error they make is that they appear to permit the pronoun (*him/her*) to refer to a referent that is within its own clause instead of disallowing a referent in its clause. For example, in a sentence like 54, 5½-year-olds allow *her* to refer to the local subject, *Sarah* (rather than the more distant subject, *Kitty*) as much as 50% of the time (Jakubowicz, 1984; Chien and Wexler, 1990).

(54) Kitty says that [Sarah should point to her].

This result makes it look like children don't know the Principle of Pronouns until quite late in development. However, recall from the beginning of section 7.4 that pronouns do not *have* to be interpreted as referring to something within the sentence. In fact, in a simple main clause like *Sarah should point to her*, the pronoun *her* cannot refer to anything within the sentence and can only refer to some individual who is either physically present when the sentence is spoken or who has been mentioned already in the conversation. In other words, language allows two different routes to finding the referent for a pronoun: one route takes the reference from a referent within the sentence, and the other route takes the reference from the discourse setting. The former is referred to as **binding** and is considered a syntactic process. The latter is not considered a syntactic process but is instead seen as something done through discourse and pragmatics.

(55) Context: Imagine a party with four people: Ben, Jerry, Dave, and Samantha. Ben says to Jerry: "Tom said she arrived late."

Whom does *she* refer to in this context? Most likely, it refers to Samantha, since she is the only female individual in the discourse. Now consider the same setting, in which instead Ben says,

(56) Tom said [he arrived late].

Whom does *he* refer to? It could refer to *Tom* (via binding), but it could also refer to *Dave*. But *Dave* is not an NP in the sentence. This shows that pronouns get their reference either through the discourse (context) or from within the sentence itself. This process of obtaining reference through discourse is referred to as **coreference**. The process of obtaining reference within the sentence is referred to as **binding** (see sidebar 7.10).

However, as we saw in section 7.4.1, the Principle of Pronouns states that in order to get reference from within the sentence, that referent must be at least one clause away. Thus, *him* in 57 cannot refer to Tom because Tom is in the same clause.

(57) Tom kissed him.

What if children are making errors in interpreting pronouns because they have not yet learned the difference between coreference (via discourse) and binding (via sentence-internal mechanisms)? What might be happening is that children allow *Tom* to pick out its referent from discourse, and they allow *him* to pick out a referent, and just by accident, they picked out the same referent from discourse (Tom). This is called **accidental coreference**. Accidental coreference is not really a violation of the Principle of Pronouns; it's just a quirk of how language works.

One way to test this is to present children with sentences in which the discourse option is not available. When we use what is called a **quantified phrase** (QP), such as *every girl, no boy, seven mice*, the QP does not pick out a particular individual. Instead, it denotes a set of individuals. Going back to our party context, imagine that Samantha says,

(58) Every boy invited him.

The quantified subject picks out a set of people, while the pronoun picks out an individual. This means that *every boy* and *him* cannot accidentally refer to the same individual, since one picks out a set and the other picks out an individual. So if children allow *him* to refer to *every boy* (meaning something like 'every boy invited himself'), then this would be a true violation of the Principle of Pronouns.

Researchers tested this hypothesis and found that in fact children at age 5;6 do provide more adultlike responses when accidental coreference is impossible (as in 59c–59d). They used sentences like the following:

(59)	a. Goofy is washing him.	NP-pronoun
	b. Goofy is washing himself.	NP-reflexive
	c. Every boy is washing him.	QP-pronoun
	d. Every boy is washing himself.	QP-reflexive

What researchers discovered was that although children (incorrectly) interpreted sentences like 59a as meaning the same thing as 59b, they correctly distinguished the meanings of 59c and 59d (Chien and Wexler, 1990; see table 7.6).

Table 7.6
Results of Chien and Wexler's (1990) study

Sentence	Type	Percent Correct
Goofy is washing him.	NP-pronoun	50%
Goofy is washing himself.	NP-reflexive	90%
Every boy is washing him.	QP-pronoun	86%
Every boy is washing himself.	QP-reflexive	85%

Source: Chien and Wexler (1990).

Taking all these facts into consideration, the consensus in the field is that both the Principle of Reflexives and the Principle of Pronouns are present in child grammar, even though there is the complication of accidental coreference for the Principle of Pronouns. Now let us examine the Principle of R-Expressions.

7.4.4 Principle of R-Expressions (Principle C)

The third principle of binding is the Principle of R-Expressions, which has to do with how proper nouns and other non-pronominals are interpreted. Do children have knowledge of this principle, and if so, at what age can we detect it?

The seminal work on this issue was done by Crain and McKee (1985), who tested sixty-two English-speaking children (mean age 4;2) using the Truth Value Judgment Task (TVJT). Children were presented with a story acted out using toy characters. A puppet then made a statement about the story, and the child was asked whether what the puppet said was right or not. If the puppet said the right thing, the child was to reward the puppet (e.g., give the puppet some carrots). Otherwise, the child was to punish the puppet (e.g., give the puppet some rocks). Crain and McKee tested children in four conditions:

(60) a. The Smurf_i ate the hamburger when he_i was inside the house.
b. When he_i was inside the house, the Smurf_i ate the hamburger.
c. *He_i ate the hamburger when the Smurf_i was inside the house.
d. When the Smurf_i was inside the house, he_i ate the hamburger.

Recall that the Principle of R-Expressions states that an R-expression cannot take a referent anywhere in the sentence. Sentence 60c is ungrammatical on the reading in which the pronoun and name are coreferential, since the pronoun *he* is structurally higher than *Smurf*, which means they are coindexed. This is therefore a violation of the Principle of R-Expressions, which explains why this sentence is ungrammatical. The other three sentences, meanwhile, are grammatical on the coreferential reading because none of them violate this principle. Example 60a involves an R-expression (*Smurf*) that is structurally higher than the pronoun *he*. In this case, the R-expression is not in violation of the principle since nothing structurally higher than it is coreferential with it. Notice that this sentence includes a subordinate clause that begins with *when*. The domain of binding for the embedded pronoun is that subordinate clause, *when he was inside the house*. Because the pronoun does not have a referent within that clause, this is not a violation of any binding principles.

As for 60b, this sentence has inverted the subordinate clause in 60a with the main clause. As such, the pronoun linearly precedes the R-expression, so you might expect a violation of the principle here. However, there is no violation because *Smurf* does not have a referent that is structurally higher than it. Remember that the subordinate clause headed by *when* is structurally a subordinate clause, so nothing within that clause is higher than the NP *Smurf*. The structure of 60b is this (omitting some parts for reasons of space):

(61)

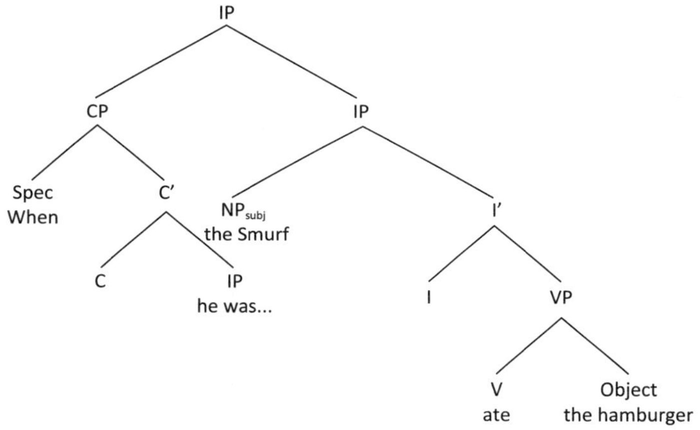

We can see that anything within the *when* clause is not structurally higher than *Smurf*, so no violation of the Principle of R-Expressions obtains. The facts are similar for 60d.

In sum, these four sentences, as remarkably similar as they are on the surface, are very different structurally. Only one of these four sentences is structurally just right to create a violation of the Principle of R-Expressions. Crain and McKee tested children on these sentences by asking them to judge their truth relative to a story. Crain and McKee found that even the youngest of their children (mean age 3;1) correctly rejected 60c nearly 80% of the time, while accepting the other three sentences. They concluded that knowledge of the Principle of R-Expressions is present from an early point in syntactic development. These results have since been replicated several times in various languages using various sentence structures (e.g., Chien and Wexler, 1990; Thornton and Wexler, 1999; Kiguchi and Thornton, 2004), showing rather conclusively that the

Principle of R-Expressions (or Principle C) is present in the grammars of children as young as age 3.

Researchers use many other protocols to test for evidence of this principle. Lukyanenko et al. (2014) used a preferential looking paradigm to test children as young as 30 months old. In this protocol, children saw a video in which two girls were standing. Children were familiarized with the names of the girls—say, Katie on the left and Anna on the right. Then they saw Katie patting herself on the head and Anna patting herself on the head. Then they saw Katie patting Anna, and Anna patting Katie. In other words, they saw both a reflexive action and a transitive action involving the verb *pat*. Then they saw both girls next to each other, with Katie patting herself, and Anna patting Katie.

Children then heard the test sentence. In the critical condition, they heard something like, "Is she patting Katie?" If *she* refers to Katie (non-adultlike interpretation), then children should look at the screen showing Katie patting herself, while if *she* refers to someone other than Katie (adultlike interpretation), then they should look at the screen in which Anna is patting Katie. Lukyanenko et al. found a statistically significant preference in eye gaze for the *Anna patting Katie* screen, showing that even as young as 30 months, children prefer referential nouns (like names) to *not* refer to structurally higher nouns within the sentence. This is taken as evidence that the Principle of R-Expressions (Principle C) is present in children as young as 30 months.

In sum, then, we know that children have knowledge of the Principle of R-Expressions intact from as young as 30 months. Notice that in some studies the sentences involved in testing the Principle of R-Expressions are often lengthy (multiple clauses), and the sentences with violations are very similar to completely acceptable sentences. So the fact that researchers have so reliably and uniformly found the same thing (that children possess knowledge of the Principle of R-Expressions) is taken as evidence that this principle is present in the grammars of children from the earliest testable ages. This is often the test for whether a principle is part of Universal Grammar (UG), and we can safely conclude that the Principle of R-Expressions is indeed part of UG.

7.4.5 Summary of the Binding Principles

This section was about the restrictions on interpreting noun phrases (NPs). We saw that there are three main types of NPs: reflexive pronouns (*herself*), non-reflexive pronouns (*her*), whose reference is inherently variable, and R-expressions (*Mary*, *the girl*), whose reference is stable. According to the principles for interpreting these NPs in adult grammar, reflexives have to refer to a structurally higher NP within their own smallest clause; non-reflexive pronouns cannot refer to a structurally higher NP within their own smallest clause but instead can refer to an NP in a higher clause of their sentence or to something mentioned within the discourse, even if it's not part of the sentence at all; and R-expressions have their own reference, so they don't take their reference from another NP inside or outside of their clause.

In terms of children's knowledge of these principles, we saw that from around age 3 children correctly interpret reflexives as needing a referent within their clause, and they correctly interpret R-expressions as having their own, independent reference. They appear to make errors in interpreting non-reflexive pronouns until as late as age 6 or 7. However, researchers believe that rather than indicating a lack of understanding of the rule for interpreting these forms, the errors arise from a misunderstanding of when pronouns can get their reference from something in the discourse.

7.5 Summary

This chapter was about how children decipher the structural relations among words in sentences. Semantic bootstrapping explains how children exploit regular correspondences between the semantic properties of lexical categories (like nouns and verbs) and their syntactic function in order to learn word order, and children seem to know the basic word order of their language quite early, even before they begin to produce multiword utterances themselves.

Beyond word order, syntax is concerned with functional categories and structure. Much of the chapter focused on these components of grammar and how and when children acquire them. The main takeaway is that although children frequently omit functional category markers (e.g., tense morphemes, agreement morphemes, determiners) in their earliest sentences, there is evidence that children's underlying syntactic representations are quite sophisticated and adultlike. In particular, the fact that children reliably produce finite verbs high up in the sentence structure (e.g., before negation) and produce nonfinite verbs much lower in the structure (e.g., after negation) tells us that even at this early stage children have some implicit understanding of the architecture of sentences, even when these architectural frameworks are abstract. Differences between children's and adults' sentences in terms of functional structure can be explained by the idea that children allow the root node of a sentence to be a projection lower than a complementizer phrase (CP), while adults always require the root node to be CP. This idea is known as the truncation hypothesis.

Finally, we looked at how different types of NPs are interpreted, the restrictions on their interpretation in terms of syntactic binding and discourse, and children's knowledge of these restrictions. We saw that while children correctly interpret reflexives and R-expressions from early on, it takes them longer to work out the conditions under which non-reflexive pronouns can derive their reference from another NP in their sentence.

The overarching theme of this chapter is that children know a great deal about the hidden structure of grammar—much more, in fact, than they appear to know on the surface.

7.6 Further Reading

Friedemann, Marc-Ariel, and Luigi Rizzi (eds.). 2000. *The Acquisition of Syntax: Studies in Comparative Developmental Linguistics*. New York: Routledge.

Guasti, Maria-Teresa. 2002. *Language Acquisition: The Growth of Grammar*. Cambridge, MA: MIT Press.

7.7 Exercises

1. Semantic bootstrapping rests on many assumptions about what the language learner brings to the task of language learning (i.e., what learners must know in advance) and what learners can infer from their environment (i.e., both linguistic and nonlinguistic input). What are some of these assumptions? Think of at least three assumptions this approach makes. Are these reasonable assumptions? Why or why not?

2. Read the following child utterances. For each utterance, state whether the sentence is finite or nonfinite or if you can't tell. If you can't tell whether the sentence is finite or nonfinite, state why you can't tell.

Utterance	Finite/Nonfinite/Can't tell?
1. What that?	
2. Captain Bob grew chickens.	
3. This writes.	
4. I like toys.	
5. We going to have some ginger ale.	
6. That spell two of them.	
7. I going take it.	
8. That's a little one.	
9. Give me the button.	
10. Where you go?	

3. In this chapter we presented an explanation for children's production of optional infinitives (OIs) in terms of a slight difference between the child's syntactic representation and the adult's: children allow the structure to be truncated, so that the highest (root) node of the tree is a phrase lower than the complementizer phrase (CP). Another possible explanation is that the child does have the full adultlike structure and always has CP as the root node, but because children have a lower cognitive capacity than adults have, they simply leave out certain morphemes when they speak. That is, their underlying representation is exactly like that of adults, but they leave stuff out in their speech. In fact, children do have a smaller working-memory capacity than adults have, so they are not able to keep track of as many concepts at once as an adult can. Perhaps inflectional marking is just one extra thing that children leave out because something has to give.

Consider the latter possible explanation for the phenomenon of OIs. What would be an advantage of accounting for this in terms of cognitive capacity rather than grammatical knowledge? What would be a disadvantage of this approach, compared to the grammatical approaches? Which approach has more advantages, in your view?

4. In this chapter we talked about children's negative utterances progressing through stages. First, children place the negative form outside the rest of the utterance, at the beginning or end of the sentence. Later they begin to place the negative word inside the sentence, between the subject and verb, but they don't seem to separate negation from the auxiliary verb that marks tense (e.g., *don't*, *can't*). Finally, they distinguish between affirmative auxiliaries (*do*, *can*) and negation (*not*). Propose a tree structure for each of the following child utterances. Which of Klima and Bellugi's stages do you think each utterance is from?

(i) No make bat
(ii) me don't have Legos pancakes
(iii) And I gonna not wear my earrings

5. Similar to negation, children's formation of questions appears to develop in stages. First, children produce questions by using only rising intonation. Later they begin to produce more varied wh-questions and sometimes invert the auxiliary with the subject. Finally, they always invert the auxiliary with the subject. Propose a tree structure for each of the following child utterances. Which of Klima and Bellugi's stages do you think each utterance is from?

(i) Eat that?
(ii) What you doing?
(iii) Where's that princess?

7.8 References

Becker, Misha, and Susannah Kirby. 2016. A-movement in language development. In Jeffrey Lidz, Joe Pater, and William Snyder (eds.), *The Oxford Handbook of Developmental Linguistics*, pp. 230–278. Oxford: Oxford University Press.

Bloom, Paul. 1990. Subjectless sentences in child language. *Linguistic Inquiry* 21: 491–504.

Borer, Hagit, and Kenneth Wexler. 1987. The maturation of syntax. In Thomas Roeper and Edwin Williams (eds.), *Parameter Setting*, pp. 123–172. Dordrecht: Reidel.

Brown, Roger. 1973. *A First Language*. Cambridge, MA: Harvard University Press.

Chien, Yu-Chin, and Kenneth Wexler. 1990. Children's knowledge of locality conditions in binding as evidence for the modularity of syntax and pragmatics. *Language Acquisition* 1: 225–295.

Choi, Soonja. 1988. The semantic development of negation: A cross-linguistic longitudinal study. *Journal of Child Language* 15: 517–531.

Cinque, G. 1999. *Adverbs and Functional Heads: A Cross-linguistic Perspective*. New York: Oxford University Press.

Conroy, Anastasia, Eri Takahashi, Jeffrey Lidz, and Colin Phillips. 2009. Equal treatment for all antecedents: How children succeed with Principle B. *Linguistic Inquiry* 40: 446–486.

Crain, Stephen, and Cecile McKee. 1985. The acquisition of structural restrictions on anaphora. In Stephen Berman, Jae-Woong Choe, and Joyce McDonough (eds.), *Proceedings of NELS 15*, pp. 94–110. Amherst, MA: Graduate Linguistic Student Association.

Crain, Stephen, and Rosalind Thornton. 1998. *Investigations in Universal Grammar*. Cambridge, MA: MIT Press.

Crain, Stephen, Rosalind Thornton, and Keiko Murasugi. 2009. Capturing the evasive passive. *Language Acquisition* 16: 123–133.

Crisma, Paula. 1992. On the acquisition of wh-questions in French. *Geneva Generative Papers* 1(2): 115–122.

Deen, Kamil. 2005. Productive agreement in Swahili: Against a piecemeal approach. In Alejna Brugos, Manuella Clark-Cotton, and Seungwan Ha (eds.), *Proceedings of the 29th Annual Boston University Conference on Language Development*, pp. 156–167. Somerville, MA: Cascadilla Press.

Deen, Kamil Ud. 2011. The acquisition of the passive. In Jill de Villiers and Tom Roeper (eds.), *Handbook of Generative Approaches to Language Acquisition*, pp. 155–188. Dordrecht: Springer.

Demuth, Katherine. 1989. Maturation and the acquisition of Sesotho passives. *Language* 65: 56–80.

Demuth, Katherine, Francina Moloi, and Malillo Machobane. 2010. Three-year-olds' comprehension, production and generalization of Sesotho passives. *Cognition* 115: 238–251.

Déprez, Viviane, and Amy Pierce. 1993. Negation and functional projections in early grammar. *Linguistic Inquiry* 24: 25–67.

Elbourne, Paul. 2005. On the acquisition of Principle B. *Linguistic Inquiry* 36: 333–366.

Fisher, Cynthia, Henry Gleitman, and Lila Gleitman. 1994. On the semantic content of subcategorization frames. *Cognitive Psychology* 23: 331–392.

Fox, Danny, and Yusef Grodzinsky. 1998. Children's passive: A view from the *by*-phrase. *Linguistic Inquiry* 29: 311–332.

Friedmann, Na'ama, and Rama Novogrodsky. 2004. The acquisition of relative clause comprehension in Hebrew: A study of SLI and normal development. *Journal of Child Language* 31: 661–681.

Gleitman, Lila. 1990. The structural sources of verb meanings. *Language Acquisition* 1: 3–55.

Gleitman, Lila, Kimberly Cassidy, Rebecca Nappa, Anna Papafragou, and John Trueswell. 2005. Hard words. *Language Learning and Development* 1: 23–64.

Gordon, Peter, and Jill Chafetz. 1990. Verb-based versus class-based accounts of actionality effects in children's comprehension of passives. *Cognition* 36: 227–254.

Goro, Takuya, and Sachie Akiba. 2004. The acquisition of disjunction and positive polarity in Japanese. In G. Garding and M. Tsujimura (eds.), *Proceedings to the 23rd WCCFL*, pp. 101–114. Somerville, MA: Cascadilla Press.

Grinstead, J., P. Lintz, M. Vega-Mendoza, J. De la Mora, M. Cantú-Sánchez, and B. Flores-Avalos. 2014. Evidence of optional infinitive verbs in the spontaneous speech of Spanish-speaking children with SLI. *Lingua* 140: 52–66.

Guasti, Maria-Teresa. 2002. *Language Acquisition: The Growth of Grammar.* Cambridge, MA: MIT Press.

Hirsh-Pasek, Kathy, and Roberta Golinkoff. 1996. The intermodal preferential looking paradigm: A window onto emerging language comprehension. In Dana McDaniel, Cecile McKee, and Helen Smith Cairns, *Methods for Assessing Children's Syntax*, pp. 105–124. Cambridge, MA: MIT Press.

Hirsh-Pasek, Kathryn, Roberta Golinkoff, Paul Fletcher, F. DeGaspe-Beaubien, and Kathleen Cauley. 1985. In the beginning: One-word speakers comprehend word order. Paper presented at the Boston University Conference on Language Development.

Hoekstra, Teun, Nina Hyams and Misha Becker. 1999. The role of the specifier and finiteness in early grammar. In David Adger, Susan Pintzuk, Bernadette Plunkett, and George Tsoulas (eds.), *Specifiers: Minimalist Approaches*, pp. 251–270. Oxford: Oxford Linguistics.

Holmberg, Anders. 2010. Verb second. In T. Kiss and A. Alexiadou (eds.), *Syntax: An International Handbook of Contemporary Syntactic Research*, pp. 342–382. Berlin: Walter de Gruyter.

Horgan, Dianne. 1978. The development of the full passive. *Journal of Child Language* 5: 65–80.

Hyams, Nina. 1986. *Language Acquisition and the Theory of Parameters.* Dordrecht: D. Reidel.

Jakubowicz, Celia. 1984. On markedness and binding principles. In Charles Jones and Peter Sells (eds.), *Proceedings of the North Eastern Linguistics Society 14*, pp. 154–182. Amherst: University of Massachusetts, Amherst, GLSA Press.

Keenan, Edward L., and Bernard Comrie. 1977. Noun phrase accessibility and universal grammar. *Linguistic Inquiry* 8(1): 63–99.

Kidd, Evan, Silke Brandt, Elena Lieven, and Michael Tomasello. 2007. Object relatives made easy: A cross linguistic comparison of the constraints influencing young children's processing of relative clauses. *Language and Cognitive Processes* 22(6): 860–897.

Kiguchi, Hirohisa, and Rosalind Thornton. 2004. Binding principles and ACD constructions in child grammars. *Syntax* 7(3): 234–271.

Klima, Edward, and Ursula Bellugi. 1966. Syntactic regularities in the speech of children. In John. Lyons and Roger. J. Wales (eds.), *Psycholinguistic Papers*, pp. 183–208. Edinburgh: Edinburgh University Press.

Krämer, Irene. 1993. The licensing of subjects in early child language. In *MIT Working Papers in Linguistics*, vol. 19: *Papers on Case and Agreement*, pp. 197–212. Cambridge, MA: MIT Press.

Lukyanenko, Cynthia, Anastasia Conroy, and Jeffrey Lidz. 2014. Is she patting Katie? Constraints on pronominal reference in 30-month-olds. *Language Learning and Development* 10(4): 328–344.

Maratsos, Michael, Dana Fox, Judith Becker, and Mary Anne. Chalkley. 1985. Semantic restrictions on children's passives. *Cognition* 19: 167–191.

McKee, Cecile. 1992. A comparison of pronouns and anaphors in Italian and English acquisition. *Language Acquisition* 2: 21–55.

Naigles, Letitia. 1990. Children use syntax to learn verb meanings. *Journal of Child Language* 17: 357–374.

O'Brien, Karen, Elaine Grolla, and Diane Lillo-Martin. 2006. Long passives are understood by young children. In David Bamman, Tatiana Magnitskaia, and Colleen Zaller (eds.), *Proceedings of the 30th Annual Boston University Conference on Language Development*, pp. 441–451. Somerville, MA: MIT Press.

Phillips, Colin. 1995. Syntax at age two: Crosslinguistic differences. In Carson Schütze, Jennifer Ganger, and Kevin Broihier (eds.), *Papers on Language Processing and Acquisition*, pp. 325–382. Cambridge, MA: MIT Press.

Pierce, Amy. 1989. On the emergence of syntax: A crosslinguistic study. PhD diss., Massachusetts Institute of Technology.

Pierce, Amy. 1992. *Language Acquisition and Syntactic Theory: A Comparative Analysis of French and English Child Grammars*. Dordrecht: Kluwer.

Pinker, Steven. 1982. A theory of the acquisition of lexical interpretive grammars. In J. Bresnan (ed.), *The Mental Representation of Grammatical Relations*, pp. 655–726. Cambridge, MA: MIT Press.

Pinker, Steven. 1984. *Language Learnability and Language Development*. Cambridge, MA: Harvard University Press.

Pinker, Steven. 1987. The bootstrapping problem in language acquisition. In Brian MacWhinney (ed.), *Mechanisms of Language Acquisition*, pp. 339–441. Hillsdale, NJ: Lawrence Erlbaum Associates.

Poeppel, David, and Kenneth Wexler. 1993. The full competence hypothesis of clause structure in early German. *Language* 69: 1–33.

Radford, Andrew. 1988. Small children's small clauses. *Transactions of the Philological Society* 86: 1-43.

Radford, Andrew. 1990. *Syntactic Theory and the Acquisition of English Syntax: The Nature of Early Child Grammars of English*. Oxford: Blackwell.

Rizzi, Luigi. 1993/1994. Some notes on linguistic theory and language development: The case of root infinitives. *Language Acquisition* 3: 371–395.

Rizzi, Luigi. 1997. The fine structure of the left periphery. In Liliane Haegeman (ed.), *Elements of Grammar: A Handbook of Generative Syntax*, pp. 281–337. Dordrecht: Kluwer.

Roeper, Thomas, and Jill de Villiers. 1994. Lexical links in the wh-chain. In Barbara Lust, Gabriella Hermon, and Jaklin Kornfilt (eds.), *Syntactic Theory and First Language Acquisition: Crosslinguistic Perspectives*, vol. 2, pp. 357–390. Hillsdale, NJ: Lawrence Erlbaum Associates.

Salustri, Manola, and Nina Hyams. 2007. The imperative as RI analogue: New data and competing theories. In Kamil Deen, Jun Nomura, Barbara Schultz, and Bonnie. D. Schwartz (eds.), *Proceedings of the Inaugural Conference of Generative Approaches to Language Acquisition—North America*, pp. 285–296. Cambridge, MA: UConn Occasional Papers in Linguistics.

Sano, Tetsuya, and Nina Hyams. 1995. Agreement, finiteness and the development of null subjects. In Mercè Gonzalez (ed.), *Proceedings of Northeast Linguistics Society 24*, pp. 543–558. Amherst: University of Massachusetts, Amherst, GLSA Press.

Snyder, William, and Nina Hyams. 2015. Minimality effects in children's passives. In Elisa Di Domenico, Cornelia Hamann, and Simona Matteini (eds.), *Structures, Strategies and Beyond: Essays in Honour of Adriana Belletti*, pp. 343–368. Linguistik Aktuell/Linguistics Today 223. Amsterdam: John Benjamins.

Thornton, Rosalind. 1990. Adventures in long-distance moving: The acquisition of complex wh-questions. PhD diss., University of Connecticut.

Thornton, Rosalind, and Kenneth Wexler. 1999. *Principle B, VP-ellipsis, and Interpretation in Child Grammar.* Cambridge, MA: MIT Press.

Valian, Virginia. 1991. Syntactic subjects in the early speech of American and Italian children. *Cognition* 40: 21–81.

Wexler, Kenneth. 1994. Optional infinitives, head movement, and the economy of derivations. In David Lightfoot and Norbert Hornstein (eds.), *Verb Movement*, pp. 305–350. Cambridge: Cambridge University Press.

Wijnen, Frank. 1997. Temporal reference and eventivity in root infinitivals. In Jeannette Schaeffer (ed.), *The Interpretation of Root Infinitives and Bare Nouns in Child Language*, pp. 1–25. MIT Occasional Papers in Linguistics 12. Cambridge, MA: MIT Working Papers in Linguistics.

Yuan, Sylvia, and Cynthia Fisher. 2009. "Really? She blicked the baby?" Two-year-olds learn combinatorial facts about verbs by listening. *Psychological Science* 20: 619–626.

Yuan, Sylvia, Cynthia Fisher, Yael Gertner, and Jesse Snedeker. 2007. Participants are more than physical bodies: 21-month-olds assign relational meaning to transitive novel verbs. Paper presented at the biennial meeting of the Society for Research in Child Development, Boston.

V Module 5: Beyond Monolingual and Typical Language Acquisition

8 Language Acquisition under Nontypical Circumstances

So far we have considered how language is acquired in the most typical case: by a child with full sensory abilities (hearing, sight) and cognitive abilities who is exposed to one (or more) language(s) from the earliest moments of life. But certainly there are cases in which language is acquired under different circumstances: at a later stage of life, with a smaller range of sensory abilities, or with a different set of cognitive abilities. How does language develop under these types of circumstances?

Studying language development under nontypical circumstances enables us to underscore the biological component of language and language acquisition. When you take away sensory abilities but have typical cognitive function and access to language (in whatever modality), language acquisition proceeds normally. If you change the biological state of the learner by changing their age (i.e., they are much older than usual) or their cognitive resources, language acquisition proceeds differently. Thus, although there is a social component to language acquisition and use, the development of human language is a fundamentally *biological* process. The studies we describe in this chapter can serve to highlight this fact.

8.1 Late First-Language Acquisition

The majority of people who acquire a language late in life (after, say, puberty) are second language learners; they already acquired a first language during infancy, and now they are learning a second (or third!) one. But occasionally it happens that a child remains unexposed to language for the first several years of life, so when they do acquire a language later on, they are acquiring a late first language. As we will see in section 8.2, this is the case for many children who are born deaf. But it can also happen when there is no hearing impairment. We will examine these cases in this section.

The idea that language is most effectively and efficiently acquired early in life, and with more difficulty later in life, is known as the **critical period hypothesis**. Critical periods (also sometimes called sensitive periods) for learning are found in many species: there is a critical window of time, usually some time between birth and sexual maturity, when an organism must acquire some species-specific ability. For some species it is navigation (e.g., indigo buntings; Emlen, 1969), for others it is locating prey (e.g., barn owls; Knudsen and Knudsen, 1990), and for certain species of songbirds it is learning birdsong (e.g., swamp sparrows; Marler and Peters, 1987; Marler, 1991). There is also evidence of a critical period for vision in a number of species, such as cats (Wiesel and Hubel, 1965; Hubel and Wiesel, 1970; Hensch, 2005).

The case of birdsong is well studied and perhaps the most famous. Each species of songbird has a species-specific song that identifies its members. This song is used in mating rituals, and without this song, a member of the species will likely not be able to find a mate and reproduce. In most songbird species young chicks learn the song through exposure to the song, but what's interesting is that this exposure must come at a particular time point in development (the critical period), and the exact timing of this critical period varies across species. For example, in swamp sparrows the critical window is from about 20 days to about 4 months of age (Marler and Peters, 1987). If the chick hears the song before this critical period, but not during the period, the chick will grow into a mature bird not having mastered its species-specific mating song. Likewise, if the chick does not hear the song before or during the critical period, but hears it after the critical period, the chick will grow into a mature bird without knowledge of its mating song. This has been tested experimentally, and no matter how much exposure a chick gets outside of the critical period, if it did not get exposure to the mating song during the critical period, it will never learn its full mating song. This idea of a critical period carries over to other species and other functions, including, according to many, human language.

Lenneberg (1967) noticed that children who lost their hearing in childhood after having been able to hear in the first few years of life had a significant advantage in learning oral language over children who had been deaf since birth, suggesting that the early window of exposure to language was critical. Even more tellingly, Lenneberg found that children who had lost their language ability due to brain trauma (resulting in aphasia, or language loss) were better at overcoming their aphasia and regaining language than adults who suffered a similar brain trauma and aphasia later in life. Lenneberg suggested that childhood was a critical time in the life of a person for learning language. In this section and in section 8.2, we will look at some additional evidence for this hypothesis.

8.1.1 Feral Children

There are many fascinating tales of so-called wild children, children who were raised by animals or in homes where they were isolated from human contact for various reasons. Some of these stories go back centuries or even millennia, as in the story told by Herodotus of the Egyptian king Psammetichus I, who sought to answer the question of which language was the oldest language, Egyptian or Phrygian. To solve this puzzle the king took two young babies from their families and had them raised by shepherds, who were instructed to feed, but not speak to, the boys. The king's expectation was that whatever language the boys started speaking spontaneously would be the answer to his question. According to the legend, the word the boys first spoke sounded like *bekos*, the Phrygian word for 'bread'. The king had his answer—Phrygian was the original language.

One does not need to reach so far back into antiquity to find stories of feral children. However, many of these tales remain unverified or turn out to be hoaxes, such as the twins Amala and Kamala, who were "found" by missionaries in the 1920s near Calcutta, India. Even the documented cases are shrouded in mystery and uncertainty. Another such case is Victor, the "wild boy of Aveyron," who was discovered in the late eighteenth century living in the forest in France. Victor was taken in by the French doctor Jean-Marc Gaspard Itard, who undertook to study and attempt to educate and rehabilitate the boy. After five years of working with Victor, Itard had been able to teach Victor some manual signs, but Victor remained unable to speak or to acquire conventions of social interaction. It remains unclear at what age Victor had been abandoned (a recent account suggests he was about 12 years old when he was discovered and likely had not been abandoned before he was 10; Frith, 2003) and whether he had a cognitive condition that could have impeded language development, such as autism, independent of his lack of exposure to language (see section 8.4.2 on the relationship between autism and language).

Another mysterious case is that of Kaspar Hauser, a boy who was locked in a dark dungeon-like cell for an unknown number of years in Nuremberg, Germany, and was discovered in the 1820s, likely at the age of 16. Unlike Victor, Kaspar Hauser was able to acquire language to a reasonable degree and was able to form social attachments. However, nothing is known about his early childhood and what kind of exposure he may have had to language prior to (or during) his imprisonment. Further adding to the mystery surrounding this individual, he was murdered five years after his discovery. (See Frith, 2003, for a detailed discussion of both Victor of Aveyron and Kaspar Hauser.)

The most recent documented case of a child kept in isolation is that of Danielle, a 7-year-old girl discovered in Florida in 2005, who appeared to be severely malnourished, incontinent, and without any language ability, having been kept by her mother in a closet in their home. Danielle was eventually adopted by another family, but it is unclear what path, if any, her language development has taken. While there are some newspaper articles about her case (DeGregory, 2008) with reports that she could produce some words, there does not seem to have been any scientific study of Danielle's rehabilitation.

8.1.2 Genie

One of the most well-known and thoroughly studied cases of a feral child in modern times is that of a girl known by the name Genie, who was kept in almost complete isolation by her deranged father from about age 20 months until she was discovered at age 13½ in 1970. During the period of her isolation Genie endured inhumane abuses that are difficult to comprehend: during her waking hours Genie remained tied to a potty-chair, and at night she slept in a crib with a mesh top. She was fed only baby food, other family members were forbidden to speak to her (though her brother and father reportedly barked at her like dogs), and because of her father's intolerance for sound Genie was beaten for making any sounds at all. Thus, she was deprived of both language input and the experience of producing vocalizations herself.

When Genie was removed from these horrific circumstances at age 13½ and taken into custody of the state, she spent several years in the care of a team of psychologists, linguists, and education specialists who attempted to rehabilitate her and also study her late development of language, cognition, and social abilities. While Genie's social and emotional development is fascinating, here we will focus on her language development in the years of her rehabilitation.

At the time of her discovery, there was no evidence that Genie had any language ability whatsoever. She neither spoke nor responded reliably to speech input. Part of her difficulty with producing speech had to do with the fact that she had been fed only baby food (mush) for more than 13 years, so the musculature of her mouth, tongue, and throat had not been able to develop normally (due to the lack of chewing and swallowing solid food). Another difficulty had to do with the fact that since she had been punished for making sounds, she had learned to suppress voluntary sound production. Sound production in speech requires a finely coordinated regulation of airflow from the lungs through the vocal tract as well as the vibration of that sound through the vocal folds. Apparently, Genie's lack of practice producing sound for all those years made it difficult for her to learn to coordinate the necessary bodily systems to produce normal-sounding speech.

But production of speech was clearly not Genie's only problem, since she did not appear to comprehend language either. Thus, researchers attempted to "teach" her language during this time. How did this process unfold? The general picture is that Genie was able to acquire *some* aspects of language, but her trajectory followed a nontypical path and the ultimate level she attained fell far short of full grammatical knowledge.

One aspect of language that Genie seemed to latch onto quickly was vocabulary. According to Susan Curtiss (personal communication, August 2001), one of the linguists working with Genie, the girl was a voracious learner of lexical items, picking up words rapidly and seeking words for all of the new things she was encountering in her expanding world. In fact, Curtiss tells a story about entering a store with Genie where spools of thread were sold. On the thread display there were dozens of different colors and shades, and Genie wanted to learn the name for each color. Of course, English does not have basic color terms for all of these different shades and hues, although new color terms can be created either through modification (orangey-yellow, very dark red) or analogy (sea green, sky blue). According to Curtiss, Genie was disappointed in these substitutes for brand-new color labels.

Despite Genie's almost insatiable appetite for new words, her lexical development was unlike that of typical children in a number of ways. As we saw in chapter 5, children typically start out by acquiring nouns, which label concrete objects, people, and animals, and a few basic verbs (along with performative words like "bye-bye"). Adjectives, in particular color words, tend to be acquired much later. But for Genie, a large number of her early lexical items consisted of color words, numbers words, and verbal expressions like "stop it" and "spit" (Curtiss et al., 1974).

Genie's phonological development, like her lexical development, exhibited some unusual patterns. For one thing, she did not produce the CVCV reduplicated forms that are so common in typical children's earliest words (*wawa* 'water', *baba* 'bottle', *mama*); instead, she reproduced the adultlike syllable structure but altered or omitted certain sounds (e.g., [ræbɪ] 'rabbit'). Another deviation from typical phonological development can be seen in some of her sound substitutions. While typical children substitute sounds for one another that share certain features (e.g., an alveolar stop is substituted for a velar stop in fronting, or a labial stop is substituted for a labial fricative in stopping; see chapter 4), some of Genie's substitutions did not appear to share any features in common. For example, she sometimes substituted [k] for /s/, as in her pronunciation of *slapping* as [klæpĩ] (Curtiss, 1977, p. 81).

On the other hand, there were aspects of Genie's phonological development that pointed to an underlying phonological *system* as opposed to either simple mimicking of sounds or random sound productions. First, Genie appeared to have acquired a rule for vowel nasalization conditioned by nasal consonants. To illustrate this rule, consider the word *funny*, which is pronounced [fʌ̃ni]. In English, a vowel before a nasal consonant becomes nasalized. Genie sometimes pronounced it this way. She also pronounced the word *can* sometimes as [kæ̃n] and sometimes as [kæ̃]. With her pronunciation [kæ̃], we don't know whether she thought that nasalization is a part of the underlying representation of the vowel (i.e., maybe the /æ/ vowel is really /æ̃/). But crucially, when she substituted [t] for /n/ in *funny*, she pronounced it [fʌti], not [fʌ̃ti], indicating that her underlying representation of the vowel did not inherently contain the nasal feature.

A second aspect of Genie's phonology that points to an identifiable underlying representation comes from her production of *s*-clusters. When she began producing words starting with *s*-clusters ([sk], [st], [sp]), she sometimes deleted the /s/ and sometimes epenthesized (inserted) a schwa [ə]. For example, she pronounced the word *stove* sometimes as [to] and sometimes as [səto] and the word *spoon* was sometimes [pu] or [pũ] and sometimes [səpu]. Other consonant clusters were sometimes produced with epenthesis, such as *crayon* [krej]/[kərej] (Curtiss et al., 1974). We saw in chapter 4 that typically developing children will often remove an *s*-cluster by deleting the [s], but we didn't see many examples of breaking up a consonant cluster by epenthesis (inserting a vowel). However, this exact process is found in various adult languages, such as Arabic. Thus, Genie here demonstrates the use of a phonological strategy for producing the least marked syllable (CV); it is not used much in the language of her environment, English, but is unquestionably part of the stock of tools human language provides for achieving certain preferred sound patterns.

In terms of her syntactic development, after about a year or two of exposure to language, Genie was eventually able to string words together to make phrases and sentences, and she generally adhered to the argument structure requirements of verbs. In other words, she generally used intransitive verbs with only one argument (1a), transitive verbs with two arguments (1b), and ditransitive verbs with three arguments (1c).

(1) a. Genie laughing.
    b. Genie love Curtiss.
    c. Grandma gave me cereal.

As her sentences became longer and more complex, many were semantically transparent. Some examples are given in 2.

(2) a. Bears have sharp claw.
    b. In hospital, shot hurt arm.
    c. Want go shopping.
    d. I like wheelchair in hospital.
    e. Willie slap my face.

But as is evident from the utterances in 1 and 2, Genie's syntax was not like that of typical, fluent language users. She used very few function words, the possessor *my* and the determiner *another* being some of the only ones, and negation was only expressed externally (i.e., outside of the clause, typically clause initial), as in 3.

(3) a. No more have.
    b. No like hospital.
    c. Not have orange record.

Such expressions are reminiscent of typically developing children's telegraphic speech (see chapter 7). But Genie's language differed from that of typical toddlers in a couple of ways. In one direction, some of her expressions were more complex than those of a typical 2-year-old, in that she would sometimes include multiple possessors, as in *Valerie mother coat* ('Valerie's mother's coat') or multiple adjectives (*Little white clear box*). In the other direction, Genie's language differed from typical toddlers who, by the time they can produce sentences containing five or six words, produce function words (articles, auxiliary verbs) at least *some* of the time. Genie never produced them.

Finally, Genie differed from typical language learners in that she plateaued at this telegraphic stage and never became fully fluent in language. Sadly, because she had been kept from language for so long, Genie was not able to acquire the full system of grammar.

8.1.3 Chelsea

Children who are deprived of language due to almost complete isolation from normal human contact, like Genie and the other children profiled above, can teach us something about the parameters of the timing of language acquisition. If a first language is not acquired

Sidebar 8.1: What Genie Tells Us about the Brain

The team of researchers who tried to study and rehabilitate Genie were interested to know which parts of her brain she was using to process language. For most right-handed people, the main language centers (Broca's and Wernicke's areas) are lateralized to the left hemisphere, so that language is mainly processed in the left brain, while ambient sounds tend to be processed more using right-hemisphere brain structures.[1] Genie was right-handed, but did she process language in her brain the same way as people who had grown up with language?

They studied this by performing a dichotic listening experiment. In a dichotic listening experiment, the subject is given two different auditory inputs, one in each ear, at the same time, and they are asked to point to the picture corresponding to one of the inputs. For example, you might hear the instructions (in both ears) "Point to the ___" followed by the word "car" in your left ear and the word "house" in your right ear. Our brains have *contralateral connections* to our bodies, which means that the left side of the body is processed by the right side of the brain, and vice versa. Simplifying somewhat, the left ear is connected to the right hemisphere, and the right ear is connected to the left hemisphere. Since language is typically processed in the left hemisphere, people in this experiment will typically point to the picture of what they hear in the right ear, which is the house in this example (right-ear input → left-hemisphere processing).

On the other hand, if instead of a word you hear a different environmental sound in each ear, for example, a piano chord in the left ear and a train whistle in the right ear, since this type of sound is processed mainly in the right hemisphere, you will point to the picture of what you heard in your left ear, the piano in this example (left-ear input → right-hemisphere processing).

Genie, in contrast, exhibited a different pattern of responses. She responded by pointing to the picture matching the word or the sound she heard in her *left* ear. That is, she responded to words the same way she (and typical right-handers) respond to ambient sounds. It was as if she was processing all sound, including language, through her right hemisphere. The researchers surmised that because she had been prevented from acquiring language during the critical period, the brain structures that would normally be used for language processing had either atrophied or been repurposed.

during early childhood, it does not appear to be learned normally at a later age.[2] However, as fascinating as these cases are, from both a humanistic and a scientific perspective, they present a number of confounding factors: these children also lacked emotional support and the experience of human touch, and it is difficult to discern whether there may have been cognitive impairments present from birth, independent of language deprivation. We know that touch is critical for normal cognitive development, as children who are deprived of touch (e.g., children raised in orphanages where the staff do not hold them) exhibit cognitive and emotional disorders (Ardiel and Rankin, 2010). Could the lack of human touch and interaction, and the emotional trauma that causes, itself impede language learning when language becomes present only later in life? To partially address this question, let's look at one case of an individual whose profound deafness prevented her from acquiring spoken language.

Curtiss (1988) interviewed a woman who was born deaf and did not receive hearing aids until she was 32 years old. This woman, called Chelsea, lived with her family in a rural area of northern California where there was no school for the Deaf[3] or other means by which her family could provide her with sign language. Although doctors misdiagnosed Chelsea as being intellectually disabled, Chelsea's family knew she was deaf and cared for her as best they could, using gestures and "home signs." Finally, at age 32, Chelsea was given a hearing aid and began to acquire oral language. Her language development followed a bizarre trajectory: while she was able to use words, it was unclear whether she associated these words with anything like a grammatical category (noun, verb), her word combinations seemed haphazard, and many were completely devoid of meaning. Some examples of her utterances are given in 4.

(4) a. The small a the hat.
    b. I Wanda be drive come.
    c. The they.
    d. The woman is bus the going.
    e. Daddy are be were to the work.

Chelsea was able to learn vocabulary words but was unable to acquire any rules for combining them into coherent phrases and sentences. In her comprehension, as well as in her production, Chelsea's performance revealed a lack of understanding of English grammatical structure. These observations help us understand that while a loving and emotionally nurturing environment is extremely important for many aspects of human development, exposure to language itself is absolutely critical early in life for language to develop normally.

8.2 Language Acquisition in Deaf Children

Having talked about how spoken language develops with exposure from early infancy (chapters 3–7) and with delayed exposure (section 8.1), we turn now to the acquisition of language in the signed modality. That is, here we discuss language development in deaf children.

What causes deafness? Deafness can be congenital (as are about half of childhood cases of severe hearing impairment; Morton, 1991) or can be acquired, usually caused by middle ear infections. Deafness can also be the result of injury or trauma to the ear. In the past, adults sometimes punished children for "bad" behavior by boxing their ears. This form of punishment involved striking children on the ears with one's fist; with enough force, the eardrum can rupture. This is why the German composer Ludwig van Beethoven lost his hearing: he had his ears boxed too many times, weakening his hearing, and eventually he became completely deaf. Thankfully this is no longer a common practice.

Hearing impairment is defined as the inability to hear sounds less than 60 decibels, which is roughly equivalent to the loudness of a baby's cry. **Functional deafness** is defined as the inability to hear sounds less than 90 decibels. If you've ever heard a baby cry, you will quickly realize why children born deaf do not acquire spoken language spontaneously—the vast majority, or perhaps all, of language sound is simply unavailable to them.

Only around 5 percent of deaf babies are born into families in which a parent is deaf (Mitchell and Karchmer, 2004). Since most hearing people do not sign fluently or know even rudimentary sign language, the vast majority of deaf babies will not be exposed to language unless their parents find a way to expose their children to American Sign Language (ASL) or give them technology that makes sound accessible to them. Hearing technology is a controversial topic we will address in section 8.2.3. In section 8.2.1 we will focus on how deaf children acquire ASL when it is available during infancy. In section 8.2.2 we consider *late* acquisition of ASL, and in section 8.2.3 we turn to the acquisition of oral language in deaf individuals.

8.2.1 Acquisition of Sign Language in Deaf Children

ASL, like other sign languages (e.g., British Sign Language, Nicaraguan Sign Language, French Sign Language, Chinese Sign Language), is a natural human language with the same types of grammatical structures and constraints we see in any human language. Importantly, ASL is not a system of mimed gestures or a manual version of English. Although signs can have a greater degree of iconicity than spoken words (e.g., the hand shape or movement of the sign may resemble the relevant object or action it signifies), ASL is a symbolic system in which there is often an arbitrary relationship between form and meaning, just like in spoken language. Importantly, it has its own syntax, morphology, lexicon, and phonology.

What are signs in sign language? Signs are composed of **sublexical features** the same way that words in spoken language are composed of individual sounds that have phonetic features (place and manner of articulation). The sublexical features of signs involve the **hand shape**, **movement**, and **place of articulation** of each sign. These distinct features allow us to identify minimal triples in sign language, analogous to the minimal pairs we find in spoken language. Some minimal triples are illustrated in figure 8.1.

Each of these sets of signs forms a minimal triple because they share two of the three features and differ only in the third. For example, CANDY, APPLE, and JEALOUS all share the same place of articulation and movement, but they differ in hand shape.

In addition to these features, signs can be modified through **inflection**. Inflection in spoken language generally involves affixation or tonal changes. In sign language, instead, inflection involves a modification of one of the sublexical features (e.g., a sign's *movement* feature might be modified). ASL has an extremely rich morphological system in which verbs can be inflected to mark temporal aspects of events—like whether an event happened continuously, habitually, repeatedly, for a long time—and more. These inflections involve the verb's sign being modified in a particular way, such as by changing one of the three phonological features mentioned above (movement, hand shape, place of articulation). Some verb signs are also articulated using different locations in space to convey subject and/or object agreement (more on this below). Sign languages also make use of **nonmanual** gestures, which can serve as grammatical markers for yes-no questions, wh-questions, markers of discourse topics, and negation, among other things.

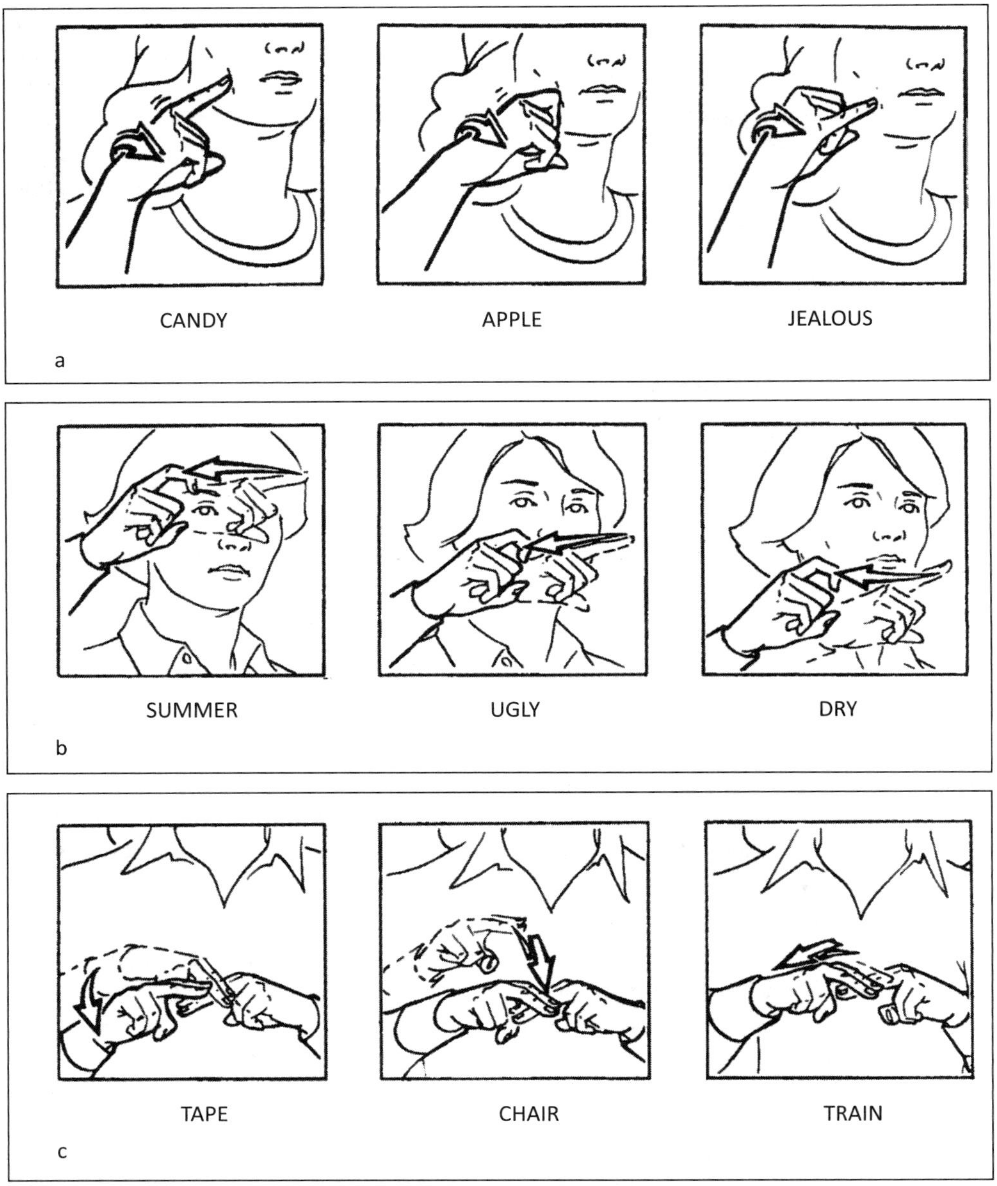

Figure 8.1
Contrastive features in ASL. *a*, Signs contrast in hand shape. *b*, Signs contrast in place of articulation (location). *c*, Signs contrast in movement. Image reprinted with permission from Poizner, Howard, Edward Klima, and Ursula Bellugi, *What the Hands Reveal about the Brain* (Cambridge, MA: MIT Press, 1987), p. 4, fig. 1.1.

Sidebar 8.2: Sign versus Spoken Language

Sign language is *formally equivalent* to spoken language, meaning it contains the same principles and hierarchical structures and the same kinds of features and mechanisms to combine words to convey meaning. But it operates through a different *modality*, namely, via bodily gestures and movements rather than vocal sounds. What are some of the differences that might arise from the different modalities? Do you expect to find a difference in the timing of the *onset* of language production because of the difference in modality? If so, in which modality do you expect to find first words earlier? Why do you expect this?

One possible difference between the signed and spoken modalities is *iconicity*. Though this is not true of all (or even most) signs, some nominal signs may, by their hand shape, resemble the objects they denote (e.g., the ASL sign for TREE involves the forearm in a vertical alignment with the fingers spread); some verbal signs can be reminiscent of an action by the way the hand moves through space. Some signs, such as the ASL signs for YOU and ME are very iconic—the sign for ME is made by the speaker pointing to him- or herself, and the sign for YOU is made by the speaker pointing to the person they're talking to. Before reading ahead, make a prediction about whether you expect sign or spoken language to be developed earlier in terms of

(a) word production

(b) word comprehension

(c) pronouns

Explain your predictions!

When deaf babies are exposed to sign language from early infancy, they acquire sign language in much the same way that hearing children acquire spoken language (Petitto, 2000). Around age 6–12 months they go through a stage of manual babbling that bears formal similarities to hearing babies' vocal babbling. Just as canonical babbling in hearing babies typically consists of repeated syllable units, deaf babies produce signs with a particular hand configuration and movement in a repetitive, cyclic fashion.

Just as ASL-acquiring infants are like hearing infants in going through a babbling stage (but using their hands), they begin to acquire first words on or around their first birthday just as hearing babies do. There is some evidence that deaf babies begin to produce their first word signs slightly earlier than hearing babies produce their first oral words, but only by about a month. One could suppose that this reflects an advantage of the iconicity of manual signs over spoken words, but there is evidence that children's early signs are *not* more iconic than adult signs. An alternative explanation is that the motor skills required to move the shoulder, elbow, and wrist develop earlier than the motor skills needed to move the vocal articulators to create discernable speech sounds (Meier et al., 2008). Researchers characterize this difference in terms of a spoken language "disadvantage" rather than a sign language "advantage."

In terms of developing their phonological systems, the earliest acquired feature is place of articulation (also called location): at about 1 year of age ASL-signing children produce the location component correctly around 75% of the time. The movement component is somewhat more challenging, being produced correctly at this age around

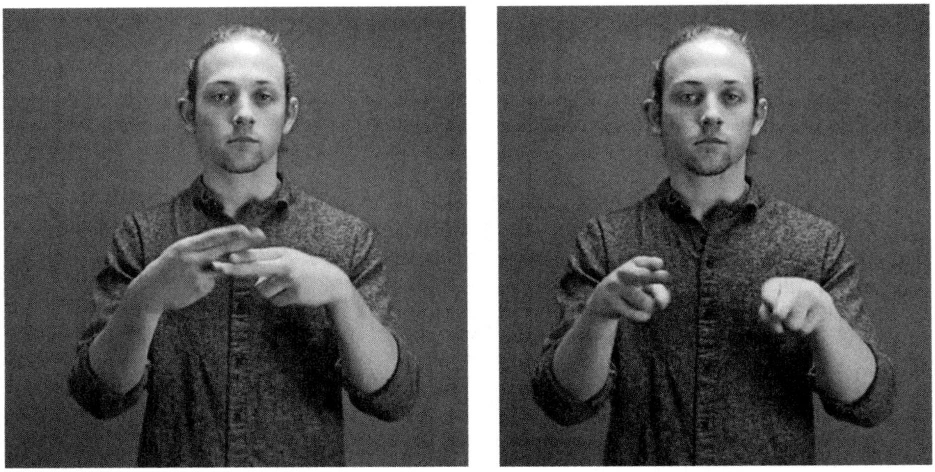

Figure 8.2
The ASL sign for the word EGG. A child might make the error of extending four fingers instead of two. Images from ASL Signbank (https://aslsignbank.haskins.yale.edu/).

50–60% of the time. Hand shape is the most difficult for young signers; it is produced correctly only about 25–40% of the time, though one study found a much higher rate of correct production (Chen Pichler, 2012). Importantly, children's errors in producing these features are systematic and regular, just like the phonological errors of hearing children (see chapter 4). An example of a hand shape error would be producing the sign for EGG (two fingers extended in the adult form) with four fingers extended (see figure 8.2; to see a video of an adult and a child signing this word, visit https://www.youtube.com/watch?v=1ld-_jscH2s).

The order of acquisition of these phonological features (location < movement < hand shape) is consistent with a markedness hierarchy, just as we have seen for oral phonological development (see chapter 4). That is, location is a less marked feature than hand shape.

When ASL-acquiring children begin combining word signs into sentences, they produce appropriate word orders. Early research seemed to expose a preference for SVO orders on the part of young signers (Hoffmeister, 1978), but more recent work has revealed that children produce various word orders, all grammatical in adult ASL. Although ASL has SVO as its default word order, other word orders are used in certain circumstances. For example, in a construction called **focus doubling**, the word that is focused for emphasis is expressed once in its typical position in the sentence and again at the end of the sentence, as in 5 (example is from Lillo-Martin and De Quadros, 2005).

(5) JOHN CAN READ CAN
'John really <u>can</u> read.'

Children acquiring ASL natively produce both default (SVO) and alternative word orders in appropriate contexts.

Agreement morphology in ASL employs something known as **spatial syntax**. This means that a signer will use a location in his or her **signing space** (roughly the semicircle of space that extends immediately in front of the signer's torso) as a referent for some entity they are talking about (e.g., a person or thing). When the signer needs to refer to that entity again, they will articulate a verbal sign as originating or ending in that location to indicate that the entity is either the agent or the patient of the verb's action, respectively. Young children acquiring ASL reportedly have difficulty doing this, and they will often instead articulate only the citation form of a sign (i.e., produce an uninflected sign), and they make these omission errors until about age 3;6 (Meier, 1982). Other researchers have found that children sometimes stack referents, using the same location in space repeatedly as a referent, even though the meaning the child is trying to convey would require multiple locations for distinct referents. Such errors can persist even as late as age 4 (Loew, 1984). On the other hand, more recent studies have found earlier correct production of agreement using spatial syntax; one study found that children may use eye gaze rather than pointing/signing to indicate agreement, and do so as early as age 2 (De Quadros and Lillo-Martin, 2007). It is unclear what accounts for these different results, but clearly more research is needed into the acquisition of agreement in native ASL.

Aspects of sign language that appear to be acquired somewhat later include the combination of nonmanual with manual gestures (e.g., furrowed brow together with a wh-word like WHAT), certain types of classifiers that relate to the shape of an object, or the shape one's hand would have when handling that object (Schick, 1990).

We have seen that there are numerous parallels between signed and spoken language in terms of their development in early childhood. The following summarizes some of the parallels we see between deaf native signers and hearing children:

(6) a. parallel developmental milestones
babbling, first words, word combinations
b. appropriate word orders
correct default word order with context-appropriate alternative orders
c. systematic mispronunciation
For example, the word EGG might be produced with four fingers extended (hand shape error)
d. omission of inflection
For example, a child might use the same location in their signing space to indicate different referents; hearing children sometimes omit the necessary subject-verb agreement morphology for their language.

Another parallel we can observe between children acquiring signed and spoken language involves **pronoun reversals**. Before proceeding, think back to the prediction you made in sidebar 8.2 about the acquisition of the pronouns YOU and ME in ASL (which involve pointing) compared to the acquisition of these pronouns in spoken language. Did you expect an advantage for the highly iconic signs?

It turns out that, once again, hearing and deaf children acquire these pronoun forms in much the same way, including making the same kinds of errors. A longitudinal study of two deaf ASL-acquiring children by Petitto (1987) revealed the following trajectory. The children Petitto studied sometimes pointed to people and objects in their environment, including the people they were interacting with, by about 10–12 months (the same age that hearing children begin pointing as a nonlinguistic gesture). Between about 12–18 months of age, the children avoided pointing to people and objects around them and, instead, referred to these entities with a lexical label (i.e., a word sign) (the same age hearing children start labeling familiar objects with words). At age 21–23 months, an interesting shift occurred: the children resumed pointing to people and objects, including themselves, but they appeared to confuse the meanings of the signs YOU and ME. That is, they produced (and interpreted) pointing to themselves to mean "you," and they produced (and interpreted) pointing to the conversation partner to mean "me." By 27 months of age, the children had corrected their error. Hearing children have also been reported to make these pronoun reversals, using the word *me* to mean their partner and the word *you* to mean themselves, at the same stage of development (Charney, 1978).

All of these examples of parallels in the native acquisition of sign and spoken language are important because they underscore the unavoidable conclusion that *language is language*. Humans, in the early years of life, will acquire a grammatical system when they are presented with accessible language input, irrespective of modality.

8.2.2 Late Acquisition of ASL

Having established some of the main patterns of acquiring ASL natively, let's return to the question of the late acquisition of a first language. At the end of section 8.1 we saw that for the woman known as Chelsea, although she received access to spoken language input at age 32, she was not able to acquire grammar. Chelsea could learn vocabulary words, but she was not able to combine those words according to the kinds of rules languages use to build hierarchical structures. What about late exposure to sign language?

In a study by Newport (1991), adult signers who had learned ASL at different ages were tested for their grammatical knowledge of ASL. These subjects had been using ASL on a daily basis for at least 30 years at the time of testing. And yet, while all subjects produced sentences with grammatical word order (all of them used SVO order), there was enormous variation in terms of their morphological ability. Only the native signers made no errors in ASL morphology. Signers who began learning ASL in middle childhood (around age 4–6 years) at a school for the Deaf made some errors in morphology. Signers who had not learned ASL until after puberty made a large number of errors in their ASL morphology. Thus, late exposure to either spoken or signed language, even in a loving family environment, is not enough to enable someone to learn a first language with native-like fluency late in life.

Further evidence comes from a unique situation that arose in Nicaragua in the 1980s, which enabled researchers to study and track the development of sign language by people at different ages. In 1977 the first school for the Deaf opened in that country's capital, Managua, and students of various ages began attending the school. The teachers at the school were instructed to teach the students oral Spanish, but many of the children attending the school used something called *Mimicas*, "home signs" or gestures developed in isolation. The deaf children at this school used their home signs to communicate with one another on the playground and on the bus (they were supposed to use oral Spanish in the classroom), and soon their shared home signs evolved into a kind of pidgin sign language, called LSN (Lenguaje de Señas Nicaragüense). And within a short time the younger children at the school (mostly younger than age 7) began to develop a full-fledged sign language, known as ISN (Idioma de Señas Nicaragüense). ISN differs from LSN in having greater morphological complexity and more arguments per verb.

Researchers who have studied the evolution of this language have noticed that both *age of entry* into the school and *year of entry* matter greatly for the complexity of the language a student ended up developing. Students who entered the school in 1983 or *later* developed a more articulated language than students who entered the school *before* 1983. Moreover, students who were *younger* (up to about age 10) when they entered the school developed a more articulated language than students who were *older* when they started school (ages 10 to 27 years). Even more stunningly, the two factors interacted, such that for older signers, year of entry mattered much less than for younger signers, and year of entry did not matter at all for these older signers in terms of their morphological development. That is, while the younger students were able to take advantage of the enriched language input that had evolved by 1983, older students were not: they struggled to acquire morphological structure even if the input contained a morphologically richer language (Senghas, 1995; Kegl et al., 1999; Senghas et al., 2004).

It seems the ability to acquire a first language begins to diminish sometime in middle childhood (perhaps between ages 4 and 7 years) and gradually dissolves in adulthood into nothing more than an ability to repeat words, without any organizing principles. Interestingly, late acquisition of a *first* language consistently results in poor (though not nonexistent) abilities to produce and understand language. When a first language has been acquired, however, whether spoken or signed, facility in acquiring a second language (again, whether spoken or signed) is significantly increased (Mayberry, 2010).

8.2.3 Acquisition of Oral Language in Deaf Children

While some deaf children acquire sign language, many others instead acquire oral language. This is partly due to the fact that 95% of deaf children are born into families in which both parents have typical hearing ability, and it is often most straightforward for these parents to give their deaf child access to oral language. The question of whether to raise a deaf child with sign or oral language is complex and deeply personal. One reason for pursuing oral language is that parents want to allow their child to communicate with all members of the family, many of whom likely do not know sign language. Another consideration has to do with resources: Is there a school for the Deaf nearby where the child could learn sign language? And would the parents themselves have the resources to learn it? Many parents also feel that the best way to give their child opportunities to succeed in mainstream (hearing, oral) society is through oral language. On the other hand, language outcomes in oral language for children using hearing technology are quite variable: while some children catch up to their peers, others do not fare as well.

Historically, instruction in oral language for deaf children involved the use of sound amplification (e.g., through hearing aids) and an emphasis on lip reading, and outcomes were often quite poor. Children forced to acquire oral language without hearing technology face many difficulties and barriers, including isolation and restricted opportunities in life. Although pedagogical techniques have been developed over the years for teaching deaf children speech through touching the mouth and throat during vocalization of different sounds, many contrasts can remain impossible for deaf children to perceive. For example, high-frequency fricatives, known as sibilants ([s], [z], [ʃ], [ʒ], for example) are often not possible to perceive. Voicing can be felt by touching the front of the throat (see chapter 3), but many phonetic features, such as nasalization, cannot be. Relatedly, while many deaf individuals become remarkably skilled at lip reading, there are many sounds and distinctions that cannot be ascertained from visual cues.

Because English literacy is based on the language *English*, deaf children forced to learn oral English without the benefit of hearing technology suffer greatly in terms of literacy. Most deaf individuals do not surpass a third-grade reading level (Ratner, 2001). To the extent that they do learn to read and write English, they often make grammatical errors, as seen in 7 (from Ratner, 2001).

(7) a. The cat under the table.
 b. Tom has pushing the wagon.
 c. Beth made candy no.
 d. Who the baby did love?

In recent years, hearing aid technology has improved from what was available generations ago, and a different type of technology, **cochlear implants** (CI), is now widely avail-

Sidebar 8.3: Lips as Cues to Speech

Which phonological features would not be determinable from the visual cues used in lip reading? Can you think of any pairs of words in English that would *look* the same when they were pronounced?

What other sorts of cues do you think could be used by someone reading lips in order to understand what someone was saying?

able. However, the use of CI remains controversial. Unlike hearing aids, which amplify the sound that is received by the natural, biological structures involved in hearing, a CI bypasses the structures within the ear that are responsible for hearing and stimulates the auditory nerve directly with an electronic signal. In the surgical procedure required for a CI, the cochlea itself is damaged and can be destroyed, resulting in a loss of any residual hearing ability—though recent improvements in CI technology may mitigate this damage. Moreover, there is significant variability in outcomes (for example, the ability to judge sound direction or distance for adults with CI), and more research is needed to understand the effectiveness of CI on language development, including reading outcomes (Marschark et al., 2010).

We mentioned above that many families choose to raise their deaf child with oral language but that the decision to do so is complex. In families of deaf babies in which other family members are deaf, there can be a feeling of betrayal if the parents opt for hearing technology, in particular CI, which is an irreversible procedure (as noted above, the cochlea can be destroyed in the implantation). Many members of the Deaf community feel that providing hearing technology to deaf children risks weakening the community because these children may choose not to associate with or support this community. As it is, the Deaf community struggles to be visible to general society and to have their language recognized for what it is—a full-fledged human language—and there is concern that fewer numbers in the community may, over time, further weaken their status and their chances for recognition and visibility. Related to this, there is a fear that sign languages may become (further) endangered and go extinct. On a deeper level, many deaf individuals do not see their deafness as a disability and take great pride in their deafness and in belonging to their community. From this perspective, it can be taken as a personal affront if a family member opts to give their child hearing technology. For an in-depth consideration of this controversy, we refer the interested reader to Paludneviciene and Leigh (2011).

It's important to remember that there is no need, in principle, to choose one route over the other—children can become bilingual in both sign and spoken language. We address the topic of bilingualism in chapter 9. For now let us mention some advantages of raising a deaf child as a bimodal bilingual speaker (*bimodal* because sign and spoken languages use different modalities). One is that knowledge of sign language can be helpful when hearing is impeded due to noisy conditions or multiple speakers, or when a device is turned off or out of battery power. Another is that having sign language as a first language can aid later learning of oral language. That is, there is evidence that acquisition of oral language as a second language (L2) in later childhood is significantly more successful than acquisition of oral language as a late first language (Mayberry, 2010). Finally, bimodal bilingual people will also have greater potential to be **bicultural**; that is, they will be able to relate to, and interact with, members of both the hearing and Deaf communities. This can be especially important for people with family members who are deaf, but for any individual who is deaf, connection to the larger Deaf community can be an important affirmation of their personal identity.

8.2.4 Summary

We have seen in these three subsections that when deaf children are provided with ASL input from birth, they acquire ASL in much the same way that hearing children acquire their native spoken language. We have also seen that if ASL input is not available from birth and hearing technologies are not provided, children struggle to learn ASL later in life: words and word order can be learned, but morphology presents significant challenges. When hearing technologies are provided, spoken language development can proceed without problems, but there is great variability in outcomes. And in the absence of good quality hearing technology, oral language learning is labored and difficult for deaf children, and literacy remains a perpetual barrier.

8.3 Language Acquisition in Blind Children

It is readily apparent how access to spoken language is impeded for children born without the ability to hear. But if spoken language primarily involves sound-based transmission, should children without access to sight be at any disadvantage for acquiring language?

Before we assume that nothing about language is lost without visual access, consider for a moment how much we use visual cues and information to figure out what someone is talking about. Not only gestures (e.g., pointing) but the eye gaze of our interlocutor, actions and events we can witness, and various entities in our environment that might not be perceivable without vision (e.g., objects at a distance) can supply important information that we rely on when inferring the meaning of someone's utterance. For example, close your eyes and imagine that your friend says to you, "The flim is glipping the zertok" or "That kurbim is so wusk!" How would you begin to figure out what a *flim*, *zertok*, or *kurbim* was, what kind of action *glipping* was, or what kind of quality *wusk* denoted?

8.3.1 Lexical and Grammatical Development

In fact, the path of language development for blind children is not terribly different from that of sighted children. Studies of the early lexicons of blind and sighted children find minor differences: sighted children have words for outside objects like *moon* and *flag*, which are presumably perceived by sight rather than touch, while blind children use a greater proportion of proper names (including *Mommy*) than common nouns. But there is much overlap in the makeup of their lexicons: in particular words for food, toys, some animals, clothing articles, and objects found in the home (Bigelow, 1987). The most striking difference between the two groups of children in their lexical composition concerns function words: the blind children studied by Bigelow used no function words (e.g., wh-words), while the sighted children in the comparison group did (Nelson, 1973).

Landau and Gleitman (1985) also found a difference in the use of function words in their study of the grammatical development of blind children compared to sighted children. In particular, Landau and Gleitman found that blind children used fewer auxiliary verbs (*be*, *have*, *do*) than sighted children. The lack of function words in the vocabularies of blind children is curious. It's not the case that sighted children can see the referents of words like *what* and *is*, as these words have no referents. Why should blind children be at a disadvantage in acquiring them?

Landau and Gleitman suggest an intriguing explanation. They point out that in a previous study of sighted children (Newport, Gleitman, and Gleitman, 1977) a positive correlation had been found between parents' use of *yes-no* questions to their children and the children's acquisition of auxiliary verbs. That is, sighted children of parents who produced more *yes-no* questions acquired auxiliary verbs earlier, and sighted children of parents who produced fewer *yes-no* questions acquired auxiliary verbs later. Why should this be? Recall from chapter 7 that in a *yes-no* question in English, the auxiliary verb gets moved to the beginning of the sentence (and if there is no auxiliary verb, *do* gets inserted and is moved to the beginning of the sentence). The beginning of a sentence is particularly salient, so children who hear a lot of *yes-no* questions in their input are likely to notice these auxiliary verbs and acquire them earlier.

Returning to the case of blind children, Landau and Gleitman looked at the speech of the blind children's parents and found that they used relatively few *yes-no* questions. Why? It is not uncommon for parents to ask their (sighted) children rhetorical questions, such as "Is that a doggie?" when looking at a picture of a dog. But it may not be reasonable to ask a blind child such questions, since they are not likely to have ready answers. The same could extend to wh-questions, which could explain Bigelow's finding that blind children did not produce any instances of wh-words or expressions (*What's that?*). Most likely their parents did not use these expressions much in speech to their blind children.

8.3.2 Acquisition of Perception Verbs

Landau and Gleitman's study looked at language development in three blind children, but one of the three children, a girl named Kelli, served as the subject of an in-depth longitudinal study. Landau and Gleitman were interested in not only general facets of blind children's grammatical development but in particular whether blind children could learn the meanings of words related to visual perception, such as the verbs *look* and *see*—and if so, how this came about.

First, Landau and Gleitman asked whether Kelli used the verbs *look* and *see* spontaneously and appropriately. A list of some of Kelli's spontaneous productions are given in 8, showing that she does use these words in normal conversation. Kelli produced all of these spontaneous utterances before she was 5 years old, and most before she was 3;6. Do her uses seem appropriate or inappropriate?

(8) a. I'm gonna come over and see Tabatha.
  b. See? It's in my lap.
  c. Look what I have!
  d. Here, Scooter, look at the harmonica.
  e. Don't see that, Sommer!

Next, Landau and Gleitman investigated Kelli's interpretation of the verb *look* by asking her to "look" at objects in different ways ("look real hard/gently" or "look with your foot"). Kelli responded to these commands by touching the objects in different manners or with different body parts. For Kelli, "looking" at something meant exploring it by touch.

But did this mean that Kelli thought that *look* and *touch* meant the same thing? Landau and Gleitman answered this question by pitting the two verbs against each other. They would tell Kelli, for example, "You can touch the table but don't look at it," to which Kelli would respond by tapping the table with her hand. Then, the researchers would tell her, "Now you can look at it," at which point Kelli's response changed from just a light tap to a more thorough haptic exploration of the table. Thus, even though Kelli's interpretation of the verb *look* necessarily involved touching, she understood a subtle difference in the lexical meanings of the words *look* and *touch*.

How was this possible? Their first hypothesis was that Kelli's mother used the verb *look* when the object in question was near Kelli or in her hand, but not when the object was far away. This is a reasonable hypothesis given that for Kelli, looking involved touching. However, when Landau and Gleitman tabulated the number of times Kelli's mother used various verbs when the object was near Kelli versus far from her, their hypothesis was not confirmed. Instead, the verbs Kelli's mother used most frequently when the object was near or in Kelli's hand were *put*, *give*, and *hold*. But Kelli didn't make the mistake of thinking that these verbs meant 'look' or 'see'. What, then, enabled Kelli to figure out that *look* and *see*, but not other verbs, meant something about perception?

Landau and Gleitman next considered the kinds of sentence frames in which perceptual and nonperceptual verbs were used by Kelli's mother. Recall from chapter 5 (section 5.4) that a sentence frame is the constellation of arguments (normally noun phrases, NPs) that are required by a given verb to satisfy its semantics: intransitive verbs take only one NP argument (a subject), transitive verbs take two NP arguments (subject and object), and so forth. Speaking more generally, a sentence frame can also include other types of phrases, like prepositional phrases (PPs) and clauses, that typically occur with a given verb. What is interesting about perceptual verbs is that many of them can occur with either a direct object NP (*I see the dog*) or PP (*Look at the tree*) or a whole clause (*I saw that you left a message*; *John heard that Mary was the winner*). But the verbs *look* and *see*, in particular, also permit a range of other uses that are not so transparently linked to visual perception. For example, when someone says, "That rock looks heavy," they are using the verb *look* to mean something like 'appear' rather than 'observe'. The verb *see*, in addition to its meaning relating to visual perception, can also mean something like 'find out', as in "Let's see if Grandma's home," in which the embedded question can be answered by a phone call rather than by visual information. It can also mean something like 'understand', as in "I see what you mean," in which there is no visual meaning.

After segregating the uses of *look* and *see* with a perceptual as opposed to a nonperceptual meaning and examining the sentence frames of these verbs when used to mean direct perception, Landau and Gleitman discovered a different distribution of these verbs with respect to whether the object the mother was talking about was near Kelli. In fact, by isolating the cases of *look* and *see* that had a perceptual meaning from their other uses and lining those up with cases in which the object was near Kelli or in her hand, they found a much stronger positive correlation.

Once they recognized that Kelli was likely using information about the sentence frames that verbs occurred in, combined with other information from the environment to learn verb meanings, Landau and Gleitman reasoned that sighted children probably employed a similar strategy. Namely, they suspected that all children, whether blind or sighted, restricted verb meanings by the argument structures of their sentences. This was the origin of the syntactic bootstrapping hypothesis (see section 5.4).

8.3.3 Summary

In this section we have seen that, in contrast to children born deaf and not exposed to sign language or given hearing technology, language development for children born blind is not disrupted to any great extent. Although blind children do not have direct visual information about their world, they are able to use information within the language itself (sounds, words, sentences), along with information about the world they obtain through their other senses, to develop language normally. The study of blind children's language development provides a good illustration of why assumptions must be tested empirically (i.e., the assumption that blind children would be at a disadvantage for language learning does not hold up) and also of how investigation of language development in nontypical populations can lead to important insights about the nature of language development more generally (in this case, the syntactic bootstrapping hypothesis).

8.4 Impaired Language Acquisition

We saw in sections 8.1 and 8.2 that language does not develop in the typical manner when language input is unavailable to the learner, either because the input is withheld (as in the case of Genie) or because the child is deaf and there is no sign language input. In this section we will look at situations in which language does not develop typically despite language being fully available in the input. These are cases of language impairment.

8.4.1 Specific Language Impairment

Language is a cognitive function, and when a child is born with a cognitive impairment such as intellectual disability, language can be one of the various affected cognitive functions. For example, both fragile X syndrome and Down syndrome are conditions arising from genetic abnormalities that yield cognitive or intellectual disability, and language is affected (to varying degrees) along with other cognitive functions, such as problem solving, mathematical cognition, and general learning abilities. However, language can be impaired in its development while other cognitive functions develop typically. The condition in which language is selectively impaired is known as **specific language impairment**, or SLI.

The diagnosis of SLI is given when the following criteria are met:

(9) a. a nonverbal IQ score in the normal range[4]
b. normal hearing
c. no diagnosis of autism or autism spectrum disorder (see section 8.4.2)
d. no seizure or other neurological disorders
e. no traumatic brain injury
f. normal motor control
g. normal social and emotional development
h. no language deprivation
i. at least one standard deviation below age norms for general language ability

Language ability is measured in terms of **expressive language** (language production) and **receptive language** (language comprehension). For some children with SLI, both expressive language and receptive language are affected; for other children, either expressive or receptive language is adversely affected, but not both. When only one of these modes of language is impaired, it is more common for expressive language to be impaired and for receptive language abilities to be spared. Language ability is assessed using standardized assessments, such as the Clinical Evaluation of Language Fundamentals (CELF) or the Test of Language Development (TOLD), when these are available for the language being studied, in combination with informal assessments. These assessments are carried out by speech-language pathologists.

Keep in mind that a language deficit is different from a speech deficit, although both types of deficit can yield problems in expressive language. Speech concerns the articulation and production of speech sounds, while language involves issues of representation of and access to underlying language structure. Some children may in fact have both types of deficits, but they need not coincide. However, children with speech deficits are more likely to be referred for evaluation and intervention compared to children without a speech deficit (Rice, Warren, and Betz, 2005); therefore, more children receiving speech-language therapy may in fact suffer specifically from speech deficits or both speech and language deficits.

While much progress has been made in the past decades in understanding the nature of SLI, many important questions remain unanswered. One such question is whether SLI represents merely a delay in the acquisition of language or represents a fundamental difference in how language is acquired and perhaps a long-term or even permanent language disability. Longitudinal studies have found that for most children with SLI, the most basic grammatical errors subside with time (e.g., missing verbal inflectional morphemes, determiners, incorrect case on pronouns; see section 8.4.4.1), but for many children language

Sidebar 8.4: Effects of SLI

Consider for a moment why it is important to provide interventions for children with SLI. How do you think SLI affects children in terms of

(a) academic achievement?

(b) social development?

(c) self-esteem?

How do you think SLI might affect the families of children with this disorder?

problems persist into the school years (Aram and Nation, 1980). Studies have also revealed that for some individuals initially diagnosed with SLI as preschoolers, language deficits persist into adolescence and adulthood (Weiner, 1974; Hall and Tomblin, 1978).

Before discussing the main traits of SLI, it is important to note that SLI is extremely common: the prevalence of SLI is 7% among kindergarteners, with a slightly higher prevalence among boys (8%) than girls (6%) (Tomblin et al., 1997). The high prevalence of SLI underscores the need for well-trained and qualified speech-language pathologists to provide interventions that can help these children improve their language abilities.

First we will look at the specific grammatical characteristics of SLI and at examples of speakers of different languages with SLI. Then we will discuss some theories about what underlies this disorder.

8.4.1.1 Grammatical Characteristics of SLI

One of the challenges in characterizing, explaining, and treating SLI is that its traits can be quite varied across individuals. As noted above, for some children it is mainly an expressive impairment, while for other children receptive language is also affected. For some children an expressive impairment can arise from both a speech and a language disorder, while for others the primary disorder may involve language but not speech. SLI can (but need not) be associated with a late onset of language production. On the other hand, late language production is not necessarily an indicator of language impairment. While most children begin using words around age 10–12 months and word combinations by 15–18 months, some children remain silent until they are 3 years old and then begin speaking in nearly full sentences.

But some characteristics are broadly associated with SLI. In terms of general language development, children with SLI typically have a late onset of language production and unusual difficulty producing language. In terms of morphosyntax, English-speaking children with SLI frequently omit function words and morphemes such as tense morphemes, determiners, and auxiliaries. There are also some phonological deficits, such as coda deletion (*bead* [bid] pronounced as [bi]) and cluster reduction (*box* [baks] as [bak]). Another phonologically related deficit involves nonword repetition. By age 4, typically developing

(TD) children do not have any difficulty repeating back simple nonwords like *wug* or *dax*. Children with SLI, however, experience some difficulty with this. Nonword repetition requires holding the made-up word in phonological working memory, so problems completing this task are considered to involve a phonological deficit.

Vocabulary deficits are associated with SLI. According to some studies, children with SLI produced their first words later than children without SLI (Paul, 1996; Rescorla, 2002; Trauner et al., 1995) and at later ages perform below their peers but similar to younger children on vocabulary measures (Rice, 2004).

The following are some examples of speech by 4-year-olds with SLI:

(10) a. Adult: The baby is drinking milk and …
Child: Dog chew bone. (target: The dog is chewing a bone.)

b. Adult: She's combing her hair. What did she do?
Child: Comb hair. (target: (She) combed her hair) (Leonard, 2014)

(11) The man got on the boat. He jump out the boat. He rocking the boat. He drop his thing. He drop his other thing. He tipping over. He fell off the boat. (Lindner and Johnston, 1992)

What do you notice about these children's speech? As you can see in these examples, some of the main grammatical categories missing from SLI speech include

(12) (a) third-person-singular *-s*
(b) past-tense *-ed*
(c) copula and auxiliary *be*
(d) nominative case (*he*, not **him*)
(e) determiners (*a/the*)

The plural *-s* marker is also often omitted. In addition to these grammatical errors, it is typical in the speech of children with SLI to find less use of *do*-support in questions and more use of demonstrative pronouns (*that*) instead of a full noun phrase; generic terms (*thing*, *stuff*); or general-purpose verbs (*do*, *make*, *put*, or *go*) instead of more specific verbs. Some of these errors are reminiscent of TD children's telegraphic speech, but for children with SLI the errors continue for many years past when TD children outgrow this stage.

These are the primary characteristics of SLI for English-speaking children, but SLI has been found in every language for which it has been searched. Many of the traits of SLI speech are consistent across these languages (omission or substitution of function morphemes, syntactic and phonological errors, difficulties with constructions involving wh-movement, such as relative clauses and wh-questions), but some of the traits are language specific. Table 8.1 provides sources for information about how SLI presents in a variety of languages.

Table 8.1
Sources for crosslinguistic features of SLI

Language	Source
Cantonese	Fletcher et al., 2005
German	Rice, Noll, and Grimm, 1997
Hebrew	Abdalla et al., 2013
Inuktitut	Crago, Allen, and Ningiuruvik, 1993
Spanish	Bedore and Leonard, 2001; Grinstead et al., 2009, 2013

8.4.1.2 Causes of and Explanations for SLI

As discussed, there are a number of points of disagreement about SLI—for example, whether it represents merely a long delay in language development or a true deviance (abnormal development). Some of the disagreement likely stems from the fact that it is a somewhat heterogeneous disorder, so that different children may present with different symptoms or developmental trajectories. Another likely source of disagreement is the fact that the etiology of SLI is unclear. Unlike disorders such as Down syndrome or Williams syndrome, which have very clear genetic causes, researchers do not yet know exactly what causes SLI. It clearly has a genetic component, however. Evidence comes from a large number of studies that found that children diagnosed with SLI are significantly more likely to have immediate family members with some type of language disorder than are children not diagnosed with SLI (Leonard, 2014). Although not all children with SLI have family members with language disorders, there are cases in which large numbers of members within the same extended family are affected by language disorders (Hurst et al., 1991).

Other evidence comes from twin studies. In most twin studies on language impairment, a significantly higher proportion of monozygotic twins (twins who share 100% of their DNA) shared a diagnosis of SLI than dizygotic twins (twins who share 50% of their DNA, just like non-twin siblings). A number of different regions in the genetic code have been associated with language-related deficiencies. Most famous, perhaps, is the FOXP2 gene, a gene located in the chromosomal region 7q31 (31st band on the long arm of chromosome 7); however, individuals with a disruption involving this particular gene may have broader oral-motor deficiencies that go beyond grammatical errors and difficulties. In addition, other chromosomal regions have been linked to deficits in nonword repetition and expressive language. Ongoing research seeks to identify the exact relationships between genetic abnormalities and observable language deficiencies in individuals with SLI.

As for the correct formal characterization of SLI, there are two primary types of accounts: those that link SLI to a problem with representing grammatical knowledge (Rice, 2004; Rice and Wexler, 1996) and those that link SLI to a problem with processing that knowledge (Leonard, et al., 1992).

Accounts based on representation point out similarities between the errors made by children with SLI and those of younger, TD children, such as optional infinitives in languages like English and German. Recall that inflections on verbs to mark tense and/or agreement are represented in the inflectional, tense, and agreement heads in the syntactic structure (see chapter 7, section 7.2). One explanation for why these markers are omitted by typical children is that it takes time for children's grammatical representation to mature and, early on, they fail to represent the full syntactic structure or perhaps certain grammatical features within the structure. This is called the optional infinitive (OI) stage. Thus, one idea about why children with SLI fail to mark tense and/or agreement at considerably later ages than typical children is that they exhibit an extended OI stage (Rice and Wexler, 1996); in other words, for children with SLI, the grammatical representation takes longer to mature than for children without SLI.

Another possible explanation for language difficulties in children with SLI is that the ability to process, access, and use linguistic knowledge is impaired, rather than linguistic knowledge itself. Such an explanation takes into account the fact that children with SLI often exhibit some (mild) cognitive limitations in domains other than language in addition to their language limitations. Many aspects of cognitive processing could give rise to the language impairments we see with SLI, but one possible culprit is working memory. Working memory is the ability to hold some piece of information in memory so that it can be manipulated somehow. For example, in a listening span task, you hear a sentence and must recall the last word (or series of words) in the sentence. Children with SLI tend to have a smaller listening span than TD children, suggesting a lower working-memory capacity. It is not hard to imagine that if auditory working memory is impaired, it will be difficult to learn word meanings (many repetitions will likely be needed); this, in turn, could have repercussions on other areas of grammatical development. Children with SLI also have some difficulties with tasks measuring phonological memory, such as nonword repetition (you hear a nonword, such as "blicket," and have to repeat it). In this task, children with SLI perform at a level significantly below age-matched controls in terms of accuracy, and their accuracy worsens (along with the gap between them and TD children) the more syllables a word has (Kamhi et al., 1988; Marton and Schwartz, 2003).

To summarize, we have seen that SLI is a developmental disorder in which language (not necessarily speech) is impaired while other cognitive faculties remain relatively spared. SLI has been found in numerous different and typologically diverse languages. While morphosyntax is often affected by SLI, children with SLI usually display great problems with phonological working memory. For some individuals with SLI, the language disorder persists well into adolescence and adulthood, while for others it seems to resolve in childhood. As yet, the exact cause of SLI is not known, although there is good evidence that it has a genetic basis.

8.4.2 Autism and Autism Spectrum Disorder

Autism spectrum disorder (ASD) refers to a developmental disability that affects social cognition, language, and communication, and it can involve repetitive gestures or other motor behaviors and/or attentional abnormalities. ASD is generally more familiar to people than SLI; however, the incidence of ASD is much lower than that of SLI. According to the Centers for Disease Control (CDC), ASD occurs in 1 in 68 children, or 1.4% of children (compare with 7% for SLI). It is 4.5 times more common in boys than in girls, so the prevalence rate is 1 in 42 for boys but 1 in 189 for girls (CDC website). Although its precise cause cannot be determined in every case, there is strong evidence that genetics plays an important role. Besides the striking asymmetry in incidence between boys and girls, twin studies have revealed that monozygotic twins have a 60% concordance rate (both twin siblings have autism if one of them does), whereas dizygotic twins had no concordance. When evaluating for a broader array of developmental disorders involving social and communicative development, monozygotic twins displayed over 90% concordance, whereas dizygotic twins had just 10% (Bailey et al., 1995). More recent discoveries of rare genetic mutations involving specific genes or genomic loci that can be determined to cause ASD further underscore the important role that genetics plays in the presentation of this disorder (Betancur, 2011; Geschwind, 2011; Devlin and Scherer, 2012).

How is ASD relevant for language? Problems and delays with language, and more broadly communication, are one of the primary behavioral deficits exhibited by people with ASD. Children with ASD often have a delayed onset of language, and as many as 50% remain completely nonverbal (Bryson, Clark, and Smith, 1988). For those children with ASD who do have language, some of the most apparent language-related abnormalities are echolalia (repeating back specific words or phrases they have heard) and idiosyncratic speech, in which a child might use an existing word or phrase of their language with an idiosyncratic meaning. An example of echolalia is given in 13, and an example of idiosyncratic speech is given in 14.

(13) Mother: Want some?
Child: Want some?

(14) Child: Go get pizza (= I'm hungry)

Other common linguistic deficits with ASD relate to pragmatics, discourse, and prosody. Pragmatics has to do with how language is used within social contexts, from taking turns to knowing the appropriate way to address someone to understanding what background information is reasonably known to the other participants in a conversation. Children with ASD have a tendency to use either overly precise or stilted speech, spell out details that are already known, or omit necessary information; they may also struggle with repairing communication breakdowns, not requesting clarification effectively or responding to such requests inappropriately (Eigsti et al., 2011).

Sidebar 8.5: Theory of Mind and Language

An aspect of cognitive development that is relevant for both language and the specific profile of children with ASD is Theory of Mind (ToM). ToM is the ability to understand that other people have minds that are different from your own and that others can therefore have beliefs or thoughts that are different from yours. This is related to the understanding that people can have beliefs that are *false*. One traditional test for ToM is a **false belief** task: A child is presented with a display containing two hiding places. A character (Sally) enters the display, hides an object in one location, and then goes away. While that character is gone, another character (Ann) moves the hidden object to another hiding place. When the original character, Sally, returns, the child is asked: Where will Sally look for the object? If the child has ToM, they will understand that Sally could not have known that the object was moved and will look for it in the original location. If the child does *not* yet have ToM, they will suppose that Sally's thoughts and beliefs are the same as their own and will look for the object in the new location.

TD children start to "pass" this task (i.e., answer that Sally will look in the original hiding place) after about age 4 or 5 years (Wimmer and Perner, 1983; Wellman, 1985; and many others). However, children with ASD continue to answer that Sally will look in the new location. One idea about the nature of ASD is that it involves a kind of "mindblindness" (Baron-Cohen, 1995), or the inability to understand and represent the contents of others' minds.

Deficits are found in other areas of grammar as well, however, including syntax and morphology. Phonological deficits appear to be less common and more often associated with children with ASD with lower IQs (so-called low-functioning) or limited to nonword repetition, a skill that taps into phonological working memory (Bishop et al., 2004). Semantic deficits have been found in some studies but not others (Eigsti et al., 2011).

Deficits in syntax seen in children with ASD include producing sentences with less syntactic complexity (Tager-Flusberg et al., 1990; Scarborough et al., 1991) and fewer function words— compared to typical children with the same mean length of utterance (MLU)— and experiencing extra difficulty interpreting transitive sentences (Prior and Hall, 1979). Closely related to syntax, errors in morphosyntax tend to involve difficulties producing past-tense verbs, lower MLU than age-matched typical children, and omission of verbal morphology that is reminiscent of SLI speech (Roberts et al., 2004). Relatedly, in languages other than English, there is further evidence that some of the syntactic difficulties of children with ASD closely match those of children with SLI: errors with clitics (a type of pronoun) in Romance languages and Greek and with wh-movement (Durrleman et al., 2016; Durrleman and Delage, 2016).

It is important to bear in mind that there is enormous variability across individuals with ASD in terms of their language outcomes. This variability is part of what led to the shift away from labeling the disorder as pure autism (with related disorders such as Asperger's) to labeling the range of expressions as ASD.

8.4.3 Hemispherectomy

As we saw in sidebar 8.1, language tends to be lateralized in the left hemisphere of the brain. This is somewhat of a simplification, since some aspects of language are processed on the right side, but as a generalization, the language centers of the brain are widely seen as residing on the left side. Moreover, functions like general auditory processing occur on the right side of the brain, as does spatial cognition, mathematics, and other skills. So the brain tends to specialize: it dedicates certain neural real estate to being predominantly for one function.

An interesting fact is that this lateralization of functions does not happen until several years after birth. The process begins early in infancy (perhaps before birth) and continues through puberty. This means that infants and toddlers have brains that are not fully lateralized yet. But because their brains are so plastic, or malleable, the manner in which the brain organizes itself can (and indeed does) change over time. This change is thought to occur so as to maximize the brain's efficiency.

We might speculate that this process of lateralization is tied to the idea of a critical period: that once the brain has specialized (picked which neural real estate will do which functions), it is impossible to reorganize itself enough to acquire a new language. That is, lateralization proceeds though childhood, and once it is done, the brain loses a lot of the plasticity that allowed lateralization to happen in the first place. This loss of plasticity results in what looks like a critical period for language—the inability of the brain to reorganize in response to new kinds of linguistic input.

In the case of Genie (see section 8.1.2), her brain may have specialized, but in a manner that is not normal, and because of this she was unable to acquire language fully after she was rescued from her home. Likewise, adults that begin acquiring a (second) language after the critical period might already have specialized brains that are not plastic enough to rejig themselves to maximally learn a new language. It should be stressed that this is a very neural view of the critical period hypothesis, and certainly not the only one with merit in the field. Nonetheless, the relationship between language acquisition and lateralization is one worth considering.

There are many ways to investigate this, but in this section we will focus on one very interesting, yet unresolved, question. It involves children who suffer from epilepsy—a brain disorder that has no cure and the source of which is not fully understood. Like ASD, epilepsy is a spectrum disorder in that patients exhibit a range of symptoms. A common one is seizures, which involve a hyperactive section of the brain causing severe neural disruption. This hyperactivity is a center of massive neural activation—so much activation that it overwhelms the system and results in a seizure. Some have described epileptic seizures as neural electric thunderstorms that begin in one part of the brain and spread across the brain during the seizure. As the thunderstorm traverses the brain, the seizure changes in its intensity, causing different parts of the person to convulse and seize up. The seizure passes once the thunderstorm subsides and neural activity returns to normal. There is no cure for epilepsy, though drugs and dietary changes may help control it.

In the 1950s, a common treatment for epilepsy was, believe it or not, hemispherectomy—removal of an entire hemisphere of the brain. This may seem like an excessive response to a disorder, but there is some logic to it. Recall that the seizures originate in one part of the brain (typically the same part in any one individual) and then spread throughout the rest of the brain. The thinking was that if that hemisphere was removed entirely, then the source of the electrical thunderstorm was removed and the seizures would be alleviated. This treatment was typically reserved for very young children. While that increases the shock factor to be sure (remove half a baby's brain?), again, there was some logic there. The thinking was that a young brain is plastic enough to recover from the removal of an entire hemisphere, so early removal of the disordered part of the brain was key.

This treatment fell out of favor in the 1960s for a variety of reasons, but doctors began to revisit the idea when no new treatments were developed in the ensuing few decades. Furthermore, the reality was that early hemispherectomy was one of the best treatments available for young infants with epilepsy. The long-term prognosis for children who had a hemispherectomy was significantly better than age-matched peers who went through drug treatment alone. So in the 1990s, the treatment began to be performed again at some specialized medical facilities.

And here is where it gets interesting for the purpose of language acquisition. Consider this: If language is lateralized in the left hemisphere (as is the case in the vast majority of right-handed people and over 60% of left-handed people as well), and if that left hemisphere is removed from the individual, what happens to language? Is it like removing language entirely from that human being? That remarkable question has actually been investigated.

Curtiss and De Bode (2003) and others since then have studied this issue, following forty-three children who had left-hemisphere removals at varying ages (some were just infants; some were as old as 14 years). They tracked these patients for several years, periodically testing them on various language tasks as well as other cognitive tasks. They were looking at how well children performed on the language tasks over time as well as whether they achieved (near) native-like competence in language.

What they found was in some ways not surprising and in other ways very surprising. Not surprisingly, they found that for the first six months or so, children were silent. There was very little in the way of language production, and language comprehension was sketchy too. But after six months post-surgery, some interesting findings began to emerge. They found that the younger the child was at the time of the procedure, the greater their final proficiency in language. So this confirms the intuition that motivated the whole hemispherectomy procedure for epileptic children in the first place: young brains are plastic and can recover better than older brains. Moreover, those older children (who had the procedure after the age of 10 years, which is thought to be around the endpoint of the critical period) generally performed very poorly on language tests, suggesting that their abilities to use language would never reach native-like levels.

But perhaps most surprisingly, they found that a small number of older children actually did reach higher levels of proficiency, on par with the younger children. One child in particular, who had the hemispherectomy at the age of 12 years, performed on par with children who had received the procedure as infants.

Why is this surprising? Well, if a child who had their left hemisphere removed at the age of 12 is still able to achieve near-native-like competence in language, doesn't this refute the idea of a critical period? Perhaps this shows that the critical period is not all that critical after all: it might be a period of heightened sensitivity to language, and not an all-or-nothing period, as the word 'critical' might suggest. And so perhaps others might be able to overcome the critical period under different conditions. In fact, some researchers refer to critical periods as sensitive periods precisely for this reason.

However, this is probably not the best understanding of this fact. First, it is not easy to extrapolate these results to normally developing children. After all, the hemisphere that was removed was the *disordered hemisphere*. That means it is possible that lateralization had not proceeded in a normal fashion with some of these children; language may have lateralized in the right hemisphere simply because the left hemisphere was far too damaged to house such a high-powered, critical function as language. If this is the case, then the critical period may still be intact, since the hemisphere that was removed was not the one that language had lateralized to.

Second, because these were inherently disordered brains, perhaps the reduction in plasticity that we think happens with the onset of adulthood may not have occurred quite as predicted. We know that the adult brain is plastic to some degree, so perhaps children with epilepsy retain plasticity for a longer time than people without epilepsy. If so, then the critical period may be longer for individuals with the disorder.

And third, as one of the authors of this study argued (Susie Curtiss—the same Curtiss who worked with Genie, by the way), this shows that language is such a *biological imperative* that the brain finds a way to maximize the chances of expression of language even in brains that are disordered. It may be that the removal of the entirety of the left hemisphere forces the brain *in some children* to reorganize to accommodate language, and rather than being evidence against the critical period, this should be seen as evidence of just how embedded language is in our biology.

In summary, the study of children who receive left-hemisphere removals seems to show that the younger the child has the procedure, the better the final outcome. This is an expected result. But more interestingly, some children are able to acquire language to a very high degree even when they receive a hemispherectomy relatively late in development. While on the face of it this might challenge the critical period, we have argued that it does not necessarily do so. Rather, one can also look at the facts as supporting the idea of a critical period for language, and the fact that some older children still acquire language shows how powerful the biological need to express language is.

8.5 Summary

In this chapter we looked at a variety of circumstances under which language acquisition can take place that are unlike the typical circumstance we have been assuming throughout this book: How is language acquired when the input is withheld until a later time point than early childhood? How is it acquired when important sensory abilities are compromised? How is it acquired when the input is fully available but cognitive faculties are impaired or when the brain is radically changed through removal of a hemisphere?

Across these cases the picture that emerges is that the biological drive to acquire language is powerful—so powerful that, even when access to the world is altered by loss of vision or audition, as long as language is there in the input children will create language. Even when that language input is less than a full-fledged language, as in the case of the pidgin sign language input to the children in Nicaragua, children will create language.

We have seen that the human brain is astoundingly resilient. When language input is not available in early childhood, people will begin learning words when it does become available; and yet, the brain's ability to create grammar is not unbounded. Acquisition of language outside the critical period is different: words are learned rapidly, but the older one gets, the more difficult it becomes to acquire grammatical morphemes and rules. Function words are either absent or used in an unprincipled way by older learners.

We also saw evidence that language occupies a distinct module in the architecture of the mind: it can be singularly impaired (as in SLI), and removing the left-hemisphere structures that support it often results in aphasia, the loss of language. On the other hand, like other modules of cognition, language interacts with other domains of cognition in important ways, as we saw in the social-communicative impairment of autism.

8.6 Further Reading

Landau, Barbara, and Lila Gleitman. 1985. *Language and Experience*. Cambridge, MA: Harvard University Press.

Leonard, Laurence. 2014. *Children with Specific Language Impairment*. Cambridge, MA: MIT Press.

8.7 Exercises

1. Consider Genie's phonological, lexical, and syntactic development after she was exposed to language. In what ways was her acquisition in each domain **similar** to that of typically developing (TD) children? In what ways was her acquisition **different** from that of typical children? List at least one similarity and one difference, for each domain of grammar, between Genie's development and the patterns we see in typical children.

2. In this chapter we discussed many ways in which sign language is formally equivalent to spoken language in its structure and features. Spoken languages have dialects and accents. Do you think sign languages likewise have dialects and accents? What do you think this would mean? That is, how might signs vary according to accent?

After you have thought about this question, watch the video found here: https://lingdept.wordpress.com/2017/09/16/do-sign-languages-have-accents-video-collaboration-between-department-of-linguistics-at-gallaudet-and-mental-floss/.

Did the information in this video surprise you? Why or why not?

3. In Landau and Gleitman's study of language development in blind children, one of their questions was whether blind children could learn the meanings of color words and use them appropriately. Color is a physical attribute that blind children have no perceptual access to. Surely it presents a nearly impossible concept for them to acquire?

In fact, Landau and Gleitman found that blind children did learn these words and used them appropriately. That is, although a blind child would not be able to tell the experimenter what color an object was, she knew that color was a feature of tangible, concrete objects (flowers could be green, but ideas could not be) different from size, texture, or shape.

Given how blind children learn the meanings of verbs like *look* and *see* (see section 8.3), what cues in language do you think blind children might use to figure out what color words mean?

4. Read the following conversations between a mother (MOT) and each of her two daughters (CHI in each conversation). One of the two sisters has SLI and the other is TD (the mother's utterances are omitted in the conversation with the TD child). Calculate the MLU and upper bound (longest utterance as counted in morphemes) for each child based on the utterances given.

Child OM, age 3;9 (SLI)	Morphemes
MOT: What holiday's coming up soon, Olivia?	
CHI: um [fi] (=free) (.) food.	
MOT: Really food?	
CHI: yeah.	
MOT: Halloween is coming up, right?	
CHI: yeah.	
MOT: What are you gonna be for Halloween?	
CHI: Belle.	
CHI: Princess Belle.	
MOT: Princess Belle. What's Lyra gonna be?	
CHI: lion.	
CHI: and you be [gos] (=ghost).	
MOT: I'm gonna be a ghost? Okay.	

CHI: and and and dad/ daddy be be be dragon.
MOT: Daddy's gonna be a dragon? I don't know that daddy has a dragon costume.
CHI: yes I have a dra/ I I have a dragon costume in my closet.
CHI: mommy I wan' you take me zoo zoo
CHI: I wan' you take me zoo (.) right now?
MOT: We're not going to the zoo today.
CHI: an' next time (.) I tell you (.) I need go zoo.
Total Morphemes

Child OM, age 4;8 (SLI)	Morphemes
CHI: what's a matter?	
MOT: I'm just trying to think about what else I should bring for lunch.	
CHI: mm go now.	
MOT: You're ready to go? ... What did you do at the science museum yesterday?	
CHI: uh ride the train!	
MOT: You did? You rode the train? Oh that's fun.	
CHI: and I scream!	
MOT: You screamed on the train? In the tunnel?	
CHI: yeah.	
MOT: What else did you do? Did you see the butterflies?	
CHI: yeah.	
CHI: and I didn't uh (.) we see a [kalepilerz] (=caterpillars)	
CHI: didn't see a caterpillars	
MOT: You didn't see any caterpillars?	
CHI: no.	
MOT: Oh. Were they hiding?	
CHI: we didn't see any dinosaurs too.	
MOT: Did you see the farm animals?	
CHI: no.	
CHI: making a new farm for the [dakas] (=alpacas)	
Total Morphemes	

Child LM, age 2;9 (TD)	Morphemes
CHI: oh I didn't mean to step on this.	
CHI: I didn't mean to step on it.	
CHI: the other one is in the kitchen.	
CHI: well that small white box?	
CHI: maybe daddy knows.	
CHI: here (.) here	
CHI: here I go.	
CHI: well these (.) these (.) my earrings need a new home	
CHI: where's (.) where's the lid?	
CHI: okay.	
CHI: at nighttime you hava (=have to) take off your clothes.	
CHI: and then put on your PJs.	
CHI: oh here's one of my other (.) earring.	
CHI: my earrings my purple earrings all match.	
CHI: yes.	
CHI: and I gonna not wear my earrings.	
CHI: because of the (.) because of the (.) the rainstorm.	
CHI: yes.	
CHI: that's why, because it's a rainy day.	
CHI: because I just can't.	
Total Morphemes	

(i) What are the MLUs for each child at the data points? How long is each child's longest utterance? Given their profiles, do these results surprise you? Why or why not?

(ii) What are some grammatical morphemes missing from OM's speech (the child with SLI)?

(iii) Is there anything that makes calculating OM's MLU particularly challenging?

(iv) Notice the ages of the two children. How would you describe their productive language, in comparison to one another, taking into consideration their ages, MLUs, and other aspects of their language?

8.8 References

Abdalla, Fauzia, Khawla Aljenaie, and Abdessatar Mahfoudhi. 2013. Plural noun inflection in Kuwaiti Arabic-speaking children with and without specific language impairment. *Journal of Child Language* 40: 139–168.

Aram, Dorothy M., and James E. Nation. 1980. Preschool language disorders and subsequent language and academic difficulties. *Journal of Communication Disorders* 13: 159–170.

Ardiel, Evan L., and Catharine H. Rankin. 2010. The importance of touch in development. *Paediatrics and Child Health* 15: 153–156.

Bailey, Anthony, Ann Le Couteur, Irving Gottesman, Patrick Bolton, Emily Simonoff, Edward Yuzda, and Michael Rutter. 1995. Autism as a strongly genetic disorder: Evidence from a British twin study. *Psychological Medicine* 25(1): 63–77.

Baron-Cohen, Simon. 1995. *Mindblindness*. Cambridge, MA: MIT Press.

Bedore, Lisa, and Laurence Leonard. 2001. Grammatical morphology deficits in Spanish-speaking children with specific language impairment. *Journal of Speech, Language, and Hearing Research* 44: 905–924.

Betancur, Catalina. 2011. Etiological heterogeneity in autism spectrum disorders: More than 100 genetic and genomic disorders and still counting. *Brain Research* 1380: 42–77.

Bigelow, Ann. 1987. Early words of blind children. *Journal of Child Language* 14: 47–56.

Bishop, Dorothy V. M., Murray Maybery, Dana Wong, Alana Maley, Wayne Hill, and Joachim Hallmayer. 2004. Are phonological processing deficits part of the broad autism phenotype? *American Journal of Medical Genetics* 128: 54–60.

Bryson, Susan E., Barbara S. Clark, and Isabel M. Smith. 1988. First report of a Canadian epidemiological study of autistic syndromes. *Journal of Child Psychology and Psychiatry and Allied Disciplines* 29: 433–445.

Charney, Rosalind. 1978. The development of personal pronouns. PhD diss., University of Chicago.

Chen Pichler, Deborah. 2012. Acquisition. In Roland Pfau, Markus Steinbach, and Bencie Woll (eds.), *Sign Language: An International Handbook*, pp. 647–686. Berlin: Walter de Gruyter.

Clahsen, Harald, and Mayella Almazan. 1998. Syntax and morphology in Williams syndrome. *Cognition* 68: 167–198.

Crago, Martha B., Shanley Allen, and Lizzie Ningiuruvik. 1993. Inflections gone askew: SLI in a morphologically complex language. Paper presented at the Sixth Congress of the International Association for the Study of Child Language, Trieste, Italy.

Curtiss, Susan. 1977. *Genie: A Psycholinguistic Study of a Modern-Day "Wild-Child."* New York: Academic Press.

Curtiss, Susan. 1988. Abnormal language development and the modularity of language. In Frederick Newmeyer (ed.), *Linguistics: The Cambridge Survey*, vol. 2: *Linguistic Theory: Extensions and Implications*, pp. 96–116. Cambridge: Cambridge University Press.

Curtiss, Susan, and Stella De Bode. 2003. How normal is grammatical development in the right hemisphere following hemispherectomy? The root infinitive stage and beyond. *Brain and Language* 86: 193–206.

Curtiss, Susan, Victoria Fromkin, Stephen Krashen, David Rigler, and Marilyn Rigler. 1974. The linguistic development of Genie. *Language* 50: 528–554.

De Quadros, Ronice M., and Diane Lillo-Martin. 2007. Gesture and the acquisition of verb agreement in sign languages. In Heather Caunt-Nulton, Samantha Kulatilake, and I-hao Woo (eds.), *Proceedings of the 31st Annual Boston University Conference on Language Development*, pp. 520–531. Somerville, MA: Cascadilla Press.

DeGregory, Lane. 2008. The girl in the window. *Tampa Bay Times*, July 31.

Devlin, Bernie, and Stephen W. Scherer. 2012. Genetic architecture in autism spectrum disorder. *Current Opinion in Genetics and Development* 22: 227–239.

Durrleman, Stephanie, and Hélène Delage. 2016. Autism spectrum disorder and specific language impairment: Overlaps in syntactic profiles. *Language Acquisition* 23: 361–386.

Durrleman, Stephanie, Theodoros Marinis, and Julie Franck. 2016. Syntactic complexity in the comprehension of wh-questions and relative clauses in typical language development and autism. *Applied Psycholinguistics* 37: 1501–1527.

Eigsti, Inge-Marie, Ashley B. de Marchena, Jillian M. Schuh, and Elizabeth Kelley. 2011. Language acquisition in autism spectrum disorders: A developmental review. *Research in Autism Spectrum Disorders* 5: 681–691.

Emlen, Stephen. 1969. The development of migratory orientation in young indigo buntings. *Living Bird* 8: 113–126.

Faust, Miriam, Lilly Dimitrovsky, and Shira Davidi. 1997. Naming difficulties in language-disabled children: Preliminary findings with the application of the tip-of-the-tongue paradigm. *Journal of Speech, Language, and Hearing Research* 40: 1026–1036.

Fletcher, Paul, Laurence Leonard, Stephanie Stokes, and Anita M.-Y. Wong. 2005. The expression of aspect in Cantonese-speaking children with specific language impairment. *Journal of Speech, Language, and Hearing Research* 48: 621–634.

Fortescue, Michael. 1984. Learning to speak Greenlandic: A case study of a two-year-old's morphology in a polysynthetic language. *First Language* 5: 101–114.

Friedmann, Naama, and Rama Novogrodsky. 2004. The acquisition of relative clause comprehension in Hebrew: A study of SLI and normal development. *Journal of Child Language* 31: 661–681.

Friedmann, Naama, and Rama Novogrodsky. 2011. Which questions are most difficult to understand? The comprehension of Wh questions in three subtypes of SLI. *Lingua* 121: 367–382.

Frith, Uta. 2003. *Autism: Explaining the Enigma*. Malden, MA: Blackwell Publishing.

Gallaudet Research Institute. 2001. *Regional and National Summary Report of Data from the 1999–2000 Annual Survey of Deaf and Hard of Hearing Children and Youth*. Washington, DC: Gallaudet Research Institute, Gallaudet University.

Geschwind, Daniel. 2011. Genetics of autism spectrum disorders. *Trends in Cognitive Science* 15(9): 409–416.

Grinstead, John, Alisa Baron, Mariana Vega-Mendoza, Juliana De la Mora, Myriam Cantú-Sánchez, and Blanca Flores. 2013. Tense marking and spontaneous speech measures in Spanish SLI: A discriminant function analysis. *Journal of Speech, Language, and Hearing Research* 56: 352–363.

Grinstead, John, Juliana De la Mora, Mariana Vega-Mendoza, and Blanca Flores. 2009. An elicited production test of the optional infinitive stage in child Spanish. In Jean Crawford, Koichi Otaki, and Masahiko Takahashi (eds.), *Proceedings of the 3rd Conference on Generative Approaches to Language Acquisition—North America*, pp. 36–45. Sommerville, MA: Cascadilla Press.

Hall, Penelope K., and J. Bruce Tomblin. 1978. A follow-up study of children with articulation and language disorders. *Journal of Speech and Hearing Disorders* 43: 227–241.

Hensch, Takao. 2005. Critical period plasticity in local cortical circuits. *Nature Reviews: Neuroscience* 6: 877–888.

Hoffmeister, Robert J. 1978. The development of demonstrative pronouns, locatives, and personal pronouns in the acquisition of American Sign Language by deaf children of deaf parents. Unpublished doctoral dissertation, University of Minnesota.

Hubel, David H., and Torsten N. Wiesel. 1970. The period of susceptibility to the physiological effects of unilateral eye closure in kittens. *Journal of Physiology* 206: 419–436.

Hurst, J., M. Baraitser, E. Auger, F. Graham, and S. Norell. 1991. An extended family with a dominantly inherited speech disorder. *Neurology* 32: 347–355.

Kamhi, A., H. Catts, D. Mauer, K. Apel, and B. Gentry. 1988. Phonological and spatial processing abilities in language-impaired children. *Journal of Speech and Hearing Disorders* 49: 169–176.

Kegl, Judy, Ann Senghas, and Marie Coppola. 1999. Creation through contact: Sign language emergence and sign language change in Nicaragua. In Michel DeGraff (ed.), *Language Creation and Language Change: Creolization, Diachrony, and Development*, pp. 179–237. Cambridge, MA: MIT Press.

Knudsen, Eric I., and Phyllis F. Knudsen. 1990. Sensitive and critical periods for visual calibration of sound localization by barn owls. *Journal of Neuroscience* 10: 222–232.

Landau, Barbara, and Lila Gleitman. 1985. *Language and Experience: Evidence from the Blind Child*. Cambridge, MA: Harvard University Press.

Lenneberg, Eric. 1967. *Biological Foundations of Language*. New York: Wiley.

Leonard, Laurence B. 2014. *Children with Specific Language Impairment*. Cambridge, MA: MIT Press.

Leonard, Laurence B., M. Cristina Caselli, Umberta Bortolini, Karla K. McGregor, and Letizia Sabbadini. 1992. Morphological deficits in children with Specific Language Impairement: The status of features in the underlying grammar. *Language Acquisition* 2: 151–179.

Lillo-Martin, Diane. 2016. Sign language acquisition studies. In Edith L. Bavin and Letitia R. Naigles (eds.), *The Cambridge Handbook of Child Language*, pp. 504–526. 2nd ed. Cambridge: Cambridge University Press.

Lillo-Martin, Diane, and Ronice De Quadros. 2005. The acquisition of focus constructions in American Sign Language and Língua de Sinais Brasileira. In Alejna Brugos, Manuella R. Clark-Cotton, and Seungwan Ha (eds.), *Proceedings of the 29th Annual Boston University Conference on Language Development*, pp. 365–375. Somerville, MA: Cascadilla Press.

Lindner, Katrin, and Judith Johnston. 1992. Grammatical morphology in language-impaired children acquiring English or German as their first language: A functional perspective. *Applied Psycholinguistics* 13: 115–129.

Loew, Ruth. 1984. Roles and reference in American Sign Language: A developmental perspective. PhD diss., University of Minnesota.

Marler, Peter. 1991. The instinct to learn. In Susan Carey and Rochel Gelman (eds.), *The Epigenesis of Mind: Essays on Biology and Cognition*, pp. 37–66. Hillsdale, NJ: Lawrence Erlbaum Associates.

Marler, Peter, and S. Peters. 1987. A sensitive period for song acquisition in the song sparrow, *Melospiza melodia*: A case for age-limited learning. *Ethology* 76: 89–100.

Marschark, Marc, Thomastine Sarchet, Cathy Rhoten, and Megan Zupan. 2010. Will cochlear implants close the reading achievement gap for deaf students? In Marc Marschark and Patricia E. Spencer (eds.), *The Oxford Handbook of Deaf Studies, Language, and Education*, pp. 127–143. Oxford: Oxford University Press.

Marton, K., and R. Schwartz. 2003. Working memory capacity and language processes in children with specific language impairment. *Journal of Speech, Language, and Hearing Research* 46: 1138–1153.

Mayberry, Rachel. 2010. Early language acquisition and adult language ability: What sign language reveals about the critical period for language. In Marc Marschark and Patricia E. Spencer (eds.), *The Oxford Handbook of Deaf Studies, Language, and Education*, pp. 281–291. Oxford: Oxford University Press.

Meier, Richard. 1982. Icons, analogues, and morphemes: The acquisition of verb agreement in American Sign Language. PhD diss., University of California, San Diego.

Meier, Richard, Claude Mauk, Adrianne Cheek, and Christopher Moreland. 2008. The form of children's early signs: Iconic or motoric determinants? *Language Learning and Development* 4: 393–405.

Mervis, Carolyn B., Byron F. Robinson, Jacquelyn Bertrand, Colleen A. Morris, Bonita P.Klein-Tasm, and Sharon C. Armstrong. 2000. The Williams syndrome cognitive profile. *Brain and Cognition* 44: 604–628.

Mitchell, Ross, and Michael Karchmer. 2004. Chasing the mythical ten percent: Parental hearing status of deaf and hard of hearing students in the U.S. *Sign Language Studies* 4: 128–163.

Morton, Newton E. 1991. Genetic epidemiology of hearing impairment. *Annals of the New York Academy of Sciences* 630: 16–31.

Nelson, Katherine. 1973. Structure and strategy in learning to talk. *Monographs of the Society for Research in Child Development* 38(1–2), serial no. 149: 136.

Newport, Elissa. 1991. Contrasting concepts of the critical period for language. In Susan Carey and Rochel Gelman (eds.), *The Epigenesis of Mind: Essays on Biology and Cognition*, pp. 111–130. Hillsdale, NJ: Lawrence Erlbaum Associates.

Newport, Elissa, Lila R. Gleitman, and Henry Gleitman. 1977. Mother I'd rather do it myself: Some effects and non-effects of maternal speech style. In Catherine Snow and Charles Ferguson (eds.), *Talking to Children: Language Input and Acquisition*, pp. 109–151. New York: Cambridge University Press.

Paludneviciene, Raylene, and Irene W. Leigh. 2011. *Cochlear Implants: Evolving Perspectives*. Washington, DC: Gallaudet University Press.

Paul, Rhea. 1996. Clinical implication of the natural history of slow expressive language development. *American Journal of Speech-Language Pathology* 5: 5–21.

Petitto, Laura Ann. 1987. On the autonomy of language and gesture: Evidence from the acquisition of personal pronouns in American Sign Language. *Cognition* 27(1): 1–52.

Petitto, Laura Ann. 2000. On the biological foundations of human language. In Karen Emmorey & Harlan Lane (eds.), *The Signs of Language Revisited: An Anthology in Honor of Ursula Bellugi and Edward Klima*, pp. 447–471. Mahwah, NJ: Lawrence Erlbaum Associates.

Prior, Margo R., and Lesley C. Hall. 1979. Comprehension of transitive and intransitive phrases by autistic, retarded, and normal children. *Journal of Communication Disorders* 12: 103–111.

Ratner, Nan Bernstein. 2001. Atypical language development. In Jean Berko Gleason (ed.), *Language Development*, pp. 215–256. 5th ed. Boston: Allyn and Bacon.

Rescorla, Leslie. 2002. Language and reading outcomes to age 9 in late-talking toddlers. *Journal of Speech, Language and Hearing Research* 45: 360–371.

Rice, Mabel. 2004. Growth models of developmental language disorders. In Mabel L. Rice and Steven F. Warren (eds.), *Developmental Language Disorders: From Phenotypes to Etiologies*, pp. 207–240. Mahwah, NJ: Lawrence Erlbaum Associates.

Rice, Mabel, Karen Noll, and Hannelore Grimm. 1997. An extended optional infinitive stage in German-speaking children with specific language impairment. *Language Acquisition* 6(4): 255–295.

Rice, Mabel L., Steven F. Warren, and Stacy K. Betz. 2005. Language symptoms of developmental language disorders: An overview of autism, Down syndrome, fragile X, specific language impairment, and Williams syndrome. *Applied Psycholinguistics* 26: 7–27.

Rice, Mabel, and Kenneth Wexler. 1996. A phenotype of specific language impairment: Extended optional infinitives. In Mabel L. Rice (ed.), *Toward a Genetics of Language*, pp. 215–238. Mahwah, NJ: Lawrence Erlbaum Associates.

Roberts, Jenny A., Mabel L. Rice, and Helen Tager-Flusberg. 2004. Tense marking in children with autism. *Applied Psycholinguistics* 25: 429–448.

Rymer, Russ. 1993. *Genie: A scientific tragedy.* New York: Harper Collins Publishers.

Scarborough, Holis, Leslie Rescorla, Helen Tager-Flusberg, Ann Fowler, and Vicki Sudhalter. 1991. Relation of utterance length to grammatical complexity in normal or language-disordered groups. *Applied Psycholinguistics* 12: 23–45.

Schick, Brenda. 1990. The effects of morphosyntactic structure on the acquisition of classifier predicates in ASL. In Ceil Lucas (ed.), *Sign Language Research: Theoretical Issues*, pp. 358–374. Washington, DC: Gallaudet University Press.

Senghas, Ann. 1995. The development of Nicaraguan Sign Language via the language acquisition process. In Dawn MacLaughlin and Susan McEwen (eds.), *Proceedings of BUCLD 19*, pp. 543–552. Boston: Cascadilla Press.

Senghas, Ann, Sotaro Kita, and Asli Özyürek. 2004. Children creating core properties of language: Evidence from an emerging sign language in Nicaragua. *Science* 305: 1779–1782.

Spaulding, Tammie J., Elena Plante, and Rebecca Vance. 2008. Sustained selective attention skills of preschool children with specific language impairment: Evidence for separate attentional capacities. *Journal of Speech, Language, and Hearing Research* 51: 16–24.

Tager-Flusberg, Helen, Susan Calkins, Tina Nolin, Therese Baumberger, Marcia Anderson, and Ann Chadwick-Dias. 1990. A longitudinal study of language acquisition in autistic and Down Syndrome children. *Journal of Autism and Developmental Disorders* 20: 1–21.

Tomblin, J. Bruce, Nancy L. Records, Paula Buckwalter, Xuyang Zhang, Elaine Smith, and Marlea O'Brien. 1997. Prevalence of specific language impairment in kindergarten children. *Journal of Speech, Language and Hearing Research* 40: 1245–1260.

Trauner, Doris, Beverly Wulfeck, Paula Tallal, and John Hesselink. 1995. Neurologic and MRI profiles of language impaired children. Technical Report CND-9513. University of California at San Diego, Center for Research in Language.

Weiner, Paul S. 1974. A language-delayed child at adolescence. *Journal of Speech and Hearing Disorders* 39: 202–212.

Wellman, Henry. 1985. *The Child's Theory of Mind.* Cambridge, MA: MIT Press.

Wiesel, Torsten N., and David H. Hubel. 1965. Extent of recovery from effects of visual deprivation in kittens. *Journal of Neurophysiology* 28: 1060–1072.

Wimmer, Heinz, and Josef Perner. 1983. Beliefs about beliefs: Representation and constraining function of wrong beliefs in young children's understanding of deception. *Cognition* 13: 103–128.

9 Acquisition of More than One Language

9.0 Introduction

Many people who live in a largely monolingual society marvel at bilingualism, the ability to speak and understand more than one language. For monolingual adults the process of learning a second language is often fraught with frustration and difficulty: we pore over vocabulary lists and struggle to memorize seemingly arcane rules and irregular patterns. Perhaps because of this, many parents worry that exposing their child to more than one language at the beginning of life will cause confusion and stunt their child's linguistic development or even their general cognitive development. But nothing could be further from the truth. For children, acquiring two languages at once is no more difficult than acquiring a single language. Moreover, bilingualism is extremely common—even the norm—in many parts of the world, and researchers estimate that more than half of the world is bilingual.

In this chapter we'll look at the acquisition of multiple languages in early and later childhood. This will allow us to see both similarities and differences in the acquisition of language when learning two languages happens simultaneously or in sequence. We'll also consider what was historically one of the primary theoretical questions about bilingualism: When in a bilingual person's development are two languages represented as separate languages? Nowadays researchers recognize that a bilingual person's languages are interconnected, and we'll look at the evidence showing that they are in fact distinct language systems and how they interact.

Later in the chapter we will discuss two potential effects of a bilingual society. One is the growing population of heritage-language speakers—people who began acquiring a language in early childhood, lost some facility with the language later, and then as older adolescents or adults are attempting to relearn the language.[1] The other topic concerns what can happen to a language that has lower prestige within a bilingual society, namely, that it can become endangered. These final two topics are often closely related: many heritage speakers are speakers of endangered languages.

Note that throughout this chapter, we use the term **bilingualism** to refer to knowing more than one language; we don't use different terms for knowing two versus three or more languages.

9.1 Bilingualism in Early Childhood: Simultaneous Bilingualism

When a child grows up learning more than one language from birth, this is referred to as **simultaneous bilingualism**. This often happens when one of the child's parents is a native speaker of one language, the other parent is a native speaker of a different language, and each parent uses his or her native language in speaking to the child. It can also happen when the parents speak the same language natively but that language is different from the language spoken in the larger community, as long as the child has sufficient exposure to both languages from an early age (for example, the parents speak their native language at home, and the child hears the community language through day care or a full-time babysitter or nanny). In simultaneous bilingual acquisition, we can't really say that one language is the first language and one is the second; rather, the child has two first languages, hence the term **simultaneous**. François Grosjean, who has conducted considerable research on bilingualism, advocates referring to the two languages as Language A and Language α (alpha), so as not to give greater importance to either language or to imply that one language is primary.

While we can talk about children being exposed to their languages from the same time point in the abstract, and in some cases this is in fact accurate (as when each parent speaks a different language to the child from the time the child is born), in practice it is not always quite so cut and dried. That is, a child who starts being exposed to a language after birth but within the first year or two of life, as long as the exposure to that language is maintained, will grow up to be bilingual. So at what point do we distinguish simultaneous bilingualism from **successive bilingualism**?

This is a point of much disagreement in the literature. We see a range of answers to this question, from a very strict and early cutoff within the first month of life (De Houwer, 1995), to a later delineation at age 2 (Deuchar and Quay, 2000), to an even later demarcation at age 3 (McLaughlin, 1978). Researchers who place the cutoff at an extremely early point do so because infants are changing and developing so rapidly within the first months of life; De Houwer's argument is that if we want to compare bilingual language acquisition to monolingual language acquisition, we need to start at the same point (she refers to bilingual acquisition before 1 month as "bilingual first language acquisition" and bilingual acquisition after 1 month but before age 2 as "bilingual second language acquisition"). For our purposes, we will take a theoretically neutral stance and refer to the acquisition of multiple languages *in early childhood* as simultaneous bilingualism.

In simultaneous bilingual development, we see the same sorts of stages and progression as in monolingual language acquisition: children babble at the same age, and while they may produce their first words slightly later than monolingual babies do, they are still well within the normal range. Bilingual toddlers combine words just like monolingual toddlers do, and at about the same age (around age 2). They sometimes mix their languages, meaning they might combine a word from one language with a word from another language, as in the Spanish-English bilingual child's expression "más juice" (more juice). But in fact, they do not appear to be confused about their languages, and they do not create mixed expressions any more than adult bilinguals do when it is culturally appropriate to mix. (This is known as **code-switching**; see section 9.1.4.)

How can we tell that bilingual babies are not "confused" by their multiple languages? Let's take a look at some of the theories about this and how ideas have evolved over the past few decades.

9.1.1 The Single-System Hypothesis

In the first half of the twentieth century, it was believed that bilingual children had only a single language system that encompassed both of their languages. Leopold (1939) studied his daughter, Hildegard, who grew up speaking both English and German, and found that while most of her words were clearly either English or German, she sometimes produced forms that appeared to be mixtures of the two languages. For example, Hildegard produced the following verbs, which contain an English stem and a German infinitive suffix.

(1) a. pour-en (to pour)
    b. practice-en (to practice)

Leopold concluded that because his daughter did not seem to understand that the *-en* suffix could only attach to German verb stems, she must have a "fused" language system.

In the 1970s Volterra and Taeschner studied two German-Italian bilinguals girls and mapped out a progression from a single language system to, ultimately, an adultlike distinction between the languages. They argued that this took place in three stages. At the first stage, before about age 1;8, bilingual children were said to have just a single grammatical system that contained both vocabulary and grammatical rules from both languages. Evidence for this stage comes from the observation that these children lacked what are called **translation equivalents** (TEs). These are words in both languages for the same concept, for example, *water* in English and *agua* in Spanish. Since children at this age typically avoid synonyms (multiple words for the same thing) due to the mutual exclusivity constraint (see chapter 5), it seems reasonable that if children don't distinguish their languages, they would have just a single word for each concept.

The second stage in Volterra and Taeschner's model is reached when bilingual children have two different lexicons, one for each language, but still just a single set of grammatical rules (around ages 1;8–2;8). What this means is that although they now have TEs, they will apply a rule of syntax taken from one language in a sentence with words of either language. For example, the rules for creating possessive constructions in German and

Italian are different. In German, like in English, the possessor comes before the thing possessed and is marked with a possessive marker (*s*, like in English).

(2) Mamas Fahrrad
mama-poss bicycle
'Mama's bicycle'

In Italian, in contrast, the word order is reversed and the possessed object precedes the possessor, separated by the preposition *di* 'of'.[2]

(3) La bicicletta di mama
the bicycle of mama
'Mama's bicycle'

Volterra and Taeschner noticed that the German-Italian bilingual children they studied used only the German word order, even if the sentences they were producing were made up of Italian words.

(4) a. Miao bua
cat boo-boo
'cat's booboo'
b. Lisa bicicletta
Lisa bicycle
'Lisa's bicycle'

Finally, stage three is reached when children clearly distinguish their two languages both in terms of vocabulary and syntactic rules. At this point, around age 2;9, the children in their study produced Italian possessives with the correct Italian word order.

(5) la stellina di mama
the star-dim. of mama
'mama's little star'

Thus, the **single-system hypothesis** offers a three-stage model of how bilingual children progress from having a single, fused grammatical system to an adultlike separation of their languages.

9.1.2 The Separate-Systems Hypothesis

A very different account of early bilinguals' grammars was proposed in the 1990s, when more and more evidence emerged showing that many bilingual children had TEs from quite early ages. In fact, some reanalyses of Volterra and Taeschner's data, as well as other data that had been used to argue for the single-system hypothesis, revealed that these children in fact did have some TEs in their early vocabularies. To be sure, there were not a large number of TEs, but one could argue that having them *at all* must mean that the child is able to represent distinct lexicons. (Note that it could also suggest that bilingual children are less beholden to the mutual exclusivity bias.)

De Houwer (1995) argued that children had separate grammatical systems from the earliest point it can be observed. Evidence for a clear separation of the two languages comes not only from the existence of TEs but also from the observation that mixed productions such as Hildegard's, in which bound morphemes of one language are used with a stem of the other language, are quite rare. Although they are attested in the speech of some children (as in Leopold's observations and in another German-English bilingual's "Clean-st du dein teeth?" literally, 'clean-2sg[G] you[G] your[G] teeth' [Gawlitzek-Maiwald and Tracy, 1996]), many bilingual children never produce these kinds of mixed utterances. Instead, they reliably attach morphemes of a language onto stems of that language (Meisel, 1986). Some examples of appropriately inflected expressions of a Spanish-English bilingual child are given here (from Deuchar and Quay, 2000).

(6) a. more juice (1;7.5)
b. dos patos 'two ducks'(1;8.4)
c. shut door (1;8.9)
d. dos niños 'two children' (1;8.13)
e. more paper (1;8.16)
f. mamá come 'mama eats' (1;10.24)
g. zapato rojo 'red shoe' (literally, 'shoe red') (1;11.12)

Sidebar 9.1

We mention that the presence of translational equivalents (TEs) in a bilingual child's lexicon could signal either that they have distinct lexicons for their two languages or that they are less beholden to the mutual exclusivity constraint—that is, perhaps they allow two words in the same language to have the same meaning, even though monolingual children don't.

Au and Glusman (1990) wanted to find out whether bilingual children would suspend mutual exclusivity across their languages. Using some novel objects (e.g., seals with feathered tails and lemurs with horns), they labeled the objects with made-up words in either English or Spanish and then tested whether the children (ages 3–6) would apply a new label to that same object or would apply it to a new object. For example, they showed children a seal with a feathered tail and called it a *theri* [θɛɹi] (with English pronunciation). Then, another experimenter came in, presented the child with both the seal and a new animal, a lemur with horns, and asked in Spanish, "Which one could be called a *mido* [miðo]?" (Spanish pronunciation). Children allowed the new label to refer to the original animal as well as the new animal, suggesting that the children suspend mutual exclusivity across languages.

What if bilingual children suspend mutual exclusivity overall? A more recent experiment with younger children used preferential looking to test this in a different way. Byers-Heinlein and Werker (2009) showed 17-month-olds who were either monolingual, bilingual, or trilingual a familiar object (shoe) and a novel object (phototube). They measured infants' looks to the unfamiliar object when given a novel label ("Look at the *nil*!") compared to a baseline condition. If babies are using mutual exclusivity, they should take the novel word (*nil*) to refer to the unfamiliar object, not the familiar one. What they found was that monolingual babies showed a strong preference for looking at the unfamiliar object when they heard the label *nil*; bilingual babies also showed a preference but not quite as strongly as monolingual babies; and trilingual babies showed no preference for the unfamiliar object.

These results suggest that bilingual babies still impose mutual exclusivity within a language but that exposure to more than two languages could temper or delay the application of mutual exclusivity for word learning.

Bilingual children also reliably produce correct word orders for each respective language (as in 6f), though reversals and alternative orders can also be found (e.g., *paper more* [1;8.23]).

Moreover, many bilingual children exhibit distinct *paths* of development in their respective languages. For example, we discussed in chapter 7 that children acquiring languages such as German, French, and Dutch go through a phase in which they produce main verbs as infinitives (so-called optional infinitives [OIs]), while children acquiring languages such as Spanish and Italian do not do this. It turns out that children who are bilingual in, say, German and Italian produce their German verbs as OIs at the same rate as monolingual German-speaking children, but their Italian verbs are not produced as infinitives. Instead, both monolingual Italian-speaking children and German-Italian bilingual children overproduce Italian imperative verbs (Salustri and Hyams, 2006). Critically, German-Italian bilingual children produce Italian imperatives at the same rate as monolingual Italian-speaking children.

Other examples of separate paths of development come from asymmetries in how German-French bilingual children acquire noun phrases and sentence structure in their two languages: Koehn (1989, 1994) reports that French-German bilingual children express the plural form of articles earlier than nouns in French, but they mark plural on nouns earlier than articles in German; Meisel (1994) reports that by age 2 years, these children consistently place German verbs in final position (see section 7.2.3.2) but French verbs in nonfinal position, as is appropriate for each language. Thus, in terms of their morphosyntactic development, bilingual children can appear to progress just like monolingual children acquiring each respective language.

Evidence for language separation at an even earlier point in development comes from studies of phonological perception in infants. A perception experiment by Bosch and Sebastián-Galles (2003) found that by 12 months of age, babies who were being raised in a Spanish-Catalán bilingual home were able to discriminate two vowels—/e/ as in *bait* versus /ɛ/ as in *bet*—which are contrastive in Catalán but not in Spanish. (Monolingual Catalán-acquiring infants could also discriminate these vowels, but monolingual Spanish-acquiring babies could not.)

9.1.3 The Interdependent Development Hypothesis

A third view about how the languages of a bilingual speaker are represented in the mind is that rather than being initially completely fused or completely separate, a bilingual's languages are distinct but interconnected (Döpke, 2000). This idea comes from the observation that bilinguals are not really "two monolinguals in one body"; rather, their languages can influence one another. For instance, bilinguals can code-switch, or switch between their two languages within a sentence or a discourse (see section 9.1.4), something monolinguals obviously cannot to. There is evidence of **cross-language influence** (CLI) in language production (performance), such as, for example, when a bilingual speaker mistakenly retrieves the word they want from the wrong language. We can also see the influence of one language on the other at the more abstract level of representation, which relates to competence. For example, although we saw evidence in the previous section for separate paths of development in bilingual children's acquisition of syntax, Paradis and Genesee (1996) identified three ways in which a bilingual learner's languages can be seen to influence each other in development: through transfer, delays, or accelerations in the acquisition of certain grammatical forms. This is the idea of **interdependent development**.

Transfer might occur if a child is dominant, or more proficient, in one of the two languages. In this case, a construction from the dominant language might show up when the child is speaking the weaker one. Yip and Matthews (2007) studied some Cantonese-English bilingual children who were dominant in Cantonese. Cantonese, like Mandarin and unlike English, does not move the wh-word to the beginning of the sentence in a wh-question; instead, the wh-word remains in its underlying position (e.g., *What did you buy?* would be, literally, *You bought what?*). Yip and Matthews found that when these children were speaking English, they produced more wh-questions by leaving the wh-word in its underlying position in the sentence rather than moving it to the beginning (e.g., "Put in where?") compared to monolingual English-speaking children.

Acceleration could be observed if a child is faster in acquiring some construction in one of their languages, compared to monolingual children acquiring that language, by virtue of getting a boost from the other language. For example, monolingual children acquiring French and German generally begin to produce verb inflections and auxiliary verbs at a younger age than monolingual English-speaking children (English-speaking children, instead, continue to produce "bare verbs" until quite late—see section 7.2). Gawlitzek-Maiwald and Tracy (1996) studied a German-English bilingual child (who was dominant in German), and when she began to produce auxiliary verbs in her German utterances, she sometimes used these German auxiliaries in her English utterances as well. For example, the researchers found utterances such as 7 (*ge-* is the German prefix for a past participle):

(7) Ich hab gemade you much better
    'I have-1sg ge-made you much better'

Thus, this bilingual child was able to begin producing auxiliaries and other more complex syntactic constructions in her English utterances at an earlier stage than English monolinguals, thanks to a boost she got from her German grammar.

Finally, a **delay** might be triggered by the fact that bilingual children simply have twice as much stuff to learn. In fact, bilingual children do have measurably smaller vocabularies in each of their languages compared to monolinguals (Pearson et al., 1993). However, there does not appear to be a significant lag for bilingual children in reaching the main milestones of language development, and their overall vocabulary is likely about the same size (if not larger) than that of a monolingual child.

Thus, there is support for the idea that a bilingual's languages develop interdependently. But it is important to remember that the idea of *interdependent* systems presupposes that the languages are in fact distinct *systems* and not part of one large, fused language representation.

9.1.4 Code-Switching

As mentioned above, bilinguals can engage in what is called code-switching: switching between two languages or dialects (codes) within a sentence or a discourse.

Code-switching is subject to both discourse constraints and grammatical constraints. In terms of discourse constraints, code-switching is appropriate in some conversational contexts but not in others. Code-switching is not appropriate if the person you are talking to does not speak both of your languages. But even if you are speaking with another bilingual speaker with whom you share both languages, code-switching is only appropriate if you are alone or if you are in the company of others who also share your languages. Engaging in code-switching when monolinguals are part of the conversation can make them feel excluded. Other constraints on the appropriateness of code-switching can come from the formality of the situation. Depending on the circumstances, it may not be appropriate to code-switch in a very formal context. On the other hand, if all participants in the conversation share the same languages and the topic of conversation warrants using phrases and expressions from both languages, it is easy to imagine a case where code-switching would be appropriate even if the context is somewhat formal (for example, an international academic conference).

Interestingly, bilingual children are highly attuned to these sociolinguistic constraints on code-switching. Children are acutely aware of the linguistic profiles of the people around them, and when they encounter a stranger they can use a variety of cues to make an educated guess as to that person's language(s), such as the location in which they are meeting the person, the way the person looks (for example, if a person looks Latin American, children may assume the person speaks Spanish [Fantini, 1978]), which of their parents the person knows (if the child's parents speak different languages), and so forth. A bilingual child can then use some of these cues to guess which language or languages would be appropriate to use with the person and whether code-switching would be appropriate. Children are also very sensitive to adults' attitudes about language mixing. Lanza (1992) studied a Norwegian-English bilingual child whose two parents held very different attitudes about this. The mother, a native English speaker, was quite opposed to code-switching and would respond when her daughter mixed Norwegian and English by pretending not to understand or by correcting her. The Norwegian-speaking father, on the other hand, would continue the conversation when his daughter mixed the languages (indicating he had understood his daughter's mixed utterance) and would code-switch with her on his own. With time, this child learned her parents' preferences and would readily code-switch with her father but avoid doing so with her mother.

Children code-switch for the same reasons adults do: they may prefer to use a word or expression in the language they first learned it in, or if they have trouble remembering a word in one language, they might use the equivalent word from the other language. Yip (2013) mentions that humor can also be a motivator for children to use an expression from one language when speaking in the other language, as in the following example from a 4-year-old Cantonese-English bilingual:

(8) Lei5 bump into my fei4 tou5laam5[3]
'You bump into my fat belly.'

Apparently the word that is translated here as "fat belly" is considered cute and amusing to Cantonese speakers.

We mentioned above that code-switching is also subject to grammatical constraints, and bilingual children (and adults) may code-switch in ways that betray an asymmetry in their linguistic competence in the two languages: that is, bilinguals may be **dominant** in one language over another. This is true regardless of whether the speaker learned their languages simultaneously or successively, and, furthermore, dominance is something that can shift over one's lifespan (Grosjean, 2013). In code-switching, a bilingual speaker may be more likely to use a word or expression from their dominant language while they are speaking in their weaker language. Moreover, when a bilingual is dominant in one of their languages, they are more likely to limit code-switching to lexical categories (nouns, verbs, adjectives) when mixing from their weaker language into their stronger language; it is much less common for speakers to use functional categories (determiners, auxiliaries) when mixing from their weaker language. On the other hand, when the mixing occurs by taking lexical items from the stronger language into the weaker language, either type of category may be used.

In summary, children demonstrate a remarkable level of awareness of their linguistic environment and which language or languages are appropriate to use with the people around them. We'll give one final illustration of this, using a multilingual family known by one of this book's authors when she lived in Los Angeles. The mother was a native speaker of Castillian Spanish, the father was bilingual in French and Italian, the parents spoke French to each other, and the nanny spoke Mexican Spanish. Their child, Pablo, spoke French as his first language, Spanish as his second, and English as his third. By the age of 3, he was able to use all three languages and did so appropriately: he used French with his parents, Spanish with his nanny, and English with the (mostly monolingual English-speaking) neighbors. According to Pablo's parents, when family friends came to visit, Pablo would speak to them in whichever language they used with his parents; he knew that most English speakers were monolingual but that many French and Spanish speakers could also speak English, so code-switching with English was okay. Since his nanny spoke only Spanish but no English, he used only Spanish with her and did not code-switch—unless he wanted to say something that he didn't want his nanny to understand: then he would use English!

9.2 Successive Bilingual Acquisition

Although we might think of simultaneous bilingual acquisition as the typical way in which a person becomes bilingual, it is actually more common for a person to become bilingual by learning a second language after early childhood—that is, after a first language has been largely acquired. This is known as **successive** or **sequential bilingualism**. There are many different reasons why someone might become bilingual after early childhood. Some common reasons include immigration, learning a new language in school, acquiring a new language for work-related reasons, marriage (or a parent's remarriage) to someone who speaks another language, and personal interest later in life.

In chapter 8 we saw that when a first language is acquired outside of early childhood, as in the case of Genie and deaf individuals who do not have access to sign language or hearing technology, language does not develop in the typical way. We also saw that as people get older and farther outside the critical period, the ability to acquire a first language declines (compare the language abilities of Genie and Chelsea, for example; see sections 8.1 and 8.2). However, there is also evidence that second-language acquisition in late childhood or adolescence (or adulthood) proceeds with far more success than late *first*-language acquisition. Thus, while second language (L2) learners may experience some challenges in learning their language, all is not lost. People can and do acquire additional languages after infancy, often with great success.

In fact, there are some advantages that older language learners have over younger language learners. One concerns vocabulary acquisition. A study of a 5-year-old Japanese girl acquiring English found that she acquired vocabulary words far more rapidly than a child acquiring English natively: while the 5-year-old L2 learner acquired seventy-five words in seven weeks, and an additional ninety-six words over the following four weeks, the English L1 learner took fourteen weeks to acquire his first seventy-five words (Rescorla and Okuda, 1984; see also chapter 5). The lexical advantage could come from the 5-year-old's more advanced cognitive abilities, or it could be advantageous to already have a set of concepts with labels in one language when learning new labels for these same concepts in another language.

In terms of building sentences, child (and older) L2 learners typically progress faster than L1-learning toddlers in certain respects. Specifically, older learners may start out building longer and more complex sentences. Think back to an introductory language class you may have taken in high school or college. You didn't start out building telegraphic expressions like "eat cookie" or "man go," right? You probably started out learning expressions like "My name is X" or "I am from Y" or "Where is the bathroom?" Older learners can learn to produce these kinds of longer strings before they build up a significant knowledge base of the language's grammar.

The apparent advantage that L2 learners have over L1 learners in early word learning and sentence construction has some important caveats. One is that even though these learners seem to have an initial advantage, they may and often do ultimately fall short of attaining native-like language proficiency. This can be seen in the fact that they take longer to master inflectional morphology (if they ever do master it), including verb endings that mark tense or agreement. In one study, Blom et al. (2006) found that child L2 learners of Dutch (ages 5–7 years) were almost as accurate as native Dutch-speaking children in applying the correct verb endings (though not quite perfect: native Dutch 4-year-olds were 100% correct, while the L2 children were 90–92% correct). Adult L2 learners, in contrast, were only 57% correct, despite having about the same amount of exposure to Dutch as the child L2 learners.

The difficulty adult L2 learners have with verbal inflection can persist even after many years of language exposure. One such documented case comes from Lardiere (1998), who studied tense marking on verbs by an adult L2 learner of English. The individual she studied, whose first language was Chinese, produced tense marking less than 50% of the time when it was required in English, even after living in an (immersive) English-speaking environment for ten years; after living in this environment for almost twenty years, her production of tense marking was still only about 70%.[4]

Another caveat is that older learners tend to struggle with attaining a native-like accent. The extent to which this is true of child L2 learners is highly variable: some child L2 learners readily adapt to the native phonology and accent of their second language, while others (especially if exposure begins after about age 7) may retain traces of a non-native accent.

Ultimately, it is important to bear in mind that people can and do learn languages later in life, and it is entirely possible to become bilingual with successive exposure to one's languages. However, outcomes for older L2 learners are much more variable than they are for native learners. Some adults appear to have a gift for learning languages and can become quite fluent in multiple languages, while others struggle even with years of exposure. This is quite different from native-language acquisition (no matter how many languages are acquired) where, barring pathology or interrupted exposure, speakers are remarkably uniform in the underlying grammatical competence they attain. In addition to age of acquisition, there may be many different factors that contribute to the difference between native and later language acquisition (naturalistic versus classroom learners, quantities and qualities of exposure, and much more). Above and beyond anything else, perhaps such variability itself is the best explanation for the variability that characterizes non-native acquisition.

9.3 Language Attrition and Heritage Language

When a child grows up speaking English in an English-speaking community, there is little doubt that their knowledge of English will be maintained throughout their life. Unless the person moves to a different country and stops using English, English will remain their dominant language and the language they will pass on to their children. The same expectation is reasonable for any language that is considered a **majority language** within a society or community—the language spoken by the majority of people and/or the language associated with high social prestige, education, economic opportunity, and political power: Mandarin Chinese in China; French in France; Spanish in Spain, Mexico, and most Latin American countries; and so forth.

But for many people, the linguistic environment changes over the course of their lives. For example, children of families who immigrate to a new country may find themselves in a community that speaks a different language from the one they grew up with for the first several months or years of their life. Or even if a child doesn't move across a national border, the language spoken in the home might be different from the language of the larger community, so they may encounter a different language as soon as they begin attending school, around age 5. Even though there is no reason why a person cannot grow up bilingual and maintain both (or all) of those languages throughout life, such an outcome is not guaranteed, in particular in situations where both languages are not equally valued within society.

A common outcome for people whose linguistic environment shifts is that their first language, the one spoken in their original country and/or in their home when they were very young, can undergo **attrition**, or loss. This can happen in the following way: On greater exposure to the larger community (e.g., through television, school, making friends with people outside the family), the child quickly realizes that the language of social currency is not the language their parents speak. Children are acutely aware of which languages, dialects, and ways of speaking are considered prestigious, or highly valued, and it is important to many children to be accepted by their peers. Thus, if their home language is not spoken by their peers, children will sometimes abandon their home language in favor of the language that *is* spoken by their peers; this is the language that gives the child social currency. When this happens, the child's first language can attrite. Attrition can involve a weakening of the language (e.g., loss of vocabulary, difficulty with more complex syntactic constructions, inability to use the language to talk about more complex topics) or even a loss of the ability to produce the language at all. Generally speaking, the ability to comprehend the language is maintained; a person who retains the ability to comprehend a language but not the ability to produce it is known as a **passive** or **receptive bilingual**.

It is important to consider what effect this type of language attrition can have on families. Although many people feel it is most important for children to acquire fluently the majority language of a community (e.g., English in the United States or the UK, French in France, German in Germany) in order to secure a job later in life, there is a serious danger of losing the home language. Losing (or lessening) the ability to speak their home language can result in an erosion of the child's emotional connection to their family. In turn, this linguistic and emotional erosion can sever or weaken the ties to an important support network that is critical during the adolescent years. Although many adolescents drift emotionally from their families toward their peer group even *with* a shared language, *not having* a shared language all but guarantees a breakdown in communication and connection with the family.

When a person whose first language has attrited gets older, they may desire to reacquire or improve their knowledge and skills in that language (depending on how much it has attrited). Such a person is then called a **heritage speaker** or **heritage language learner**. Essentially, heritage speakers are dominant in their *second* language, and they are attempting to relearn their *first* language.

There are a number of interesting differences between heritage learners and second language learners in terms of their path of acquisition. Compared to second language learners, heritage learners typically have very good **pronunciation**. For example, studies of speakers with Korean or Spanish as their heritage language and English as their dominant language revealed that these speakers were significantly more native-like in their pronunciation of stop consonants in Korean or Spanish (respectively) compared to English-speaking L2 learners of these languages (Au et al., 2002). This was true even if the heritage speakers did not use their heritage language on a regular basis. On the other hand, the heritage speakers were not exactly like native speakers in their pronunciation; thus, they appear to be "somewhere in between" L2 learners and native speakers in their pronunciation ability.

Heritage learners also tend to have a reasonably large **lexicon**, but it generally consists of very common words and words used in domains related to home life (e.g., words for objects found in the home or actions associated with home, such as cooking) or religious practices. Where heritage learners tend to struggle (though not more than second language learners) is in **syntax** and **morphosyntax**. For example, many studies have found that heritage speakers make errors in gender, case, and agreement marking on noun phrases in languages that mark such features (Spanish marks gender, Arabic marks agreement, and Russian marks both gender and case). For example, Spanish heritage speakers often make errors in marking feminine gender, especially overextending masculine gender to nouns that do not end in *-a* (e.g., *la nuez* 'the-fem. nut', *la mano* 'the-fem. hand' [Montrul et al., 2008]). Complex syntactic constructions, such as relative clauses, are also known to be difficult for heritage speakers.

One of the major challenges for studying heritage-language acquisition is that the circumstances under which someone becomes a heritage speaker or learner can vary quite dramatically. The age at which a person reduced the use of their first language, the amount and type of input they received in that language, and the degree to which they have maintained use of the language since childhood, among other factors, can vary widely across individuals and accordingly affect their proficiency in the language. A further complication is that the heritage speaker's parents (and other adults who provide the input for their L1) may themselves undergo some degree of attrition in their native language. There is clear evidence that even adults can lose some knowledge of their native language once they emigrate from their home country (Pavlenko and Malt, 2011); thus, the input to children who become heritage speakers may differ in important ways from the input to children growing up in the country where this language is the majority language.

9.4 Language Endangerment and Language Revitalization

In this chapter we have been looking at how multiple languages are acquired in childhood and whether they are acquired simultaneously or sequentially. In the last section we discussed how a language can be lost in an individual speaker when that speaker's family moves to a new language community or when the family's home language is not the majority language of the community where they live. Languages can also be lost on a larger scale—whole languages can cease to be. This happens when their speakers stop speaking them.

If a Spanish-speaking family immigrates to the United States and the children in the family stop speaking Spanish, the Spanish language will attrite (perhaps temporarily) for these individual speakers (and this can have potentially negative effects on the individual and family; see the previous section), but the language itself will not disappear. Spanish is spoken worldwide by an estimated 560 million people. But consider what this would mean for a language that has only a few hundred speakers, all of them aging: the very survival of such a language depends on young people acquiring it and maintaining it. If that does not happen, it will die out.

As many as half of the languages alive today, over 3,500 languages, will be extinct by the end of this century. These are called **minority languages**—they are spoken by a relatively small number of people within a community, or they are associated with low prestige and a lack of political and economic power. Many of these languages are spoken by indigenous peoples—that is, people whose ancestors occupied a particular region longer than the colonial peoples who moved in and took control. When a language is associated with low prestige, people will often gravitate to a higher-prestige language. That language comes to be used in more and more domains of life (e.g., in legal matters, in school, in formal interactions, and eventually in informal interactions and in the home) until the low-prestige language is no longer used on a regular basis and parents stop passing it down to their children. In this circumstance, the language is **endangered** and at risk of becoming extinct.

This is a somewhat oversimplified account of how languages become endangered; there are a variety of causes of endangerment, and often the reasons for a language losing prestige are complex. Particularly in situations of colonial rule, there can be varying degrees of **coercion**, in which an indigenous language may be banned from schools, and children (and adults) may be physically punished or humiliated for speaking the language. Even without explicit coercion, speakers may see the colonial language as the path to economic success and therefore discourage their children from speaking the indigenous language, even though it forms an integral part of their history and culture. Perhaps less visibly, but no less significantly, children can be ridiculed or shunned if they do not speak the majority language just like native speakers of that language do. Such peer pressure can be an even stronger influence over children and adolescents than parental urging.

What is important to recognize is that this situation is shockingly common and widespread. Of the roughly seven thousand languages spoken in the world today, only about 6%,

Table 9.1
Selected languages and their relative level of endangerment

Language	Where spoken	Number of speakers	Endangered?
Chamorro	Guam	60,000	vulnerable
Tok Pisin	Papua New Guinea	120,000	no
Corsican	Corsica	160,000	definitely endangered
Icelandic	Iceland	330,000	no
Basque	Spain, France	700,000	vulnerable
Kurux	India	1,751,489	vulnerable

fewer than four hundred languages, have over a million speakers each. These relatively few languages dominate the world, being spoken by 94% of the earth's population. That means that the other 94%, well over six thousand languages, are spoken by a tiny fraction of the world's population, only 6% (Anderson, 2012). Now, having a small number of speakers is not in itself a death sentence for a language. Researchers believe that for most of human history, in fact, people lived in small communities that had a language different from the languages of neighboring communities, and people were most likely bilingual in whichever languages they needed for trade and other interactions with nearby groups (Nettle and Romaine, 2000). Such a state has been maintained to a large extent into modern times in the country of Papua New Guinea, a mountainous island just north of Australia that is home to the world's greatest linguistic diversity. Even there, currently, smaller languages are threatened by larger languages such as English and Tok Pisin, but they may have existed for centuries with only a few thousand speakers or even a few hundred speakers.

Some further examples of languages and their numbers of speakers and endangerment status are given in table 9.1 (data from the UNESCO Atlas of the World's Languages in Danger).[5]

Even though a relatively small number of speakers is not *necessarily* a sign of endangerment, a language with very few speakers that is *also* considered low prestige compared to another language in the community and is used in few social domains faces an uphill battle. One of the central criteria for judging a language's level of endangerment is called **intergenerational transmission**. This is the likelihood of a language being passed down naturally from parent to child in exactly the way we have discussed throughout this book. Given that languages die when people stop acquiring them, efforts to **revitalize** endangered languages are relevant for the field of language acquisition.

9.4.1 Is Language Revitalization Important?

In thinking about global communication, we might catch ourselves musing, "Wouldn't it be better if everyone in the world spoke only one language?" On a superficial level it might make certain things easier: international business and trade might operate more smoothly;

we might not have to worry about misunderstandings due to unfamiliarity with someone's language; we wouldn't have to spend so much time taking foreign language classes. But as soon as we dig just a little bit deeper, we can see that there are some important reasons to maintain and revitalize languages that are at risk of dying. The question of whether language revitalization is important, and whether it is a positive direction, is extremely complex, and a full treatment of this question goes far beyond the scope of this book. Here we give just a flavor of the considerations related to this question. First, we'll consider some things that are lost when a language dies, and then we'll consider what is gained when language diversity is preserved.

9.4.1.1 Scientific Knowledge, Including Linguistic Knowledge

Many indigenous peoples possess a wonderfully rich body of knowledge about their environments, including scientific knowledge about plant and animal species, medicinal uses of plants, and the ecosystems their people have lived in for centuries, if not longer. This knowledge is encoded in these people's language, and if the language stops being used, the information will stop being shared. While the information itself could be written down in English (or another widely spoken language), this information might then benefit an elite scientific community but not the people who gathered it. Additionally, if we want to preserve this information for humanity's benefit, it would be most efficient for the people who hold the knowledge themselves to record it in the manner most efficient for them—namely, in their own language.

Scientific knowledge includes not only ecological science but also linguistic scientific knowledge. We pointed out in chapter 2 that until recently linguists thought that the OVS word order was impossible because it had never been found in a language… until it was found. A great number of remarkable discoveries have been made about human language because of a single exemplar. One such case noted by Anderson (2012) is that in the Native American language Wichita, the verb does not agree with the subject but with the possessor of the subject (so you would get the equivalent of *My house am small* or *Your house are small* instead of *My/your house is small*). Some of the most complex morphological and syntactic systems that linguists have found have been in small languages spoken in remote places by few people. As linguists, we want to know what the possibilities of human language are: If UG offers a blueprint, what is that blueprint? When languages disappear before they can be studied in depth, the scope and richness of that blueprint is diminished.

9.4.1.2 Cultural Knowledge

People use language to talk about their culture, and sometimes language itself encodes certain properties of a culture. For example, a language might have a complex system of honorifics (i.e., using special words when speaking to someone higher or lower than you in a social setting), or women and men might speak differently because of cultural norms. Religious and historical knowledge is encoded in language. In principle, it is possible to write all of this information down in a majority language, like English or Spanish, but is this the same as preserving the experience of these cultural forms for the speakers themselves? The grandchildren of people who expressed reverence using a particular verb form rather than another might be able to read about it in a textbook, but is this the same as actually using the form themselves? We believe there is a big difference.

9.4.1.3 Identity

This is the least tangible of the things that are lost when a language dies, but it is not the least important. People's identities are complex and shaped by a multitude of factors related to cultural upbringing, socialization, religious affiliation if they have one, education, political views, and many other things, including language. For those of us whose native language is a world language like English, it may not be part of our consciousness that our language is part of our identity: we take English for granted. But imagine you were told you were never to speak your language again. How would this make you feel? Some of you reading this book may have had this experience. Or you may have voluntarily put yourself in a position of not being able to use your native language, such as in a foreign-exchange program. That is hardly a coercive situation, but still—how did you feel? Chances are, when your language is "taken away" you feel a bit lost, at least for a while. Now add to this scenario the idea that the loss of your language is not temporary but that the people telling you to stop using it are also harming or killing your relatives or forcing them to abandon their ways of life. This type of experience is utterly traumatic, and it is no accident that the loss of a group's indigenous language has been linked to increased suicide rates (Hallett et al., 2007) and other negative mental health effects, even when other social factors are controlled for. On the flip side, language maintenance and revitalization has been associated with improved mental and physical health for indigenous speakers (Whalen et al., 2016).

A world language like English might be your ticket to economic opportunity, but at what cost?

9.4.1.4 Autonomy

Finally, let's consider what is gained if speakers of a minority language are able to keep their language alive. In addition to *not* losing all of the things that would be lost if the language dies (sections 9.4.1.1–9.4.1.3 above), people gain a sense of pride, autonomy, and sovereignty when they maintain their language. Autonomy of a community means having the power to determine on their own terms whether and how their language is used, preserved, taught in schools, and so forth. Keeping the language alive means the community is able to make these determinations. Once the language dies, it is significantly more difficult to bring it back, and the "option" to keep using the language is often no longer realistic.

In our view, the question of why and whether language revitalization is important really comes down to the speakers themselves: Do they want to keep their language alive? So much has been taken from indigenous peoples by colonial powers: their land and way of life, their language, and their autonomy. It seems to us that if an indigenous group wants to preserve their language, the decision should be theirs. In that case it is the duty of the linguistic scientific community to help them in whatever way the group deems appropriate.

9.4.2 How Are Languages Revitalized?

We have talked about how and why languages become endangered and some reasons (we believe) it is important to revitalize them. The next question is, *how* are languages revitalized?

Many people are familiar with the success story of Modern Hebrew. Hebrew was the ancient language of the Jewish people, and although it had been used as a vernacular (an everyday, spoken language) in the Middle East until about 2000 years ago, in the intervening centuries it became relegated to religious texts and rites. Thus, although it still existed in written form (and was known to religious leaders and educated men in the Jewish community), by the end of the nineteenth century it was essentially a dead language—it was no longer a living language acquired naturally by children. Around the end of that century, however, Jews began returning to Palestine from various different countries; because they lacked a common language, Hebrew was revived over the course of a few decades, and now it is spoken fluently by seven million people.[6]

The success of Hebrew revival must be seen as an anomaly, however. In fact, the sociopolitical, linguistic, and religious contexts that made its rebirth possible formed a kind of perfect storm that is rarely encountered. Yet certain other languages have seen some measure of revitalization in recent decades, so let us examine some of them briefly.

The Maori language of New Zealand has increased its number of speakers from about ten to twenty thousand in the 1970s, when it was mainly used for traditional functions but not in day-to-day interactions, to about sixty to seventy thousand today. One tool that has been instrumental in rebuilding this language is the concept of **language nests** (called Kohanga Reo in Maori), which were established starting in the early 1980s. People recognized that the best hope for the language's survival lay in the children of the community, so they established preschools that would offer an immersive environment for children to acquire Maori. As the children grew, the community followed up with primary schools (Kura Kaupapa Maori) that likewise offered Maori immersion. The idea of using language nests as a means of providing language exposure to the community's young children has been adopted by other communities. Immersion preschools are now found in Hawai'i, the Cherokee Nation (Oklahoma), the Qualla Boundary of the Eastern Band of Cherokee Indians (North Carolina), British Columbia (teaching various First Nations languages), and elsewhere.

While exposing very young children to a language is an important ingredient in language revitalization, one must also ensure that those children will continue speaking it in everyday contexts outside of school and throughout their lives. The only way for a lan-

Sidebar 9.2

> Some minority language communities have not only language nests but also language immersion schools for the primary (and in some cases secondary) school grades. What are some of the challenges in establishing and running these schools? What kinds of curricular materials are needed to run a primary school, and where do you think the teachers get their materials from? Do you think it is easy or difficult to find teachers for these schools? What are some challenges for the students in terms of learning the language and maintaining it?

guage to survive from generation to generation is if it achieves intergenerational transmission. This means that adults in the community must also use the language on a regular basis. In some communities the language has skipped a generation because children thirty or forty years ago were prevented from speaking their language (this was widespread through the use of boarding schools in Australia and North America), so that many of today's parents no longer even have knowledge of the language themselves. Adult language learners face a number of challenges in gaining fluency in a new language, but adult language acquisition is not impossible. One method that has been used successfully in this context is called the Master-Apprentice Program (Hinton, 2002, 2013). This method has a simple design: An elder fluent speaker (the master) meets up with a person who wants to learn the language (the apprentice), and the two people try to engage in basic conversation with each other while doing everyday activities. Gradually, the apprentice learns more and more of the language and can use it in a wider range of contexts and activities. Although this approach does not quickly produce dozens or hundreds of new speakers at a time, it has been quite successful in creating proficient speakers of severely endangered languages.

In order to achieve intergenerational transmission, the children who go through a language nest and then an immersion primary school must grow up and pass the language on, naturalistically, to their own children. One way to encourage the use of a language into adulthood is to provide job incentives that require the language. A good example of a community that has done this is the Basque Country. Basque, a language isolate that is spoken in the Pyrenees mountains of northern Spain and southwestern France, is a mildly endangered language (it is considered "vulnerable" on the UNESCO atlas of endangered languages).[7] It is spoken by about seven hundred thousand people, but it is a minority language to Spanish (in Spain) and French (in France). In Spain (where the largest part of the Basque Country lies), the Basque language was outlawed for many decades during the dictatorship of Francisco Franco. It could not be taught in schools, and people were punished (usually by fines or imprisonment) for speaking it. Many parents chose not to pass the language on to their children, and by the 1970s the number of speakers had dropped dramatically (Lasagabaster, 2001).

In addition to the establishment of Ikastolak (Basque language schools), some of which operated in secret even during Franco's regime, the new democratic government of Spain made the Basque language co-official with Spanish within the regions of the Basque

Country, and many municipal jobs now have a language requirement. University teaching jobs, for example, require high proficiency in Basque, regardless of which subject one teaches. Having job access tied to language proficiency can be a powerful motivator to get people to continue using a language. This approach has produced impressive results: the number of Basque speakers increased by over 35% in the period from the mid-1980s to the mid-2000s, and enrollment in Basque language schools has increased by about 20% in the first decade of the twenty-first century (Lasagabaster, 2001; Valadez et al., 2015).

Another motivator for language use is more ineffable but probably not less important: coolness. We mentioned in section 9.3 that children of immigrants often gravitate toward the majority language (and away from their home language) because the majority language is considered cool. This effect is most noticeable among youths and adolescents, but it is precisely this age group that must keep using a language if they hope to pass it on to their children once they become adults. Is it possible to make a minority language seem cool? This is exactly what some hip-hop artists are working to do with indigenous languages in Guatemala and Ecuador. For example, the group Balan Ajpu creates rap music using a combination of Spanish and Tz'utujil, a Mayan language. While it is too early to know exactly how this movement will affect the number of speakers of indigenous languages, it is likely to have a positive effect. Researchers credit pop artists with helping to increase interest in languages such as Basque and Catalán (Coluzzi, 2007).

Other factors that can play a role in increasing the prestige and visibility of minority languages include giving the language official status (as has happened for Maori, Hawaiian, and Basque, for example) and having university-level programs that teach the language (as is available for Navajo, for example). Ultimately, the grassroots-level impetus for using the language, visibility of the language in everyday contexts, and a strong desire on the part of the community to speak and maintain the language will determine whether it lives or dies. The Irish language, for example, is recognized as the official language of the Republic of Ireland, and it is taught in schools at every level through university. But until a critical mass of Irish people decide to use the language for everyday interactions in all domains of life and pass it on to their children, Irish will falter under the weight of English desirability.

For those of us whose native language is spoken by millions of people around the world, we have the luxury of not worrying about our language going extinct anytime soon. But we can be allies to those speakers who are in danger of losing their language, the language of their ancestors. Having support from the majority community is an important ingredient in the success of revitalizing an endangered language (Crystal, 2000).

9.5 Summary

In this chapter we looked at some aspects of bilingualism—how multiple languages are acquired simultaneously or in sequence—and how languages can be lost in an individual (creating heritage speakers) or in a community (creating endangered or extinct languages) when they do not hold high prestige within a society. Although the bulk of this book is about first-language acquisition in children, it is important to remember that as humans we are able to acquire and lose languages throughout our lives: this is what linguist François Grosjean calls "the wax and wane of languages."

9.6 Further Reading

Grosjean, François. 2010. *Bilingual: Life and Reality.* Cambridge, MA: Harvard University Press.

Grosjean, François, and Ping Li. 2013. *The Psycholinguistics of Bilingualism.* Malden, MA: Wiley-Blackwell.

Harrison, K. David. 2007. *When Languages Die: The Extinction of the World's Languages and the Erosion of Human Knowledge.* Oxford: Oxford University Press.

Montrul, Silvina. 2010. Current issues in heritage language acquisition. *Annual Review of Applied Linguistics* 30: 3–23.

Nettle, Daniel, and Suzanne Romaine. 2000. *Vanishing Voices.* Oxford: Oxford University Press.

9.7 Exercises

1. The following table lists the proportion of words in a bilingual child's vocabulary that are translation equivalents (TEs): pairs of words in the two languages that mean the same thing (from Deuchar and Quay, 2000). The third column means the number of words that form pairs; for example, the number 2 means two words that form a pair together (in this case, the words were English *bye* and Spanish *tatai*). The far-right column means the percentage of the child's total vocabulary that involves TEs. Thus, at age 0;11 the child has three words, two of which form an equivalent pair.

Age	Total number of words	Number of words with equivalents	Percentage of words with equivalents
0;10	0	0	0
0;11	3	2	67
1;0	3	2	67
1;1	7	2	29
1;2	8	2	25
1;3	17	2	12
1;4	32	2	6
1;5	58	14	24
1;6	83	18	22
1;7	125	35	28
1;8	202	68	34
1;9	253	96	38
1;10	330	146	44

Questions:

(i) At what point is there clear evidence that this child has TEs in her lexicon? On what basis do you make this judgment? (There are several possible answers.)

(ii) Given your answer to (a), at what point do you think this child represents her lexicons as distinct systems? Can you tell? Explain your answer.

2. In section 9.1.2 we talked about one experiment that showed that at 12 months of age, bilingual babies exposed to both Spanish and Catalán could discriminate two vowels (/e/ and /ɛ/) that are contrastive in Catalán but not in Spanish, similar to babies exposed only to Catalán but unlike babies exposed only to Spanish. This experiment also revealed that at 8 months of age, the bilingual babies could *not* discriminate these vowels, although babies exposed only to Catalán *could* discriminate them. What does this finding suggest about the separate-systems hypothesis? Why might these 8-month-old babies exposed to both Catalán and Spanish fail to discriminate these sounds?

3. We mentioned in this chapter that many factors can complicate the study of heritage speakers' language, such as the age at which speakers lessened their use of the first language and the amount and type of input they received in that language. In your own words, explain what these factors are (if there are others besides these two examples, name them) and how they complicate the study of heritage-language acquisition. If you have personal experience as a heritage speaker, reflect on your own experience: Can you determine which factors have most strongly affected your level of proficiency in your language? If you are not a heritage speaker but you know someone who is, interview that person to learn about their experience and perspective.

4. In chapter 5 we talked about the mutual exclusivity constraint, and question 4 at the end of that chapter asked you to think about what this constraint implies for bilingual learners. If you completed that exercise earlier, does reading this chapter cause you to rethink your answer? Why or why not? If you did not complete that exercise, how do you think mutual exclusivity would play out in a bilingual learner?

9.8 References

Anderson, Stephen R. 2012. *Languages: A Very Short Introduction.* Oxford: Oxford University Press.

Au, Terry, and Mariana Glusman. 1990. The principle of mutual exclusivity in word learning: To honor or not to honor? *Child Development* 61: 1474–1490.

Au, Terry, Leah Knightly, Sun-Ah Jun, and Janet Oh. 2002. Overhearing a language during childhood. *Psychological Science* 13: 238–243.

Bosch, Laura, and Núria Sebastián-Gallés. 2003. Simultaneous bilingualism and the perception of a language-specific vowel contrast in the first year of life. *Language and Speech* 46: 217–243.

Byers-Heinlein, Krista, and Janet Werker. 2009. Monolingual, bilingual, trilingual: Infants' language experience influences the development of a word-learning heuristic. *Developmental Science* 12: 815–823.

Coluzzi, Paolo. 2007. *Minority Language Planning and Micronationalism in Italy*. Bern: Peter Lang.

Crystal, David. 2000. *Language Death*. Cambridge: Cambridge University Press.

De Houwer, Annick. 1995. Bilingual language acquisition. In Paul Fletcher and Brian MacWhinney (eds.), *The Handbook of Child Language*, pp. 219–250. Cambridge, MA: Blackwell Publishers.

Deuchar, Margaret, and Suzanne Quay. 2000. *Bilingual Acquisition: Theoretical Implications of a Case Study*. Oxford: Oxford University Press.

Döpke, Susan (ed.). 2000. *Cross-linguistic Structures in Simultaneous Bilingualism*. Amsterdam: John Benjamins.

Fantini, Alvino. 1978. Bilingual behavior and social cues: Case studies of two bilingual children. In Michel Paradis (ed.), *Aspects of Bilingualism*, pp. 283–301. Columbia, SC: Hornbeam.

Gawlitzek-Maiwald, Ira, and Rosemarie Tracy. 1996. Bilingual bootstrapping. *Linguistics* 34: 901–926.

Grosjean, François. 2013. Bilingualism: A short introduction. In François Grosjean and Ping Li (eds.), *The Psycholinguistics of Bilingualism*, pp. 5-25. Malden, MA: Wiley-Blackwell Publishing.

Hallett, Darcy, Michael J. Chandler, and Christopher E. Lalonde. 2007. Aboriginal language knowledge and youth suicide. *Cognitive Development* 22: 392–399.

Hinton, Leanne. 2002. *How to Keep Your Language Alive*. Berkeley, CA: Heydey Books.

Hinton, Leanne. 2013. *Bringing Our Languages Home: Language Revitalization for Families*. Berkeley, CA: Heydey Books.

Koehn, Caroline. 1989. Der Erwerb der Pluralmarkierungen durch bilinguale Kinder (Französisch-Deutsch): Eine empirische Untersuchung. MA thesis, University of Hamburg.

Koehn, Caroline. 1994. The acquisition of gender and number morphology within DP. In Jürgen Meisel (ed.), *Bilingual First Language Acquisition: French and German Grammatical Development*, pp. 29–52. Philadelphia: John Benjamins.

Lanza, Elizabeth. 1992. Can bilingual two-year-olds code-switch? *Journal of Child Language* 19: 633–658.

Lardiere, Donna. 1998. Dissociating syntax from morphology in a divergent L2 end-state grammar. *Second Language Research* 14: 359–375.

Lasagabaster, David. 2001. Bilingualism, immersion programmes and language learning in the Basque Country. *Journal of Multilingual and Multicultural Development* 22(5): 401–425.

Leopold, Werner. 1939. *Speech Development of a Bilingual Child*, vol. 1: *Vocabulary Growth in the First Two Years*. New York: AMS Press.

McLaughlin, Barry. 1978. *Second-Language Acquisition in Childhood*. Hillsdale, NJ: Lawrence Erlbaum Associates.

Meisel, Jürgen. 1986. Word order and case marking in early child language. Evidence from simultaneous acquisition of two first languages: French and German. *Linguistics* 24: 123–183.

Meisel, Jürgen. 1994. Getting FAT: Finiteness, Agreement and Tense in early grammars. In Jürgen Meisel (ed.), *Bilingual First Language Acquisition: French and German Grammatical Development*, pp. 89–130. Philadelphia: John Benjamins.

Montrul, Silvina, Rebecca Foote, and Silvia Perpiñán. 2008. Gender agreement in adult second language learners and Spanish heritage speakers: The effects of age and context of acquisition. *Language Learning* 58: 503–553.

Nettle, Daniel, and Suzanne Romaine. 2000. *Vanishing Voices*. Oxford: Oxford University Press.

Oller, D. Kimbrough, et al. 1997. Development of precursors to speech in infants exposed to two languages. *Journal of Child Language* 24: 407–425.

Paradis, Johanne, and Fred Genesee. 1996. Syntactic acquisition in bilingual children. *Studies in Second Language Acquisition* 18: 1–25.

Pavlenko, Aneta, and Barbara C. Malt. 2011. Kitchen Russian: Cross-linguistic differences and first language object naming by Russian-English bilinguals. *Bilingualism: Language and Cognition* 14: 19–45.

Pearson, Barbara, Sylvia Fernandez, and D. Kimbrough Oller. 1993. Lexical development in bilingual infants and toddlers: Comparison to monolingual norms. *Language Learning* 43: 93–120.

Rescorla, Leslie, and Sachiko Okuda. 1984. Lexical development in second language acquisition: Initial stages in a Japanese child's learning of English. *Journal of Child Language* 11: 689–695.

Salustri, Manola, and Nina Hyams. 2006. Looking for the universal core of the RI stage. In Vincent Torrens and Linda Escobar (eds.), *The Acquisition of Syntax in Romance Languages*, pp. 159–182, Amsterdam: John Benjamins.

Valadez, Concepción, Feli Etxeberria, and Nahia Intxausti. 2015. Language revitalization and the normalization of Basque: A study of teacher perceptions and expectations in the Basque Country. *Current Issues in Language Planning* 16: 60–79.

Volterra, Virginia, and Traute Taeschner. 1978. The acquisition and development of language by bilingual children. *Journal of Child Language* 5: 243–264.

Whalen, Douglas, Margaret Moss, and Daryl Baldwin. 2016. Healing through language: Positive physical health effects of indigenous language use. *F1000 Research* 5: 852.

Yip, Virginia. 2013. Simultaneous language acquisition. In François Grosjean and Ping Li (eds.), *The Psycholinguistics of Bilingualism*, pp. 119–144. Malden, MA: Wiley-Blackwell Publishing.

Yip, Virginia, and Stephen Matthews. 2007. *The Bilingual Child: Early Development and Language Contact*. Cambridge: Cambridge University Press.

Appendix A: English IPA Symbols

Consonants

	Bilabial	Labiodental	Dental	Alveolar	Post-alveolar	Palatal	Velar	Glottal
Stop	p b			t d			k g	ʔ
Nasal	m			n			ŋ	
Tap/Flap				ɾ				
Fricative		f v	θ ð	s z	ʃ ʒ			h
Approximant	w			ɹ		j	w	
Lateral approximant				l				

When two symbols are listed within a table cell, the sound on the left is **voiceless** and the sound on the right is **voiced**.

Additional sounds: t͡ʃ as in *church* and d͡ʒ as in *judge* are affricates.

Vowels

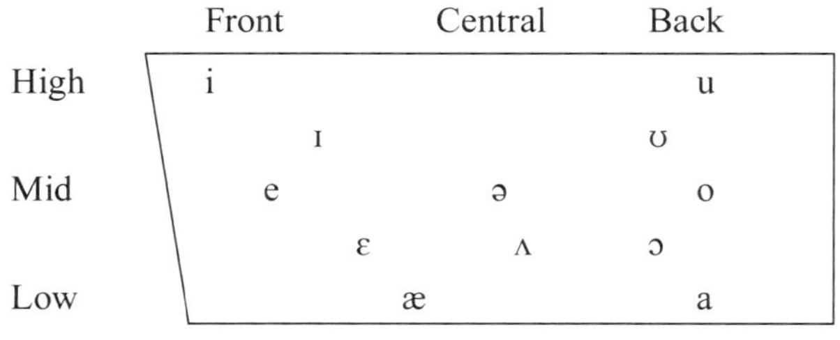

Sounds toward the center of the vowel space (ɪ, ɛ, æ, ə, ʌ, ʊ, ɔ) are **lax**, while sounds at the outer edges (i, e, u, o, a) are **tense**.

Additional sounds: aɪ (alternately aj) as in *hide*, aʊ (alternately aw) as in *house*, and ɔɪ (alternately ɔj) as in *boy* are diphthongs.

Note: The sounds listed on this page correspond to the main phonemes found in Main-stream American English. Other dialects, including British English, African American

English, Australian English, and others may have slightly different sounds. In particular, there may be variation in the vowels.

THE INTERNATIONAL PHONETIC ALPHABET (revised to 2015)

CONSONANTS (PULMONIC)

	Bilabial	Labiodental	Dental	Alveolar	Postalveolar	Retroflex	Palatal	Velar	Uvular	Pharyngeal	Glottal
Plosive	p b			t d		ʈ ɖ	c ɟ	k ɡ	q ɢ		ʔ
Nasal	m	ɱ		n		ɳ	ɲ	ŋ	ɴ		
Trill	ʙ			r					ʀ		
Tap or Flap		ⱱ		ɾ		ɽ					
Fricative	ɸ β	f v	θ ð	s z	ʃ ʒ	ʂ ʐ	ç ʝ	x ɣ	χ ʁ	ħ ʕ	h ɦ
Lateral fricative				ɬ ɮ							
Approximant		ʋ		ɹ		ɻ	j	ɰ			
Lateral approximant				l		ɭ	ʎ	ʟ			

Symbols to the right in a cell are voiced, to the left are voiceless. Shaded areas denote articulations judged impossible.

CONSONANTS (NON-PULMONIC)

Clicks	Voiced implosives	Ejectives
ʘ Bilabial	ɓ Bilabial	ʼ Examples:
ǀ Dental	ɗ Dental/alveolar	pʼ Bilabial
ǃ (Post)alveolar	ʄ Palatal	tʼ Dental/alveolar
ǂ Palatoalveolar	ɠ Velar	kʼ Velar
ǁ Alveolar lateral	ʛ Uvular	sʼ Alveolar fricative

VOWELS

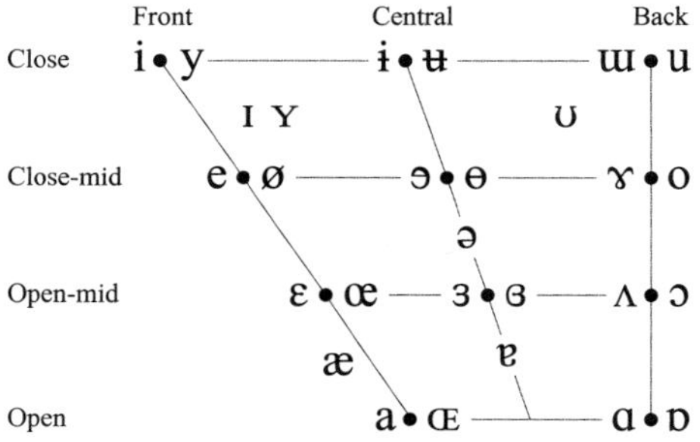

Where symbols appear in pairs, the one to the right represents a rounded vowel.

OTHER SYMBOLS

ʍ Voiceless labial-velar fricative

w Voiced labial-velar approximant

ɥ Voiced labial-palatal approximant

ʜ Voiceless epiglottal fricative

ʢ Voiced epiglottal fricative

ʡ Epiglottal plosive

ɕ ʑ Alveolo-palatal fricatives

ɺ Voiced alveolar lateral flap

ɧ Simultaneous ʃ and x

Affricates and double articulations can be represented by two symbols joined by a tie bar if necessary. t͜s k͡p

SUPRASEGMENTALS

ˈ Primary stress

ˌ Secondary stress

ˌfoʊnəˈtɪʃən

ː Long eː

ˑ Half-long eˑ

˘ Extra-short ĕ

| Minor (foot) group

‖ Major (intonation) group

. Syllable break ɹi.ækt

‿ Linking (absence of a break)

DIACRITICS Some diacritics may be placed above a symbol with a descender, e.g. ŋ̊

̥	Voiceless	n̥ d̥	̤	Breathy voiced	b̤ a̤	̪	Dental	t̪ d̪
̬	Voiced	s̬ t̬	̰	Creaky voiced	b̰ a̰	̺	Apical	t̺ d̺
ʰ	Aspirated	tʰ dʰ	̼	Linguolabial	t̼ d̼	̻	Laminal	t̻ d̻
̹	More rounded	ɔ̹	ʷ	Labialized	tʷ dʷ	̃	Nasalized	ẽ

̜	Less rounded	ɔ̜	ʲ	Palatalized	tʲ dʲ	ⁿ	Nasal release	dⁿ
̟	Advanced	u̟	ˠ	Velarized	tˠ dˠ	ˡ	Lateral release	dˡ
̠	Retracted	e̠	ˤ	Pharyngealized	tˤ dˤ	̚	No audible release	d̚
̈	Centralized	ë	~	Velarized or pharyngealized	ɫ			
̽	Mid-centralized	e̽	̝	Raised	e̝	(ɹ̝ = voiced alveolar fricative)		
̩	Syllabic	n̩	̞	Lowered	e̞	(β̞ = voiced bilabial approximant)		
̯	Non-syllabic	e̯	̘	Advanced Tongue Root	e̘			
˞	Rhoticity	ɚ a˞	̙	Retracted Tongue Root	e̙			

TONES AND WORD ACCENTS

LEVEL			CONTOUR		
e̋ or	˥	Extra high	ě or	˩˥	Rising
é	˦	High	ê	˥˩	Falling
ē	˧	Mid	e᷄	˦˥	High rising
è	˨	Low	e᷅	˩˨	Low rising
ȅ	˩	Extra low	e᷈	˧˦˧	Rising-falling
↓	Downstep		↗	Global rise	
↑	Upstep		↘	Global fall	

Typeface: Doulos SIL

For interactive IPA charts, please visit these websites: http://www.internationalphoneticalphabet.org/ipa-sounds/ipa-chart-with-sounds/ and http://www.ipachart.com/.

Appendix B: Methods in Child Language Acquisition

Introduction

Scientific inquiry begins with observation. We need to observe how something in the world is before we can inquire about it. Once we observe something, we can ask questions about how it came to be that way, or why it is the way it is, as opposed to any number of other ways. We then formulate a hypothesis about the observation, and we test it in the regular scientific method.

As with any other scientific endeavor, data is the lifeblood of the field of child language acquisition, so the field has developed many different ways to obtain that data. In this appendix, we will briefly describe the numerous tools that researchers have at their disposal to gather data that fit their research questions. The selection of a particular method over any other method depends very much on the research question at hand and the population to be studied (in this case, the ages of the children). If you are setting about designing an empirical research study, we recommend that you use this appendix as a starting point and then follow up with additional resources, some of which are provided at the end of this appendix.

We organize the methods into three broad modules. The first module contains the standard method to collect naturalistic data; the second and third modules contain methods to gather experimental data. The second module addresses experimental production data, and module 3 addresses experimental judgment (comprehension) data.

Module 1: Naturalistic Data

The standard method to gather naturalistic data was pioneered by Roger Brown and his students in the 1960s (see chapter 6 for more on Brown). We begin with a discussion of what we mean by **naturalistic data**, followed by a brief discussion of the history of this method.

What Is Naturalistic Data?

Naturalistic language data is simply recordings of spontaneously produced speech (or sign) data, produced in natural, everyday contexts. Typically, when people say *naturalistic data*, they are thinking of **longitudinal** audio/video recordings. *Longitudinal* means that data have been collected over a period of time so that you have data from the same participant(s) from time A to time B. Typically, longitudinal data targets just a few participants, since the volume of data generated makes large participant numbers prohibitive. Longitudinal data can be compared to **cross-sectional** data, in which a larger number of participants are targeted and data typically are collected in just one session. This is schematized in figure B.1, where each black dot represents one data collection point. We see that in figure B.1a, there are three participants, for each of whom data is collected eight times (over some period of time). But in figure B.1b, there are sixteen participants, for each of whom data is collected just once.

For child language data to be naturalistic, the following criteria must be met:

(i) The language data must be spoken/signed; that is, the researcher is not measuring the child's response to something, but collecting language the child produces him- or herself.

(ii) The child's productions must not be manipulated by the researcher. That is, the researcher does not try to control the quality or kind of language produced by the child. So if you present a picture that is designed to elicit certain sentence patterns, even though the response will be spoken and creative language, this does not count as naturalistic data (it is elicited production data; see module 2).

Note that the line between a naturalistic and manipulated interaction can be thin. Does providing toys count as manipulation? It depends on the intent behind providing the toys. If the toys are there to stimulate speech of any kind, then the data is still naturalistic. But

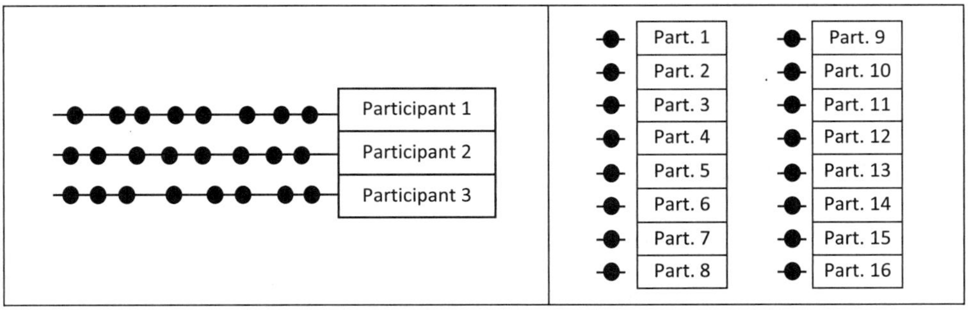

Figure B.1
Left: Longitudinal data: few participants, but many data collection points across time. *Right*: Cross-sectional data: many participants with data collected at one time point each.

if the toys are there to elicit certain structures, words, or sounds from the child, then it is manipulated and therefore not naturalistic.

(iii) The data must be recorded in some way. This was classically done with audio recorders, although video is now more common. Note that diary studies involve a kind of naturalistic data collection—an adult, usually the child's parent, records the speech of the child in a journal or diary.

How Naturalistic Data Is Collected

Roger Brown is credited with having developed the modern technique for naturalistic data collection, and he established a series of fundamental principles for naturalistic data collection and data analysis.

When you think about language development, it is not a discrete process, but rather a continuous process. That is, development happens day by day, hour by hour, in fits and starts. It does not happen all at once at predictable time points. To track this development, we need a process of **sampling**. Sampling is a method whereby one gathers data at regularly spaced intervals over a period of time. Sampling data every few weeks (for about an hour per session) gets an incremental snapshot of the development of the participant child. When you put these snapshots together, you get a broad image of the child's development.

Brown and his students collected data from three children at regular intervals of 2–4 weeks. Note that there is a trade-off between the frequency of data collection (how dense your final data set is) and the effort required to manage large volumes of data. The denser your data collection, the more data that is generated, and the harder it becomes to manage. Nevertheless, new recording technologies have led a number of researchers to begin collecting massive amounts of naturalistic data through wearable microphones. In these projects children and/or parents wear a small microphone that records their speech throughout the day. Such recordings provide quite comprehensive and ecologically valid language data, potentially including the child's own speech as well as the input from parents and/or siblings (for example, the HomeBank repository; see VanDam et al., 2016).

The collected data then needs to be transcribed for later analysis, and transcription by hand is enormously time consuming. One widely used standard format for transcription in the language acquisition field is the CHAT (Codes for the Human Analysis of Transcripts) format. This format allows a researcher to analyze the transcripts using the CLAN (Computerized Language Analysis) program. The CHAT format and CLAN program were developed by psycholinguist Brian MacWhinney, and they are available at the CHILDES website (http://childes.talkbank.org), where examples of natural language transcripts from a variety of languages, including Brown's recordings, are freely available. Some researchers and speech-language pathologists use other transcription formats, such as SALT (Systematic Analysis of Language Transcripts).

New developments in automated speech processing have made it possible to tackle the challenge of transcribing high-density recordings, such as those in the HomeBank project. However, automated speech-to-text processing is not error-free, and such transcripts need to be checked by human researchers to ensure that the resulting corpus is accurate.

Module 2: Production Data

Naturalistic data is very versatile and useful. It is typically the first data set that one gathers in a project, before one has any hypotheses about the data. For example, if you are investigating the acquisition of a language that has never been studied before, you probably want to start with naturalistic data. Because you don't know how children acquire this particular language, you can't really design any experiments.

However, if you already know something about the developmental sequence in a language, then you might want to target a particular research question. For example, if you are interested in finding out whether children can produce plural morphology on nouns correctly, then rather than gather naturalistic data and hope children produce plural nouns, you might create a context in which they have to produce plural nouns. Or if you are interested in finding out whether children can produce past-tense verbs, you might create a context in which children must produce past-tense verbs. This is no longer naturalistic, since you are manipulating the context. Such manipulations move the method from being labeled 'naturalistic' to 'experimental'.

Elicited Production

The basic idea behind this method is that the researcher creates a context in which the child must respond with a particular targeted structure. One way to elicit the production of the past tense, for example, is to ask the child to tell you what happened at some past event. For example, the experimenter might ask the child what happened at the child's last birthday party (if it was not too long ago) or on their last vacation. This is a simple manipulation, but if the researcher has a list of potential past events, then a significant number of past-tense verbs might be elicited. The researcher needs to be sure not to model the very forms they are trying to elicit from the child.

Let's say we're interested to see if children can produce plural nouns. We want to know whether they produce plurals at similar rates for regular nouns (e.g., *bird-birds*) as for irregular nouns (e.g., *sheep-sheep*), and with the different phonological versions of plural (referred to as allomorphs)—that is, [s] as in *buckets*, [z] as in *birds*, and [əz], as in *glasses*. In order to do that, we need to create a context in which children must produce a plural noun. In one of the most famous experiments in all of linguistics, Jean Berko (now known as Jean Berko Gleason) ran what came to be known as the wug test (described in chapter 6 of this book). In this experiment, Berko Gleason presented children with an odd-shaped character like the one in figure B.2 and said to children, "This is a wug." Then a second

Figure B.2
A wug. *Source*: https://commons.wikimedia.org/wiki/File:Wug.svg.

one was presented (so children were now looking at two wugs), and the experimenter said, "Now there's another one. There are two of them. There are two … ," and the child had to finish the sentence. This is a clever way to produce an obligatory context for children to produce a plural noun. Moreover, this was clever because this got away from the issue of children having already heard some plural forms in their input. Because these are nonsense words, children could not possibly rely on previous exposure to *wugs* to be able to produce the plural noun. Rather, the only way to produce the plural noun would be to take the singular, which they had just heard, and apply the plural rule (add 's') to that noun. Berko Gleason investigated numerous other nonsense nouns, some ending in voiced stops, some ending in voiceless stops, and some ending in sibilants, so as to investigate the three allomorphs of the plural *-s* in English. This is a classic example of an elicited production task because a context was provided to children to produce exactly the phenomenon that the investigators were targeting.

We find more sophisticated uses of the elicited production task when trying to elicit more complex sentence patterns. For example, Demuth et al. (2010) were interested in eliciting passive sentences from Sesotho-speaking children. They presented children with pictures of events that involved one person acting upon another (e.g., a boy pulling a girl in a wagon). If children were asked to describe the picture, they might tend to use an active sentence ("The boy is pulling the girl"). So instead of asking children *What is the boy doing?* or *What is happening here?*, Demuth et al. asked questions like *What happened to the girl?* With such questions, it is far more natural to respond with a passive, like *She was pulled by the boy.* In fact, with this manipulation, the researchers found that passives were produced by children as young as 3 years old an astounding 98% of the time.

Yet another complex pattern that could be elicited using this technique is wh-questions. The following is adapted from Yoshinaga (1996), a study that elicited wh-questions in Japanese and English (though the pictures are courtesy of Nozomi Tanaka and Ryoko Hattori). It has been found that children (and adults) tend to find subject wh-questions (*Who is pushing the girl?*) easier to understand than object wh-questions (*Whom is the boy pushing?*). This is parallel to the findings for relative clauses, as discussed in chapter 7. But does this asymmetry carry over to production? Using the elicited production technique,

Figure B.3
Left: Picture to elicit subject wh-questions. *Right*: Picture to elicit object wh-questions.

we could easily test whether children are able to produce these two kinds of wh-questions with the same degree of success. Figure B.3 (*left*) shows a picture designed to elicit subject wh-questions, and figure B.3 (*right*) is designed to elicit object wh-questions.

The protocol to elicit questions from these pictures is also quite simple. The researcher introduces the child to a puppet—let's call her Momo. Momo knows all about the pictures we are going to see, but we need to ask her to figure out what's happening in each picture. The child is then shown the first picture (figure B.3 [*left*]) and the researcher says, "Someone is pushing the girl, but we can't see. Momo knows. Can you ask her?" The child is expected to ask, "Momo, who is pushing the girl?" For object wh-questions, the protocol is very similar, with the researcher saying, "The boy is pushing someone. But we can't see. Momo knows. Can you ask her?" and the child is expected to ask, "Momo, who is the boy pushing?"

Note that there are clues in the protocol that might help the child, for example, the placement of *someone* (either in subject position or object position), but, crucially, the experimenter must not say things like "We can't see who is pushing the girl" or "Ask Momo who is pushing the girl," since these prompts would model the structure for the child.

Elicited Imitation

Slightly more controlled than the above two methods is the elicited imitation task. Here the experimenter presents the child with a sentence that they must repeat verbatim. Typically this is done in a complex experimental setup in which the imitation is actually motivated. For example, a puppet will tell the child a secret, and the child must convey the secret to the experimenter without a second puppet being able to hear. This gives the child a reason to imitate the first puppet. The reasoning behind this method is that it has been found that children generally cannot imitate structures that they have not acquired yet. So if a child has not yet acquired tense, then they will imitate sentences without tense marking.

If the child has not acquired the passive voice yet, then they will imitate passive sentences (like *The girl was pulled by the boy*) in the active voice (e.g., *The girl pulled the boy*). This task, therefore, can be used to assess whether a particular structure has been acquired by children.

Care needs to be taken with this method, however, since it is quite tricky to implement correctly. The conditions to motivate the imitation (the puppet scenarios described above) need to be done carefully, or else children might balk at the idea of imitating. Moreover, in some cultures it is deemed rude to imitate others, and some children might already have learned the taboo against this. Nonetheless, this method is widely used and holds potential for a host of possible applications.

Priming

In the 1980s, researchers discovered that when you hear a particular grammatical pattern, you are more likely to use it in your own speech in the very next utterance (or few utterances). For example, if you see a picture of a man giving a woman a flower, there are at least two ways that picture might be described, referred to as the double object and dative patterns.

(1)	a. The man gave the woman a flower	Double object
	b. The man gave a flower to the woman	Dative

What is important is that both these sentences describe the picture equally well, and the choice between these two patterns is essentially arbitrary. But research with adults has shown (e.g., Bock, 1986) that if a participant sees the picture of the man giving the woman a flower and hears one of these sentence patterns (say, the dative pattern), then when they are asked to describe a subsequent picture (say, of a girl throwing a ball to a boy), participants are more likely to use the same pattern that they heard previously. The way we describe this is that the first exposure to the pattern (the dative sentence *The man gave a flower to the woman*) is called the **prime**, and this activates an abstract pattern in the mind of the participant. When that participant is then asked to describe a new picture, because the primed pattern is still active in their mind, they are more likely to use that very same pattern, though with different words. This works whether the prime is the double object or the dative, and it works with many other syntactic patterns too.

But crucially, a pattern cannot be activated if a participant does not have knowledge of that pattern. If a child, for example, has not acquired the dative pattern, then priming them with the dative pattern will not increase the likelihood of them producing the dative. So priming is a useful technique to gauge whether a structure has been acquired or not.

Messenger et al. (2012) applied this technique to the acquisition of the passive. They showed children a picture of an event with two people and then described the picture using either an active or a passive sentence (this was the prime). They then asked the children to describe a subsequent picture of a different event with different people. The researchers coded how often children produced passives to describe the second picture. When primed with the active, unsurprisingly, children overwhelmingly produced active descriptions of the second picture. But when primed with the passive, children were significantly more likely to describe the second picture with a passive sentence. This shows that children have knowledge of the passive pattern, or else there would not be a priming effect at all.

Module 3: Comprehension Data

It is often the case that production data is the first kind of data that one obtains in a new project. However, an important fact about human language is that comprehension always exceeds production (except perhaps in the case of certain aphasic patients). Even in normal adults, we typically comprehend far more than we ever say. For example, you probably know what the word *arduous* means, but when was the last time you actually said it? If someone were recording everything you said for the last year, and they had access to only the words that you produced, they might well conclude that you do not know the word *arduous*. To get at what you actually know, they would have to conduct some kind of comprehension experiment. The same reasoning applies to children: just because children don't say certain things, or say other things incorrectly, does not necessarily mean that they do not have knowledge of those phenomena. So comprehension experiments are often seen as more representative of a child's true knowledge. This harks back to the discussion of competence and performance that we discussed in chapter 2: just because a child fails to *perform* does not mean they do not have competence.

Grammaticality/Acceptability Judgment

The classic experiment that linguists perform on adults is a grammaticality judgment task. Here you present participants with a sentence and ask them if it is grammatical or ungrammatical. For example, in sentence 1 below, the sentence is grammatical, and adults are expected to circle the Y response. Sentence 2 is ungrammatical and adults are expected to circle the N response. Sentence 3 is perhaps ungrammatical, though it could be grammatical in certain contexts. 'The girl' has been topicalized, or moved to the front of the sentence, and this sentence would make sense in a discourse like the following: *The girl, the boy kissed; his mom, the boy hugged.*

		Grammatical?	
(1)	The boy kissed the girl	Y	N
(2)	Kissed the boy the girl	Y	N
(3)	The girl, the boy kissed	Y	N

So it is unclear what an adult would say about sentence 3. For this reason, many researchers now use a more articulated scale, called a Likert scale, in such judgment tasks. For example, researchers often provide a scale from 1 to 7, where 1 is totally unacceptable and 7 is totally acceptable. To emphasize that the question is whether a sentence could be acceptable given the right context, such tasks may be referred to as **acceptability judgment tasks**.

Years ago it was thought impossible to ask young children to perform a task like this. Making a judgment about the "well formedness" (grammaticality or acceptability) of a sentence requires you to think about language as an object—that is, to make a **metalinguistic** judgment. You need to step back for a moment, not think about the *meaning* of the sentence but consider the *form* of the sentence, and make a judgment about whether that form is consistent with the rules that you have internalized about that language. Not all children (or adults) can easily do this, and simply asking a child out of the blue whether a sentence sounds OK to them typically produces a blank stare or a judgment about the meaning of the sentence.

Nonetheless, many researchers have successfully implemented such tasks with young children by first modeling lots of examples of ungrammatical sentences (e.g., *Kissed the boy the girl*) and pointing out that these sentences don't sound right. After much training of this sort, children from the age of 4 years can often succeed on these tasks (some researchers report success as young as age 3). Often, instead of asking children whether a sentence is grammatical or ungrammatical, children are asked to judge whether a puppet said something in a silly way or an OK way. Alternatively, sometimes children are given a scale using smiley faces, where on the bad end of the scale is a sad face and on the good end of the scale is a happy face, and in between are gradations of smiley faces (a little like the pain scale that pediatricians use with children to indicate the level of pain they might be feeling).

Truth Value Judgment Task (TVJT)

One of the most interesting methodological developments in the last thirty years was the invention of the Truth Value Judgment Task (TVJT). First described by Crain and McKee (1985) in their now-famous Principle C experiment (see chapter 7, section 7.4 for details), the TVJT is a method that allows us to measure children's comprehension abilities at very young ages in a wide variety of phenomena. One major advantage of the TVJT over grammaticality judgment tasks is that in this method, children do not need to access metalinguistic knowledge in order to respond correctly. Rather, all the child does is listen to a sentence and then make a judgment about whether that sentence was true or false given a story or scenario they were presented with. In fact, this is essentially what we do in everyday language—listen to people and assess the truth of their utterances. So this is a very natural task in that sense.

The key elements of the TVJT are an appropriate context, a puppet, and a setup in which the puppet is said to be learning the language and therefore prone to making errors. The task at hand for the child is to watch the contexts with the puppet, listen as the puppet describes the context, and decide whether the puppet is right or wrong. For example, consider the following scenario (taken from O'Brien et al., 2006) at the end of which the puppet (Gobu) is asked to describe something that happened in the story:

EXPERIMENTER: Bart, the gorilla, and the cheetah were relaxing in the jungle one day, when Bart found a bunch of bananas.

BART: Hey, cool! Look what I found!

GORILLA: Would you mind sharing some of those with me?

BART: No way, dude, these are mine, all mine! Hee, hee. If you want some, you're gonna have to chase me.

CHEETAH: I could chase him, but I'm not all that fond of bananas.

GORILLA: Well bananas are my favorite, so watch out Bart, here I come!!!! (Gorilla chases Bart)

EXPERIMENTER: Gobu, can you tell me something about the story?

GOBU: Well, let's see. In that story, the gorilla was chased by Bart.

In this setup, the child listens to the puppet's statement and has to judge whether the puppet is right or wrong. You and I can read that story and say clearly that the puppet is wrong—Bart was chased by gorilla (the opposite of what the puppet said). But children might interpret the passive sentence (which is what the test item is) incorrectly. They often understand passive sentences as if they were active sentences, so a child that does not know the passive might hear that test sentence and interpret it as *The gorilla chased Bart.* If that is the case, the child will say that Gobu is correct. But if the child understands the passive, then the child will say that Gobu is incorrect.

By recording whether the child accepted or rejected the puppet's statement, we can see whether they understand the passive voice or not. And the child does not have to do anything other than listen to the stories and then judge whether the puppet was right or wrong.

This method has been used to investigate a wide variety of phenomena, including passives, Principle C, quantifier scope, and many others. But while the TVJT can be applied to children as young as 3 years of age, it does present challenges.

One challenge is that the contexts and test sentences have to be very carefully constructed. The context must felicitously motivate the test sentence, or else children will misunderstand the task or realize that something is off. Children can be overly generous in accepting sentences in this type of task, particularly if there is any ambiguity. This is referred to as the **principle of charity** (Crain and Thornton, 1998), which states that if the child is at all unclear about the truth of the statement, the child will accept the statement. Because of this researchers must take care to minimize ambiguity, but they must also ensure that children are willing to reject the puppet's utterance under some circumstances (that is, ensure that children won't accept the puppet's sentences no matter what).

Another challenge is that test sentences must be testing a specific hypothesis: there must be a reasonable alternative interpretation that the researcher thinks the child might adopt, so that the child's answer is meaningful either way. In the example above, if the child says the puppet is right, we know they have parsed the puppet's sentence (incorrectly) as if it were in active voice, and if the child says the puppet is wrong, we know they have parsed the puppet's sentence (correctly) as passive.

The benefits of this method (e.g., ecological validity, applicability to 3-year-olds, fun and exciting for children) more than outweigh these challenges (see Crain and Thornton, 1998, for a full discussion of this method). As such, the TVJT has become a go-to method for language acquisitionists in recent decades.

Picture Selection

Related to the TVJT is the picture selection task. The idea here is very simple: present children with two pictures, then present them with a linguistic stimulus, and they pick the picture that matches the linguistic stimulus. For example, in figure B.4, there are two pictures, one of a banana and one of an apple. A simple picture selection task would present the two pictures to a child and ask "Where is the banana?" If the child knows the word *banana*, the child should pick the picture on the left. If they don't, they might pick randomly.

This is a simple task, used in a wide variety of experiments ranging from those testing phonological knowledge all the way up to testing complex relative clauses. The ease of creating the items, coupled with the fun and interest that such an experiment poses, makes this a very popular method.

However, a word of warning. A picture selection task seems easy to us adults, but depending on what pictures and linguistic prompts one uses, it can be a somewhat cognitively demanding task. Imagine you use a picture of a girl chasing a boy and another picture of a boy chasing a girl, and the sentence is "Point to where the girl is being chased by the

Figure B.4
Picture selection task: "Which picture shows ... banana?"

boy!" Now the picture selection task is essentially two TVJT tasks in one. The child must listen to the linguistic stimulus and then evaluate the truth of that statement relative to the first picture. The child must then do the same with the second picture. And then the child must decide which picture has the more truthful picture. Nonetheless, the picture selection task remains a popular and reliable method that language acquisitionists employ in their experiments.

Intermodal Preferential Looking Paradigm / Eye Tracking

Perhaps the biggest advancement in experimental methodology came with the technological development of the intermodal preferential looking paradigm. This method, developed by Kathy Hirsh-Pasek and Roberta Golinkoff in the 1980s, is based on the finding that human beings in general prefer to match the sensory input from one modality to sensory input from other modalities. We like to match what we see to what we hear, so when we hear language, we like to look at things that match what we are hearing. In this sense, if you hear the word *monkey*, and in front of you sit a monkey, a lion, and an ostrich, you will likely look longer at the monkey. This preference is strong in young children as well.

Hirsh-Pasek and Golinkoff capitalized on this fact by designing a clever experiment in which very young children could be tested on some complex language patterns. The beauty of this method is that you can test children who can't even talk, since this preference for matched sensory input is in place from before the onset of spoken language.

To carry out this kind of study, a parent sits on a chair with their infant in their lap, and the parent wears headphones so they cannot hear what was happening (see figure B.5). The infant has two screens in front of them, one on the left and one on the right, with a hidden sound speaker in the middle. First, the child is introduced to the characters they will see in the videos; for example, they'll see Big Bird and Cookie Monster on each screen, plus the character's label (*Look! It's Big Bird!*). Children then see a video on one screen depicting an action, say, of Big Bird pushing Cookie Monster. That screen then goes blank, and on the other screen, the opposite action appears: Cookie Monster pushing Big Bird. Then both screens go blank, and finally both screens come back on, each showing a different action (Big Bird pushing Cookie Monster on one, Cookie Monster pushing Big Bird on the other), and the child hears a linguistic stimulus that matches only one of the screens: "Look! Big Bird is pushing Cookie Monster."

If infants of about 1 year of age (as were the participants in this experiment) understand simple transitive sentences (even though they are not producing such sentences yet), then the expectation is that they will look at the screen that matches the sentence. If they do not, then they should find both screens equally interesting and look at them equally.

Hirsh-Pasek and Golinkoff indeed found a significant difference in children's preference for the matching screen, showing that at the age of about 1 year, before they are pro-

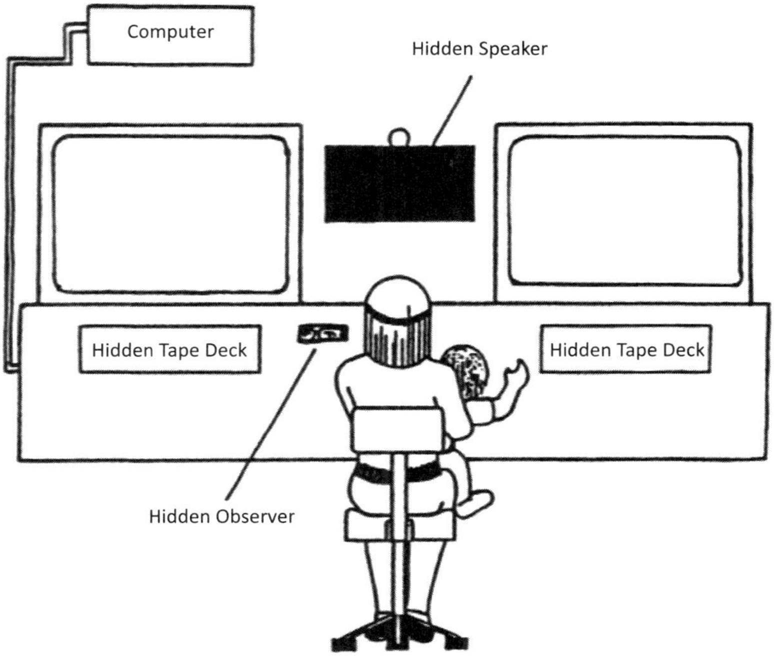

Figure B.5
The intermodal preferential looking paradigm setup. From Hirsh-Pasek and Golinkoff (1996). Reprinted with permission.

ducing sentences, children understand that the first noun in a sentence is the do-er of an action and the second noun is the one that is acted upon. Showing this kind of knowledge can only really be done with a method of this kind, since all other methods discussed so far are too challenging for children of this age to complete.

In recent years, with the advancement of technology, this method has morphed into what we call the **visual world eye-tracking paradigm**. The concept is the same: present the child with a visual scene along with a linguistic stimulus, and then code where the child looks. The difference is one of resolution. In the intermodal preferential looking paradigm, there are usually just two visual stimuli (areas of interest), and the coding is typically done with a regular video camera (roughly 30 frames per second, which is about 33 ms windows). But with eye tracking, there are typically four visual stimuli (four areas of interest), and the coding is done at smaller resolutions, typically 20 ms windows (though that could go even lower). Thus, the equipment exists nowadays to use eye tracking to test infants on a wide variety of phenomena with remarkable accuracy.

Act Out Task

Unlike the intermodal preferential looking or eye tracking methodology, the **act out task** is far less technical in requirements. The idea for this method is very simple: The researcher presents the child with a test sentence that the child is supposed to act out (using toys or props of some kind). If the child has knowledge of the target structure, they will act the test sentence out in a way consistent with the sentence's meaning. But if they do not have knowledge of the target structure, they will act the test sentence out in another way.

Continuing with our example from the acquisition of the passive voice, we know that children sometimes interpret passive sentences as if they are active. A simple way to see this is to present the child with a passive sentence and some toys and ask them to act out the sentence. The toys might be a truck and a car, and the test sentence might be *The car was smashed by the truck*. In this situation, if the child understands the passive sentence, then they will grab the truck and smash it into the car. But if the child interprets the passive as an active, then they will grab the car and smash it into the truck. So by simply observing the child's actions, we are able to infer something about their underlying understanding of the test sentence. By acting the sentence out *incorrectly*, the child is revealing that the passive voice is a challenge for them.

The act out task can be used to test a variety of different structures. For example, if we are testing the acquisition of relative clauses, we might provide children with a bunch of toy farm animals and give children a test sentence such as *The sheep jumped over the cow that the pig pushed*. In this sentence, the child should take the pig and show it pushing the cow, and then take the sheep and have it jump over that cow. This is a challenging task, since it requires that you work backward from the end of the sentence in order to correctly act it out. Needless to say, it has been found that children have trouble acting out such sentences (Tavakolian, 1978).

But in an interesting plot twist, Hamburger and Crain (1982) found that when the context is tweaked just slightly, children are able to correctly act out such complex sentences. Specifically, they noticed that in the original experiment that tested the sentence *The sheep jumped over the cow that the pig pushed*, there was only one cow present in the context. Hamburger and Crain wondered how children are supposed to interpret a relative clause like *the cow that the pig pushed* when there was no reason for the speaker to use a relative clause. (Typically, we use a relative clause to distinguish one entity from another similar one: Which cow? The one that the pig pushed.) Hamburger and Crain remedied the situation by simply adding a cow, and they then acted out one cow being pushed by the pig and the other cow being pushed by another animal. When children saw this and were asked to act out *The sheep jumped over the cow that the pig pushed*, they acted out the correct sequence impeccably well.

This shows that the act out task can be used to test some pretty complex sentence structures, but care needs to be taken to make the contexts and available props appropriate to the test sentence. Any infelicity in the setup will lead to unreliable results.

Methodologies for Infant Studies

The methodologies described in these sections look at the linguistic knowledge of children from about the age of 1 year and older. But there are a number of methods for studying comprehension in tiny infants. Naturally, all of these measure comprehension, since infants do not produce language yet.

Two methods that work well with very young babies, including newborns, are the **high-amplitude sucking** (HAS) procedure and measuring heart rate. The insight behind these methods is that babies get excited when they perceive something new in their environment, and excitation can lead to an increase in heart rate and/or in their rate of sucking. For heart rate measurement, a heart rate monitor is attached to the baby's body and a baseline heart rate is established. This is the baby's heart rate when they are not excited (e.g., the experimenter presents some stimulus repeatedly, such as the syllable [ba], and waits until the baby **habituates**, or gets bored, at which point their heart rate goes to baseline). Then the experimenter presents the new stimulus (e.g., another version of [ba] or a different syllable, such as [pa]), and if the infant perceives a difference at the shift, they will **dishabituate**, and their heart rate will increase significantly above the baseline. But crucially, this only happens if the child can detect the difference between the baseline stimulus and the new stimulus.

Figure B.6 shows what results from one heart rate monitoring item might look like. The figure shows the heart rate of a child as they are taken through a single item. The line in the graph shows the overall heart rate of the child, and on the left side of the graph (labeled A), the child has a baseline heart rate. The researcher has, at this point, introduced the baseline stimulus, saying the syllable [ba] to the child, and the child has habituated to that syllable. The researcher then introduces a new stimulus, saying the syllable [pa]. In figure B.6, you can see where this was done because the child's heartbeat becomes faster—that is, the child dishabituates. The heart rate of the child increases noticeably, to a peak at point B. Once the child begins to adjust to the new stimulus, the heart rate begins to fall back down, until the child reaches a new state of habituation (region labeled C).

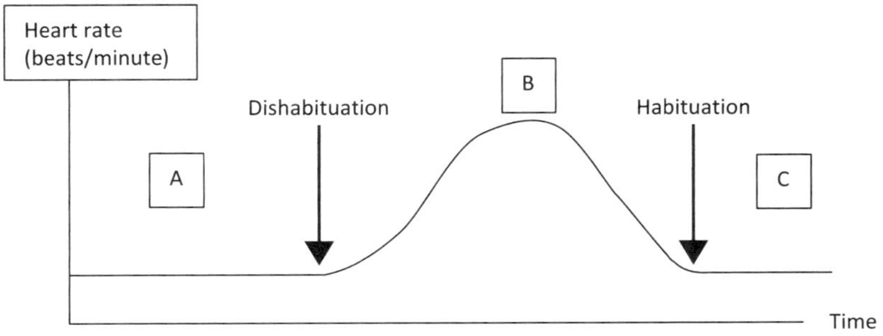

Figure B.6
Measuring habituation using heart rate.

The same principle works for sucking rate: infants suck for nourishment and for comfort, so by placing an electrode inside a pacifier nipple, one can measure the intensity with which the infant is sucking. Just as with the heart rate method, if the baby perceives a change in stimulus, their sucking rate will go up compared to the baseline rate.

This method is very useful in probing children's perceptive abilities at various ages. For example, we might want to test to see if young infants can discriminate between two very similar sounds in English, such as [f] and [v] (where the two sounds differ in voicing only), [p] and [t] (where the two sounds differ in place of articulation only), or [g] and [ŋ] (where the two sounds differ in nasality). To do this, you would habituate the child to one sound, and then switch to the other sound and see if the child disambiguates or not. This technique has been used widely to test at what age children can discriminate between foreign sounds that are not in their own language (e.g., Werker and Tees, 1984; see chapter 3).

Both of these methods (heart rate monitoring and suck rate monitoring) are based on the idea of habituation: when babies get habituated to some continually presented stimulus, they get bored, but if they perceive a change, they get surprised or excited, which is called dishabituation. As we just saw, dishabituation can be measured by changes in involuntary physical responses such as heart rate and sucking rate.

Another way to test infant speech perception is by measuring how long babies look in a particular direction. This method is known as the **headturn preference procedure**. When infants reach about 4 months of age, they are able to control their neck muscles and thus turn their heads from side to side. Infants will naturally turn their head toward a sound source they perceive. In some cases they will maintain that posture longer when the sound coming from that source is surprising to them, and this type of response shows a **novelty preference**. Researchers can show that babies discriminate two types of sound stimuli by familiarizing them with one type of sound until the infant no longer looks in the direction of the sound and then changing to a new sound. If the infant perceives the second sound as new and different, they will look toward its source (if not, they will continue to look elsewhere). In the experiment by Saffran et al. (1996), in which babies listened to two minutes of strings of nonsense syllables (e.g., *bidakupadotigolabubidaku*; see chapter 2), after the initial presentation babies were presented with two-syllable strings that had occurred in the input string. But these two-syllable strings had either occurred within one of the nonsense words in the presentation (*bida* is part of *bidaku*) or they had occurred across a word boundary (*kupa* is the last syllable of *bidaku* plus the first syllable of *padoti*). Babies looked longer at the source of *kupa* than *bida*, indicating that *kupa* was more surprising to them.

In some cases, however, babies maintain a longer orientation toward the source of the sound they prefer to listen to rather than the sound that is new or surprising. This type of response indicates a **familiarity preference**. We saw in chapter 3 that babies prefer rhythmic and phonotactic patterns that are coincident with patterns in their own language (Jusczyk et al., 1992; Mattys and Jusczyk, 2001). In these cases, headturn is not an indication of discrimination due to habituation but rather discrimination based on a preference for familiarity.

Brain-Based Methods

New technologies that allow researchers to observe the brain activity of someone performing a cognitive task can give us information about how language is stored and processed in the brain. Two common methods include **electroencephalography (EEG)**, which measures electrical activity in the brain, and **functional magnetic resonance imaging (fMRI)**, which measures changes in blood flow in different areas of the brain.

In an EEG study, a subject wears a skullcap with sensors that detect minute changes in electrical voltages in the brain in response to linguistic stimuli (also called **events**; thus, this technique is sometimes described as measuring **event-related potentials**, or ERPs). Because the sensors are placed on the scalp, this is a noninvasive technique that can be used with children at any age, including newborns (Molfese and Molfese, 1997; Espy et al., 2004). We know from EEG/ERP studies with adults that within 100 to 300 ms after hearing a language stimulus, adults begin to process word category information; around 300 to 500 ms after the stimulus, they integrate some syntactic (argument structure) and semantic information, and about 500 to 700 ms after the stimulus, more complex syntactic processing takes place. This is measured by comparing the electrical activity in brain regions while processing grammatical versus deviant sentences. Differences in the voltage when the brain processes different kinds of syntactic or semantic violations tells us when it is processing that type of grammatical information. An example of a semantic violation for the normal sentence *The goose was fed* would be *The straight-edge was fed*; an example of a syntactic violation is *The goose was in the fed* (these examples are English translations of German sentences used by Hahne, Eckstein, and Friederici, 2004).

There are fewer studies using ERP measures of children's language processing than adults'. However, some studies suggest that children show similar types of ERP responses to language input, in particular to semantic violations (Friederici and Oberecker, 2008), but their responses are slightly delayed and of longer duration than those of adults; children do not show a strong ERP response to syntactic violations before age 7 (Hahne et al., 2004).

While EEG/ERP provides remarkably precise information about *when* the brain responds to a linguistic stimulus (it has very good **temporal resolution**—measurements are in milliseconds), it tells us relatively little about *where* language processing happens in the brain. This is because electrical responses are quite dispersed over the cerebral cortex. In contrast, fMRI analysis involves scanning the brain to measure changes in blood flow in different regions as someone experiences different kinds of linguistic stimuli. The principle behind the technique is that areas of the brain that are active in a task will receive more blood flow during that task. This type of technique has somewhat poor temporal resolution (the timing of responses is measured in seconds, not milliseconds) but excellent **spatial resolution**—we can measure in millimeters where brain activity is happening. Like EEG, this technique is noninvasive and safe to use with subjects of all ages. This allows researchers to detect changes in where the brain processes language across the lifespan. For example, a verb-generation study by Szaflarski et al. (2006) found that between the ages of 5 and 25 years, there is increased lateralization of language in right-handers (that is, language is processed more and more in the left hemisphere), but from age 25–70 years, right-handed adults show less and less lateralization (language processing is less concentrated in the left hemisphere). (See also Vannest et al., 2009.)

Further Reading

Crain, Stephen, and Rosalind Thornton. 1998. *Investigations in Universal Grammar.* Cambridge, MA: MIT Press.

McDaniel, Dana, and Helen Smith Cairns. 1996. *Methods of Assessing Children's Syntax.* Cambridge, MA: MIT Press.

References

Berko (Gleason), Jean. 1958. The child's learning of English morphology. *Word* 14: 150–177.

Bock, J. Kathryn. 1986. Meaning, sound, and syntax: Lexical priming in sentence production. *Journal of Experimental Psychology: Learning, Memory, and Cognition* 12(4): 575–586.

Bornstein, Marc, O. Maurice Haynes, Kathleen Painter, and Janice Genevro. 2000. Child language with mother and with stranger at home and in the laboratory: A methodological study. *Journal of Child Language* 27: 407–420.

Bornstein, Marc, Kathleen Painter, and Jaihyun Park. 2002. Naturalistic language sampling in typically developing children. *Journal of Child Language* 29: 687–699.

Brown, Roger. 1973. *A First Language.* Cambridge, MA: Harvard University Press.

Crain, Stephen, and Cecile McKee. 1985. The acquisition of structural restrictions on anaphora. In Stephen Berman, Jae-Woong Choe, and Joyce McDonough (eds.), *Proceedings of the North East Linguistic Society (NELS) 15*, pp. 94–110. Amherst, MA: Graduate Linguistic Student Association.

Crain, Stephen, and Rosalind Thornton. 1998. *Investigations in Universal Grammar.* Cambridge, MA: MIT Press.

Demuth, Katherine, Francina Moloi, and Malillo Machobane. 2010. Three-year-olds' comprehension, production and generalization of Sesotho passives. *Cognition* 115(2): 238–251.

Espy, Kimberley Andrews, Dennis Molfese, Victoria Molfese, and Arlene Modglin. 2004. Development of auditory event-related potentials in young children and relations to word-level reading abilities at age 8 years. *Annals of Dyslexia* 54: 9–38.

Friederici, Angela, and Regine Oberecker. 2008. The development of syntactic brain correlates during the first years of life. In Angela Friederici and Guillame Thierry (eds.), *Early Language: Bridging Brain and Behavior*, pp. 215–231, Amsterdam: John Benjamins.

Hahne, Anja, Korinna Eckstein, and Angela Friederici. 2004. Brain signatures of syntactic and semantic processes during children's language development. *Journal of Cognitive Neuroscience* 16(7): 1302–1318.

Hamburger, Henry, and Stephen Crain. 1982. Relative acquisition. In Stan Kuczaj (ed.), *Language Development: Syntax and Semantics*, pp. 245–274. Hillsdale, NJ: Lawrence Erlbaum Associates.

Hirsh-Pasek, Kathy, and Roberta Golinkoff. 1996. The intermodal preferential looking paradigm: A window onto emerging language comprehension. In Dana McDaniel, Cecile McKee, and Helen Smith Cairns (eds.), *Methods for Assessing Children's Syntax*, pp. 105–124. Cambridge, MA: MIT Press.

Jusczyk, Peter, Kathy Hirsh-Pasek, Deborah Kemler Nelson, Lori Kennedy, Amanda Woodward, and Julie Piwoz. 1992. Perception of acoustic correlates of major phrasal units by young infants. *Cognitive Psychology* 24: 252–293.

Leopold, Werner. 1948. The study of child language and infant bilingualism. *Word* 4(1): 1–17.

MacWhinney, Brian. 2000. *The Child Language Data Exchange System*. Mahwah, NJ: Lawrence Erlbaum Associates.

Mattys, Sven L., and Peter W. Jusczyk. 2001. Phonotactic cues for segmentation of fluent speech by infants. *Cognition* 78: 91–121.

Messenger, Katherine, Holly P. Branigan, and Janet F. McLean. 2012. Is children's acquisition of the passive a staged process? Evidence from six- and nine-year-olds' production of passives. *Journal of Child Language* 39: 991–1016.

Molfese, Dennis, and Victoria Molfese. 1997. Discrimination of language skills at five years of age using event related potentials recorded at birth. *Developmental Neuropsychology* 13: 133–156.

O'Brien, Karen, Elaine Grolla, and Diane Lillo-Martin. 2006. Long passives are understood by young children. In David Bamman, Tatiana Magnistkaia, and Colleen Zaller (eds.), *Proceedings of the 30th Boston University Conference on Language Development*, pp. 441–451. Somerville, MA: Cascadilla Press.

Saffran, Jenny, Richard Aslin, and Elissa Newport. 1996. Statistical learning by 8-month-olds. *Science* 274: 1926–1928.

Szaflarski, Jerzy P., Scott K. Holland, Vincent J. Schmithorst, and Anna W. Byars. 2006. fMRI study of brain mapping in children and adults. *Human Brain Mapping* 27: 202–212.

Tavakolian, Susan. 1978. Children's comprehension of pronominal subjects and missing subjects in complicated sentences. In Helen Goodluck and Lawrence Solan (Eds.) *University of Massachusetts Occasional Papers in Linguistics: Papers in the Structure and Development of Child Language 4*, pp. 37–83, Amherst: University of Massachusetts Graduate Linguistic Student Association.

Valian, Virginia. 1991. Syntactic subjects in the early speech of American and Italian children. *Cognition* 40: 21–81.

VanDam, Mark, Anne S. Warlaumont, Elika Bergelson, Alejandrina Cristia, Melanie Soderstrom, Paul De Palma, and Brian MacWhinney. 2016. HomeBank: An online repository of daylong child-centered audio recordings. *Seminars in Speech and Language* 37(2): 128–142.

Vannest, Jennifer, Prasanna R. Karunanayaka, Vincent J. Schmithorst, Jerzy P. Szaflarski, and Scott K. Holland. 2009. Language networks in children: Evidence from functional MRI studies. *American Journal of Roentgenology* 192: 1190–1196.

Werker, Janet, and Richard Tees. 1984. Cross-language speech perception: Evidence for perceptual reorganization during the first year of life. *Infant Behavior and Development* 7: 49–63.

Yoshinaga, Naoko. 1996. Wh-questions: A comparative study of their form and acquisition in English and Japanese. PhD thesis, University of Hawai`i.

Notes

Chapter 1

1. Linguists use the * symbol at the beginning of a sentence to indicate that it is ungrammatical or not well formed.

Chapter 2

1. Note that *is* in sentences 2a–b is not technically an auxiliary verb, because auxiliary verbs are verbs that take another verb as their complement (as in, *The man is leaving*). We'll still refer to *is* as an auxiliary to illustrate the rule of question formation in English, because even when *is* is the only verb in the sentence, it still functions like an auxiliary verb in a relevant way: namely, *is* still inverts with the subject in a question, as we will see in the examples that follow.

2. Ages are reported in the format years;months or years;months.days. Thus, 3;2 means three years and two months. If days are reported, 3;2.15 would mean three years, two months and fifteen days.

3. This study was based on publicly available transcripts of conversations with children compiled in an online database known as CHILDES (MacWhinney, 2000). This database contains transcripts of speech data from dozens of children in many different languages. It is available at http://childes.talkbank.org.

4. Though see O'Grady (2005) for an uncommonly sophisticated exception to this trend.

5. On the other hand, it raises the question of how OSV, OVS, and VOS languages are possible. Recall that such languages are extremely rare but not nonexistent.

6. **Mass nouns** are nouns like *cereal*, *pudding*, and *rice*, which denote things that can't be counted. They are distinguished from **count nouns** (denoting things that can be counted), like *book*, *car*, and *dog*.

Chapter 3

1. Changes in vowel length do not create different sound categories in English, but they do in many other languages, such as Finnish and Japanese.

2. If you need a refresher on IPA symbols, please see appendix A.

Chapter 4

1. Braunwald (1978) reported that one child at age 11 months used the word [baʊwaʊ] to mean *dog* but also in reference to the sound of barking, the sound of an airplane or a car engine, birds, or any outside noise she could hear from inside the house.

2. Aboriginal languages of Australia are particularly known for lacking fricatives. For more information about the phonological systems of Australian languages, see Butcher (2006).

3. This ranking was obtained from cmloegcmluin.wordpress.com/2012/11/10/relative-frequencies-of-english-phonemes.

4. We assume that the other constraints (*CodaVoice, IdentVoice, and so forth) are still operational, but if we can't determine how they would be ranked in a given example, we can omit them from the tableau. In this case, since this particular child avoids all codas, we can't tell whether his IdentVoice constraint should be ranked higher or lower than his *Coda constraint.

Chapter 5

1. Not all words are referential, although all words have meaning of some kind. Function words in particular (*the*, *and*, *if*) do not refer to entities, but they have meaning. However, here were are focusing on nouns, which generally do have this property.

Chapter 6

1. As we will discuss, not all languages mark these grammatical features with inflectional morphology.

2. Some languages are "isolating," meaning that each individual morpheme is a separate word (like Mandarin); some languages are "agglutinating," meaning that derivational and inflectional morphemes are stuck onto a root word (like Turkish); some languages are "fusional," meaning that morphemes (including the root) can change their pronunciation when they are combined (like Cherokee); and some languages are "polysynthetic," meaning that many different morphemes can be added to a single root (like Inuit). Many languages use more than one of these morphological strategies. For example, English is mostly isolating but has some agglutinating morphology as well.

3. Note that the verb *be* is contractible in some environments but not others. For example, it can contract when it is phonotactically permitted (e.g., *that's*, but not **this's*) and when syntactically permitted (e.g., *Fido is bigger than Fluffy is*, but not **Fido is bigger that Fluffy's*). In addition, the verb *be* functions as a copula when the phrase that follows it is a noun phrase (NP), adjective phrase (AP), or prepositional phrase (PP) (*Fido is a dog/large/in the yard*) but as an auxiliary when the phrase that follows it is a verb phrase (VP) (*Fido is barking*).

4. In some Latin American dialects of Spanish, the second-person plural is the same as the third-person plural, but in Castillian Spanish on the Iberian Peninsula, second-person plural has a different form.

Chapter 7

1. This sentence is analyzed as follows:

Sasa	hivi	ni-ta-kul-a	ki-tu	ki-kubwa	kwa samamu		ni-na-ja-a	sana
Now	right	1s-fut-eat-IND	7-thing	7-big	because		1s-pres-hungry-IND	very

'Right now, I am about to eat something big because I am very hungry'

where 1s = first person singular, fut = future tense, IND = indicative mood, 7 = the class of the noun or the agreement an adjective has with its associated noun, and pres = present tense.

2. Thanks to Elaine Lau for help with this example.

3. As an interesting side note, these constructions are grammatical in many German dialects. In fact, the way you ask, *What do you think is in the box?* is precisely *Was glaubst du was in dem Kasten ist?* 'What do you think what is in the box?'

4. Their claim was actually about what they referred to as **extraction**, which is a cover term for several kinds of syntactic processes, the most prominent of which is relative clauses.

5. They also considered genitive RCs and object-of-comparison RCs. These are rare and complex patterns that need not concern us here.

6. We use the expression "in its clause," recognizing that a more standard definition would specify "in its local domain." Here we are limiting our examples to those involving clausal domains (not prepositional domains) and to a language in which the local domain is the clause (English). There is crosslinguistic variation in how the notion of locality is defined for binding purposes.

Chapter 8

1. The picture is actually far more complicated than described here, since it is not true that no language processing happens in the right hemisphere. But for our purposes this description will suffice.

2. Certainly, many people can and do learn a *second* language after early childhood. Second-language acquisition is a large and dynamic field of study in its own right and is not covered in this book. However, consider differences you have probably noticed in yourself or in others who have learned a second language late in life: typically, the process of second-language acquisition is effortful and conscious to at least some degree, and even the most successful second-language learners often speak with an accent.

3. The word *deaf* is written with a capital *D*, Deaf, to indicate members of the Deaf community—people who sign as their primary mode of communication. The word *deaf* with a lowercase *d* means unable to hear.

4. Researchers use a lower bound of 85 when the test has a mean score of 100 and a standard deviation of 15. However, other researchers use a different lower bound (e.g., 70 or 75) depending on the test used (Spaulding, Plante and Vance, 2008).

Chapter 9

1. There are many competing definitions of the term **heritage speaker**, as well as many varying conditions that can lead to heritage speaker status. For our purposes we adopt one of the definitions that is used by researchers and is theoretically neutral.

2. The construction in 2 is one way to form the possessive in German. The other way is actually similar to the Italian construction: the possessed object comes first, then the preposition *von* 'of, from', and then the possessor. Using this rule, the possessive noun phrase in 2 would be *Das Fahrrad von Mama,* literally 'the bicycle of mama'.

3. The numbers in the words in this example indicate tones.

4. These numbers reflect all types of tense marking, including *-s* on main verbs and inflected auxiliaries such as *is* and *does*. If one looks only at main verbs (e.g., *eats*, *runs*) and discards the auxiliary verbs, this L2 learner's production of tense was found in only about 5% of obligatory contexts.

5. The UNESCO (United Nations Educational, Scientific and Cultural Organization) interactive map can be found at http://www.unesco.org/languages-atlas/.

6. It is not quite accurate to say that the Jews who emigrated from various countries in Europe and repopulated Palestine lacked a common language; a large number of these emigrés spoke Yiddish. However, for various historical reasons Yiddish lacked the prestige that Hebrew held, so it was disfavored as a common tongue, or lingua franca.

7. A *language isolate* is a language that is not related to any other living languages. Unlike Spanish or French, Basque is not a Romance language—it is not even in the Indo-European family. Thus, even though Basque is spoken in a region where Spanish and French are also spoken, it is not at all related to these languages.

Index

Agent, 168–170
Allomorph, 144
Allophones, 63
American Sign Language (ASL), 233–238. *See also* Sign language
Antecedent, 204, 210
Argument, 124, 146, 169, 197, 199, 205, 230, 239, 244
Argument structure, 11, 124–126, 171
ASL. *See* American Sign Language; Sign language
Assimilation processes, 87–88, 93–94
Attrition, 275–276
Autism spectrum disorder, 227, 246, 251–253
Auxiliary verb, 21–26, 31, 46–47, 108, 110, 138, 139, 141, 143, 149, 172, 178, 181, 190–194, 218, 230, 242–243, 248, 271
Axiom, 181, 189

Babbling, 81–82, 235, 237, 266
Bare VP hypothesis, 185
Bilingualism
  simultaneous, 266
  successive (sequential), 266, 274
Binding, 203–216
Blind children, 242–245
Boostrapping
  prosodic, 71, 123
  semantic, 167–170
  syntactic, 120, 123, 124, 129, 169
By-phrase. *See* Passives

Categorical perception, 62–65, 73
Categorization (in word learning), 114–115, 121, 129
C-command, 207–210
Chelsea, 230–232
Child Language Data Exchange System (CHILDES), 137, 149, 295
Chomsky, Noam, 8, 17, 21, 25, 28, 30, 32–33, 138
Clause, 70–72, 123, 126
Cochlear implants, 240
Coda, 71, 95
Coda deletion, 89
Code-switching, 267, 270, 272–273
Coindexation, 203, 209, 213
Competence, 30–31, 300
Consonant assimilation, 87
Consonant cluster (reduction), 88–89
Constructivism, 39–45
Continuity, 31–32
Coreference, 203, 212
Covert contrast, 90
Crisma's Effect, 188–189
Critical period hypothesis, 225–226, 231, 253–256

Developmental Problem of Language Acquisition (DPLA), 9–10, 15, 39
Domain-general learning mechanisms, 15, 17, 24, 39–41, 43, 47
Domain-specific learning mechanisms, 39, 47

Endangered languages, 278–279, 282–284
Entrenchment, 44–46

Faithfulness constraints, 96–98
Fast-mapping, 110–111, 154
Features
lexical, 108–109, 111, 114–117
phonological, 83, 87, 93
syntactic, 186–187
Feral children, 226–227
Finite verb/finiteness, 178–180, 184, 186–187, 189, 204, 216
Fronting, 86–87
Functional categories, 172
Functional deafness, 232
Functional structure, 11–12, 171–177

Generalization, 5, 19, 83
Genie, 227–231
Gliding, 86–87
Grammaticality judgment task, 300–301

Habituation, 61, 308
Head
of a phrase, 174–175
of a relative clause, 201–202
Head-turn preference procedure, 72, 308
Hearing impairment, 232
Hemispherectomy, 253–256
Heritage language, 275, 277
High-amplitude sucking technique, 61, 307

Induction, 17–21, 25, 35, 39, 105, 111
Infant-directed speech, 67–70
Inflection. *See* Morphemes, inflectional
Input, 4–5, 9, 12, 15–18, 20, 23–26, 27–28, 32, 34–36, 38–39, 41, 44–49, 85, 115, 129, 143, 153–154, 177, 188, 197, 200, 228, 231, 238–239, 245, 253, 277
Interdependent development hypothesis (bilingualism), 270–272
Intergenerational transmission, 279, 283
Intermodal preferential looking paradigm, 125, 304–305

Lexicon, 29–30, 40–41, 66, 95, 107, 109–111, 114–115, 120–121, 242, 267–269, 277
Logical Problem of Language Acquisition (LPLA), 3–4, 8–9, 15, 33, 35, 39, 47
Low-pass filtered speech, 55–57, 68–69

MacArthur-Bates Communicative Development Inventory (MBCDI), 106–107, 110
Manner of articulation, 63, 83
Mapping problem, 112–115, 120, 129
Markedness, 83, 85, 96–98
Markedness constraints, 96–98
Maturation, 32
Meaning (word), 11, 81, 105, 108–129
Mean length of utterance (MLU), 141
Medial wh-questions, 197
Minimal pairs, 63, 82
Morphemes, 135
bound, 136, 146
derivational, 135–136
free, 136
inflectional, 135–138, 141–148
Morphological errors
of commission, 148, 150–151, 155
of omission, 148–150
overregularization, 148, 151–153, 155
Mutual exclusivity, 118–120, 129, 269

NATURE, 16
Negation, 172, 184, 186–187, 190–191, 230
external, 190
internal, 190
Negative evidence, 7–9, 109
Noun Phrase Accessibility Hierarchy (NPAH), 202
Null subjects, 187–188

Onset, 71
Optimality Theory, 95–98
Optional infinitives (OIs), 178, 180, 184, 186–189, 250, 270
Overextension, 108–109
Overregularization, 148, 151–153

Passives, 197–201, 297, 299, 302, 306
actional and non-actional, 199–200
by-phrase, 197, 200
frequency of, 198
Patient, 169
Perceptual assimilation, 65, 85
Performance, 30–31
Phoneme, 11, 63, 79, 82–84

Phonological rules, 90–94
Phonotactics, 35–36, 72–73
Phrase, 70–73, 123
Place of articulation, 63, 83, 93
Positive evidence, 7
Poverty of the stimulus, 17, 24–28, 39, 46–47
Preemption, 44
Preferential looking paradigm, 125, 215, 269, 304–305
Principle of Pronouns (Principle B), 205, 211–213
Principle of reference, 116, 117, 129
Principle of Referring Expressions (Principle C), 206, 213–215
Principle of Reflexives (Principle A), 205–206
Principles and parameters, 32–35
Projection, 174, 182, 216
Pronouns, 5, 6, 12, 42, 129, 157, 167, 201, 203, 205–207, 210–216, 235, 237–238, 246, 252,
 non-reflexive, 203, 210, 215–216
 reflexive, 203, 210, 212–213, 215
 relative, 201
 resumptive, 201
Prosody, 56–57, 69–72, 123

Quantified phrase, 212

Reduplication, 89
Referentiality, 116–117, 129
Referring expressions (R-expressions), 205–207, 213–216
Relative clauses, 32, 201–202, 248, 277, 297, 303, 306
 head of, 201
 relativizer, 201
Reversible sentences, 199, 202
Rime, 71

Semantic bootstrapping. *See* Bootstrapping
Semantics, 11, 128. *See also* Meaning (word)
Separate-systems hypothesis (bilingualism), 268
Sign language, 69, 81–82, 107, 233–235. *See also* American Sign Language
Single system hypothesis (bilingualism), 267–268
Small clause hypothesis. *See* Bare VP hypothesis
Specifier, 174, 185, 187, 193, 195, 196
Specific language impairment (SLI), 245–250
Speech segmentation, 36, 66–67, 70–73
Statistical tracking, 35–38. *See also* Transitional probability
Stopping, 86–87, 89, 93
Structure-dependent distributional learning, 170
Structure-dependent rule, 22–24, 33–34, 47
Subject-auxiliary inversion (SAI), 193
Substitution processes, 84, 86–87
Syllabic processes, 88–89
Syllable, 35, 71, 73, 81, 88–89
Syntactic bootstrapping, 123–129
Syntax, 11

Taxonomic constraint, 121
Telegraphic speech, 6
Tense, 31, 179
Theory of Mind, 252
Transfer, 271
Transitional probability, 36–38, 72–73
Truncation hypothesis, 180–184
Truth Value Judgment Task (TVJT), 154, 301–303

Universal grammar, 15, 16–19, 24, 28–29, 33–35, 38, 45–48, 105, 128–129

Verb learning, 120–126
Vocabulary spurt, 107
Vocalization, 86–87
Voice onset time (VOT), 58–63, 73, 83
Voicing, 58–59, 62–63, 87

Weak syllable deletion, 89
Wh-movement, 34
Whole object constraint, 117–119, 120, 129
Wh-questions, 192–197
Wug test, 143–145, 296–297

X-bar structures, 174, 185

Yes-no questions, 191